Introduction to
Multivariate Statistical Analysis in Chemometrics

Kurt Varmuza
Peter Filzmoser

CRC Press
Taylor & Francis Group
Boca Raton London New York

CRC Press is an imprint of the
Taylor & Francis Group, an **informa** business

CRC Press
Taylor & Francis Group
6000 Broken Sound Parkway NW, Suite 300
Boca Raton, FL 33487-2742

Printed in the United States of America on acid-free paper
10 9 8 7 6 5 4 3 2 1

International Standard Book Number-13: 978-1-4200-5947-2 (Hardcover)

Library of Congress Cataloging-in-Publication Data

Varmuza, Kurt, 1942-
Introduction to multivariate statistical analysis in chemometrics / Kurt Varmuza and Peter Filzmoser.
p. cm.
Includes bibliographical references and index.
ISBN 978-1-4200-5947-2 (acid-free paper)
1. Chemometrics. 2. Multivariate analysis. I. Filzmoser, Peter. II. Title.

QD75.4.C45V37 2008
543.01'519535--dc22 2008031581

Visit the Taylor & Francis Web site at
http://www.taylorandfrancis.com

and the CRC Press Web site at
http://www.crcpress.com

Contents

Preface

This book is the result of a cooperation between a chemometrician and a statistician. Usually, both sides have quite a different approach to describing statistical methods and applications—the former having a more practical approach and the latter being more formally oriented. The compromise as reflected in this book is hopefully useful for chemometricians, but it may also be useful for scientists and practitioners working in other disciplines—even for statisticians. The principles of multivariate statistical methods are valid, independent of the subject where the data come from. Of course, the focus here is on methods typically used in chemometrics, including techniques that can deal with a large number of variables. Since this book is an introduction, it was necessary to make a selection of the methods and applications that are used nowadays in chemometrics.

The primary goal of this book is to effectively impart a basic understanding of the methods to the reader. The more formally oriented reader will find a concise mathematical description of most of the methods. In addition, the important concepts are visualized by graphical schemes, making the formal approach more transparent. Some methods, however, required more mathematical effort for providing a deeper insight.

Besides the mathematical outline, the methods are applied to real data examples from chemometrics for supporting better understanding and applicability of the methods. Prerequisites and limitations for the applicability are discussed, and results from different methods are compared.

The validity of the results is a central issue, and it is confirmed by comparing traditional methods with their robust counterparts. Robust statistical methods are less common in chemometrics, although they are easy to access and compute quickly. Thus, several robust methods are included.

Computation and practical use are further important concerns, and thus the $\mathbb{R}$ package chemometrics has been developed, including data sets used in this book as well as implementations of the methods described. Although some programming skills are required, the use of $\mathbb{R}$ has advantages because it is freeware and is continuously updated. Thus interested readers can go through the examples in this book and adapt the procedures to their own problems. Feedback is appreciated and it can lead to extension and improvement of the package.

The book cover depicts a panoramic view of Monument Valley on the Arizona Utah border, captured by Kurt Varmuza from a public viewing point in August 2005. The picture is not only an eye-catcher, but may also inspire thoughts about the relationship between this fascinating landscape and chemometrics. It has been reported that the pioneering chemometrician and analytical chemist D. Luc Massart (1941–2005) mentioned something like "Univariate methods are clear and simple, multivariate methods are the Wild West." For many people, pictures like these are synonymous with the Wild West—sometimes realizing that the impression is severely influenced by movie makers. The Wild West, of course, is multivariate: a vast, partially unexplored country, full of expectations, adventures, and fun (for tourists), but also with a harsh climate and wild animals. From a more prosaic point of view, one may see peaks, principal components, dusty flat areas, and a wide horizon. A path—not always easy to drive—guides visitors from one fascinating point to another. The sky is not cloudless and some areas are under the shadows; however, it is these areas that may be the productive ones in a hot desert.

Kurt Varmuza
Peter Filzmoser

Acknowledgments

We thank our institution, the Vienna University of Technology, for providing the facilities and time to write this book. Part of this book was written during a stay of Peter Filzmoser at the Belarusian State University in Minsk. We thank The Austrian Research Association for the mobility program MOEL and the Belarusian State University for their support, from the latter especially Vasily Strazhev, Vladimir Tikhonov, Pavel Mandrik, Yuriy Kharin, and Alexey Kharin.

We also thank the staff of CRC Press (Taylor and Francis Group) for their professional support.

Many of our current and former colleagues have contributed to this book by sharing their software, data, ideas, and through numerous discussions. They include Christophe Croux, Wilhelm Demuth, Rudi Dutter, Anton Friedl, Paolo Grassi, Johannes Jaklin, Bettina Liebmann, Manfred Karlovits, Barbara Kavsek-Spangl, Robert Mader, Athanasios Makristathis, Plamen N. Penchev, Peter J. Rousseeuw, Heinz Scsibrany, Sven Serneels, Leonhard Seyfang, Matthias Templ, and Wolfgang Werther. We are especially grateful to the last named for bringing the authors together.

We thank our families for their support and patience. Peter, Theresa, and Johannes Filzmoser are thanked for understanding that their father could not spend more time with them while writing this book.

Many others who have not been named above have contributed to this book and we are grateful to them all.

Authors

Kurt Varmuza was born in 1942 in Vienna, Austria. He studied chemistry at the Vienna University of Technology, Austria, where he wrote his doctoral thesis on mass spectrometry and his habilitation, which was devoted to the field of chemometrics. His research activities include applications of chemometric methods for spectra–structure relationships in mass spectrometry and infrared spectroscopy, for structure–property relationships, and in computer chemistry, archaeometry (especially with the Tyrolean Iceman), chemical engineering, botany, and cosmo chemistry (mission to a comet). Since 1992, he has been working as a professor at the Vienna University of Technology, currently at the Institute of Chemical Engineering.

Peter Filzmoser was born in 1968 in Wels, Austria. He studied applied mathematics at the Vienna University of Technology, Austria, where he wrote his doctoral thesis and habilitation, devoted to the field of multivariate statistics. His research led him to the area of robust statistics, resulting in many international collaborations and various scientific papers in this area. His interest in applications of robust methods resulted in the development of $\mathbb{R}$ software packages. He was and is involved in the organization of several scientific events devoted to robust statistics. Since 2001, he has been a professor in the Statistics Department at Vienna University of Technology. He was a visiting professor at the Universities of Vienna, Toulouse, and Minsk.

1 Introduction

1.1 CHEMOINFORMATICS–CHEMOMETRICS–STATISTICS

CHEMOMETRICS has been defined as "A chemical discipline that uses statistical and mathematical methods, to design or select optimum procedures and experiments, and to provide maximum chemical information by analyzing chemical data." In shorter words it is focused as "Chemometrics concerns the extraction of relevant information from chemical data by mathematical and statistical tools." Chemometrics can be considered as a part of the wider field CHEMOINFORMATICS which has been defined as "The application of informatics methods to solve chemical problems" (Gasteiger and Engel 2003) including the application of mathematics and statistics.

Despite the broad definition of chemometrics, the most important part of it is the application of multivariate data analysis to chemistry-relevant data. Chemistry deals with compounds, their properties, and their transformations into other compounds. Major tasks of chemists are the analysis of complex mixtures, the synthesis of compounds with desired properties, and the construction and operation of chemical technological plants. However, chemical/physical systems of practical interest are often very complicated and cannot be described sufficiently by theory. Actually, a typical chemometrics approach is not based on first principles—that means scientific laws and rules of nature—but is DATA DRIVEN. Multivariate statistical data analysis is a powerful tool for analyzing and structuring data sets that have been obtained from such systems, and for making empirical mathematical models that are for instance capable to predict the values of important properties not directly measurable (Figure 1.1).

Chemometric methods became routinely applied tools in chemistry. Typical problems that can be successfully handled by chemometric methods are

- Determination of the concentration of a compound in a complex mixture (often from infrared data)
- Classification of the origins of samples (from chemical analytical or spectroscopic data)
- Prediction of a property or activity of a chemical compound (from chemical structure data)
- Recognition of presence/absence of substructures in the chemical structure of an unknown organic compound (from spectroscopic data)
- Evaluation of the process status in chemical technology (from spectroscopic and chemical analytical data)

Similar data evaluation problems exist in other scientific fields and can also be treated by multivariate statistical data analysis, for instance, in economics (econometrics), sociology, psychology (psychometrics), medicine, biology (chemotaxonomy),

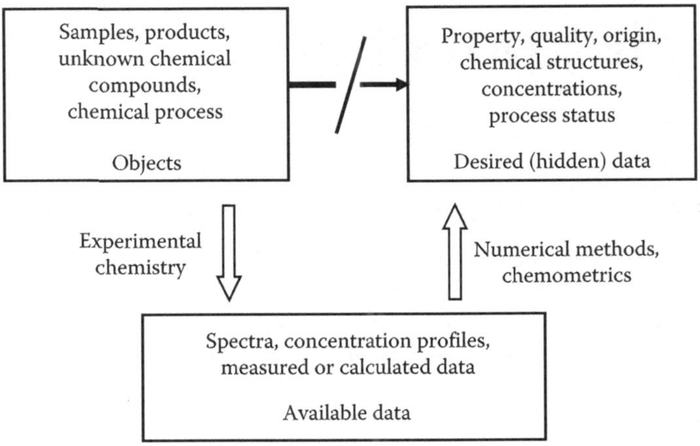

FIGURE 1.1 Desired data from objects can often not be directly measured but can be modeled and predicted from available data by applying chemometric methods.

image analysis, and character and pattern recognition. Recently, in bioinformatics (dealing with much larger molecules than chemoinformatics), typical chemometric methods have been applied to relate metabolomic data from chemical analysis to biological data.

MULTIVARIATE STATISTICS is an extension of univariate statistics. Univariate statistics investigates each variable separately or relates a single independent variable x to a single dependent variable y. Of course, chemical compounds, reactions, samples, technological processes are multivariate in nature, which means a good characterization requires many—sometimes very many—variables. Multivariate data analysis considers many variables together and thereby often gains a new and higher quality in data evaluation. Many examples show that a multivariate approach can be successful even in cases where the univariate consideration is completely useless.

Basically, there are two different approaches in analyzing multivariate statistical data. One is data driven and the statistical tools are seen as algorithms that are applied to obtain the results. Another approach is model-driven. The available data are seen as realizations of random variables, and an underlying statistical model is assumed. Chemometricians tend to use the first approach, while statisticians usually are in favor of the second one. It is probably the type of data in chemometrics that required a more data-driven type of analysis: distributional assumptions are not fulfilled; the number of variables is often much higher than the number of objects; the variables are highly correlated, etc. Traditional statistical methods fail for this type of data, and still nowadays some statisticians would refuse analyzing multivariate data where the number of objects is not at least five times as large as the number of variables. However, an evaluation of such data is often urgent and no better data may be available. Successful methods to handle such data have thus been developed in the field of chemometrics, like the development of partial least-squares (PLS) regression. The treatment of PLS as a statistical method rather than an algorithm resulted in several improvements, like the robustification of PLS regarding to outlying objects. Thus, both approaches have their own right to exist, and a combination of them can be of great advantage. The book may hopefully narrow the gap between the two approaches in analyzing multivariate data.

1.2 THIS BOOK

The book is at an introductory level, and only basic mathematical and statistical knowledge is assumed. However, we do not present "chemometrics without equations"—the book is intended for mathematically interested readers. Whenever possible, the formulae are in matrix notation, and for a clearer understanding many of them are visualized schematically. Appendix 2 might be helpful to refresh matrix algebra.

The focus is on multivariate statistical methods typically needed in chemometrics. In addition to classical statistical methods, also robust alternatives are introduced which are important for dealing with noisy data or with data including outliers. Practical examples are used to demonstrate how the methods can be applied and results can be interpreted; however, in general the methodical part is separated from application examples.

For practical computation the software environment $\mathbb{R}$ is used. $\mathbb{R}$ is a powerful statistical software tool, it is freeware and can be downloaded at http://cran.r-project.org. Throughout the book we will present relevant $\mathbb{R}$ commands, and in Appendix 3 a brief introduction to $\mathbb{R}$ is given. An $\mathbb{R}$-package "chemometrics" has been established; it contains most of the data sets used in the examples and a number of newly written functions mentioned in this book.

We will follow the guidance of Albert Einstein to "make everything as simple as possible, but not simpler." The reader will find practical formulae to compute results like the correlation matrix, but will also be reminded that there exist other possibilities of parameter estimation, like the robust or nonparametric estimation of a correlation.

In this chapter, we provide a general overview of the field of chemometrics. Some historical remarks and relevant literature to this subject make the strong connection to statistics visible. First practical examples (Section 1.5) show typical problems related to chemometrics, and the methods applied will be discussed in detail in subsequent chapters. Basic information on univariate statistics (Section 1.6) might be helpful to understand the concept of "randomness" that is fundamental in statistics. This section is also useful for making first steps in $\mathbb{R}$.

In Chapter 2, we approach multivariate data analysis. This chapter will be helpful for getting familiar with the matrix notation used throughout the book. The "art" of statistical data analysis starts with an appropriate data preprocessing, and Section 2.2 mentions some basic transformation methods. The multivariate data information is contained in the covariance and distance matrix, respectively. Therefore, Sections 2.3 and 2.4 describe these fundamental elements used in most of the multivariate methods discussed later on. Robustness against data outliers is one of the main concerns of this book. The multivariate outlier detection methods treated in Section 2.5 can thus be used as a first diagnostic tool to check multivariate data for possible outliers. Finally, Section 2.6 explains the concept of linear latent variables that is inherent in many important multivariate methods discussed in subsequent chapters.

Chapter 3 starts with the first and probably most important multivariate statistical method, with PRINCIPAL COMPONENT ANALYSIS (PCA). PCA is mainly used for mapping or summarizing the data information. Many ideas presented in this chapter, like the selection of the number of principal components (PCs), or the robustification of PCA, apply in a similar way to other methods. Section 3.8 discusses briefly related methods for summarizing and mapping multivariate data. The interested reader may consult extended literature for a more detailed description of these methods.

Chapters 4 and 5 are the most comprehensive chapters, because multivariate calibration and classification belong to the most important topics for multivariate analysis in chemometrics. These topics have many principles in common, like the schemes for the evaluation of the performance of the resulting regression model or classifier (Section 4.2). Both chapters include a variety of methods for regression and classification, some of them being standard in most books on multivariate statistics, and some being of more specific interest to the chemometrician. For example, PLS regression (Section 4.7) is treated in more detail, because this method is closely associated with the developments of the field of chemometrics. The different approaches are listed and the algorithms are compared mathematically. Also more recently developed methods, like support vector machines (Section 5.6) are included, because they are considered as very successful for solving problems in chemometrics.

The final methodological chapter (Chapter 6) is devoted to cluster analysis. Besides a general treatment of different clustering approaches, also more specific problems in chemometrics are included, like clustering binary vectors indicating presence or absence of certain substructures.

Chapter 7 finally presents selected techniques for preprocessing that are relevant for data in chemistry and spectroscopy.

1.3 HISTORICAL REMARKS ABOUT CHEMOMETRICS

An important root of the use of multivariate data analysis methods in chemistry is the pioneering work at the University of Washington (Seattle, WA) guided by T. L. Isenhour in the late 1960s. In a series of papers mainly the classification method "learning machine," described in a booklet by N. J. Nilsson (Nilsson 1965), has been applied to chemical problems. Under the name pattern recognition—and in a rather optimistic manner—the determination of molecular formulae and the recognition of chemical structure classes from molecular spectral data have been reported; the first paper appeared in 1969 (Jurs et al. 1969a), others followed in the forthcoming years (Isenhour and Jurs 1973; Jurs and Isenhour 1975; Jurs et al. 1969b,c; Kowalski et al. 1969a,b; Liddell and Jurs 1974; Preuss and Jurs 1974; Wangen et al. 1971). At about the same time L. R. Crawford and J. D. Morrison in Australia used multivariate classification methods for an empirical identification of molecular classes from low-resolution mass spectra (Crawford and Morrison 1968, 1971).

The Swedish chemist, Svante Wold, is considered to have been the first to use the word chemometrics, in Swedish in 1972 (Forskningsgruppen för Kemometri) (Wold 1972), and then in English two years later (Wold 1974). The American chemist and mathematician Bruce R. Kowalski presented in 1975 a first overview of the contents and aims for a new chemical discipline chemometrics

(Kowalski 1975), soon after founding the International Chemometrics Society on June 10, 1974 in Seattle together with Svante Wold (Kowalski et al. 1987).

The early history of chemometrics is documented by published interviews with Bruce R. Kowalski, D. Luc Massart, and Svante Wold who can be considered as the originators of modern chemometrics (Esbensen and Geladi 1990; Geladi and Esbensen 1990). A few, subjectively selected milestones in the development of chemometrics are mentioned here as follows:

- Kowalski and Bender presented chemometrics (at this time called pattern recognition and roughly considered as a branch of artificial intelligence) in a broader scope as a general approach to interpret chemical data, especially by mapping multivariate data with the purposes of cluster analysis and classification (Kowalski and Bender 1972).
- A curious episode in 1973 characterizes the early time in this field. A pharmacological activity problem (discrimination between sedatives or tranquilizers) was related to mass spectral data without needing or using the chemical structures of the considered compounds (Ting et al. 1973). In a critical response (Clerc et al. 1973) it was reported that similar success rates—better than 95%—can be obtained for a classification of the compounds whether their names contain an even or odd number of characters, just based on mass spectral data! Obviously, in both papers the prediction performance was not estimated properly.
- The FORTRAN program ARTHUR (Harper et al. 1977)—running on main frame computers at this time—comprised all basic procedures of multivariate data analysis and made these methods available to many chemists in the late 1970s.
- At the same time S. Wold presented the software soft independent modeling of class analogies (SIMCA) and introduced a new way of thinking in data evaluation called "soft modeling" (Wold and Sjöström 1977).
- In 1986 two journals devoted to chemometrics have been launched: *Journal of Chemometrics* by Wiley, and the *Journal of Chemometrics and Intelligent Laboratory Systems* (short: ChemoLab) by Elsevier; both are still the leading print media in this field.
- *Chemometrics: A Textbook* published in 1988 by D. L. Massart et al. (1988) was for a long time the Bible (*Blue Book*) for chemometricians working in analytical chemistry.
- A tremendous stimulating influence on chemometrics had the development of PLS regression, which now is probably the most used method in chemometrics (Lindberg et al. 1983; Wold et al. 1984).
- In the late 1980s H. Martens and T. Naes (1989) broadly introduced the use of infrared data together with PLS for quantitative analyses in food chemistry, and thereby opened the window to numerous successful applications in various fields of chemical industry, at present time.
- Developments in computer technology promoted the use of computationally demanding methods such as artificial neural networks, genetic algorithms, and multiway data analysis.

- D. L. Massart et al. and B. G. M. Vandeginste published a new Bible for chemometricians in two volumes, appeared in 1997 and 1998 (Massart et al. 1997; Vandeginste et al. 1998), the *New Two Blue Books.*
- Chemometrics became well established in chemical industry within process analytical technology, and is important in the fast growing area of biotechnology.

1.4 BIBLIOGRAPHY

Recently, INTRODUCTORY BOOKS about chemometrics have been published by R. G. Brereton, *Chemometrics—Data Analysis for the Laboratory and Chemical Plant* (Brereton 2006) and *Applied Chemometrics for Scientists* (Brereton 2007), and by M. Otto, *Chemometrics—Statistics and Computer Application in Analytical Chemistry* (Otto 2007). Dedicated to quantitative chemical analysis, especially using infrared spectroscopy data, are *A User-Friendly Guide to Multivariate Calibration and Classification* (Naes et al. 2004), *Chemometric Techniques for Quantitative Analysis* (Kramer 1998), *Chemometrics: A Practical Guide* (Beebe et al. 1998), and *Statistics and Chemometrics for Analytical Chemistry* (Miller and Miller 2000).

A comprehensive two-volume *Handbook of Chemometrics and Qualimetrics* has been published by D. L. Massart et al. (1997) and B. G. M. Vandeginste et al. (1998); predecessors of this work and historically interesting are *Chemometrics: A Textbook* (Massart et al. 1988), *Evaluation and Optimization of Laboratory Methods and Analytical Procedures* (Massart et al. 1978), and *The Interpretation of Analytical Chemical Data by the Use of Cluster Analysis* (Massart and Kaufmann 1983). A classical reference is still *Multivariate Calibration* (Martens and Naes 1989). A dictionary with extensive explanations containing about 1700 entries is *The Data Analysis Handbook* (Frank and Todeschini 1994).

SPECIAL APPLICATIONS of chemometrics are emphasized in a number of books: *Chemometrics in Analytical Spectroscopy* (Adams 1995), *Multivariate Pattern Recognition in Chemometrics, Illustrated by Case Studies* (Brereton 1992), *Chemometrics: Chemical and Sensory Data* (Burgard and Kuznicki 1990), *Chemometrics—From Basics to Wavelet Transform* (Chau et al. 2004), *Experimental Design: A Chemometric Approach* (Deming and Morgan 1987), *Chemometrics in Environmental Analysis* (Einax et al. 1997), *Prediction Methods in Science and Technology* (Hoeskuldsson 1996), *Discriminant Analysis and Class Modelling of Spectroscopic Data* (Kemsley 1998), *Multivariate Chemometrics in QSAR (Quantitative Structure–Activity Relationships)—A Dialogue* (Mager 1988), *Factor Analysis in Chemistry* (Malinowski 2002), *Multivariate Analysis of Quality—An Introduction* (Martens and Martens 2000), *Multi-Way Analysis with Applications in the Chemical Sciences* (Smilde et al. 2004), and *Chemometric Methods in Molecular Design* (van de Waterbeemd 1995).

In the EARLIER TIME OF CHEMOMETRICS until about 1990, a number of books have been published that may be rather of historical interest. *Chemometrics—Applications of Mathematics and Statistics to Laboratory Systems* (Brereton 1990), *Chemical Applications of Pattern Recognition* (Jurs and Isenhour 1975), *Factor Analysis in*

Chemistry (Malinowski and Howery 1980), *Chemometrics* (Sharaf et al. 1986), *Pattern Recognition in Chemistry* (Varmuza 1980), and *Pattern Recognition Approach to Data Interpretation* (Wolff and Parsons 1983).

Relevant collections of papers—mostly conference PROCEEDINGS—have been published: *Chemometrics—Exploring and Exploiting Chemical Information* (Buydens and Melssen 1994), *Chemometrics Tutorials* (Jonker 1990), *Chemometrics: Theory and Application* (Kowalski 1977), *Chemometrics—Mathematics and Statistics in Chemistry* (Kowalski 1983), and *Progress in Chemometrics Research* (Pomerantsev 2005).

The four-volume *Handbook of Chemoinformatics—From Data to Knowledge* (Gasteiger 2003) contains a number of introductions and reviews that are relevant to chemometrics: *Partial Least Squares (PLS) in Cheminformatics* (Eriksson et al. 2003), *Inductive Learning Methods* (Rose 1998), *Evolutionary Algorithms and their Applications* (von Homeyer 2003), *Multivariate Data Analysis in Chemistry* (Varmuza 2003), and *Neural Networks* (Zupan 2003).

Chemometrics related to COMPUTER CHEMISTRY and chemoinformatics is contained in *Design and Optimization in Organic Synthesis* (Carlson 1992), *Chemoinformatics—A Textbook* (Gasteiger and Engel 2003), *Handbook of Molecular Descriptors* (Todeschini and Consonni 2000), *Similarity and Clustering in Chemical Information Systems* (Willett 1987), *Algorithms for Chemists* (Zupan 1989), and *Neural Networks in Chemistry and Drug Design* (Zupan and Gasteiger 1999).

The native language of the authors of this book is GERMAN; a few relevant books written in this language have been published (Danzer et al. 2001; Henrion and Henrion 1995; Kessler 2007; Otto 1997). A book in FRENCH about PLS regression has been published (Tenenhaus 1998).

Only a few of the many recent books on MULTIVARIATE STATISTICS and related topics can be mentioned here. In earlier chemometrics literature often cited are *Pattern Classification and Scene Analysis* (Duda and Hart 1973), *Multivariate Statistics: A Practical Approach* (Flury and Riedwyl 1988), *Introduction to Statistical Pattern Recognition* (Fukunaga 1972), *Principal Component Analysis* (Joliffe 1986), *Discriminant Analysis* (Lachenbruch 1975), *Computer-Oriented Approaches to Pattern Recognition* (Meisel 1972), *Learning Machines* (Nilsson 1965), *Pattern Recognition Principles* (Tou and Gonzalez 1974), and *Methodologies of Pattern Recognition* (Watanabe 1969). More recent relevant books are *An Introduction to the Bootstrap* (Efron and Tibshirani 1993), *Self-Organizing Maps* (Kohonen 1995), *Pattern Recognition Using Neural Networks* (Looney 1997), and *Pattern Recognition and Neural Networks* (Ripley 1996).

There are many STATISTICAL TEXT BOOKS on multivariate statistical methods, and only a (subjective) selection is listed here. Johnson and Wichern (2002) treat the standard multivariate methods, Jackson (2003) concentrates on PCA, and Kaufmann and Rousseeuw (1990) on cluster analysis. Fox (1997) treats regression analysis, and Fox (2002) focuses on regression using $\mathbb{R}$ or S-Plus. PLS regression is discussed (in French) by Tenenhaus (1998). More advanced regression and classification methods are described by Hastie et al. (2001). Robust (multivariate) statistical methods are included in Rousseeuw and Leroy (1987) and in the more recent book by Maronna et al. (2006). Dalgaard (2002) uses the computing environment $\mathbb{R}$ for introductory statistics, and statistics with S has been described by Venables and Ripley (2003). Reimann et al. (2008) explain univariate and multivariate statistical methods and provide ℝ tools for many examples in ecogeochemistry.

1.5 STARTING EXAMPLES

As a starter for newcomers in chemometrics some examples are presented here to show typical applications of multivariate data analysis in chemistry and to present some basic ideas in this discipline.

1.5.1 Univariate versus Bivariate Classification

In this artificial example we assume that two classes of samples (groups A and B, for instance, with different origin) have to be distinguished by experimental data measured on the samples, for instance, by concentrations x_1 and x_2 of two compounds or elements present in the samples. Figure 1.2 shows that neither x_1 nor x_2 is useful for a separation of the two sample groups. However, both variables together allow an excellent discrimination, thereby demonstrating the potential and sometimes unexpected advantage of a multivariate approach. A naive univariate evaluation of each variable separately would lead to the wrong conclusion that the variables are useless. Of course in this simple bivariate example a plot x_2 versus x_1 clearly indicates the data structure and shows how to separate the classes; for more variables—typical examples from chemistry use a dozen up to several hundred variables—the application of numerical methods from multivariate data analysis is necessary.

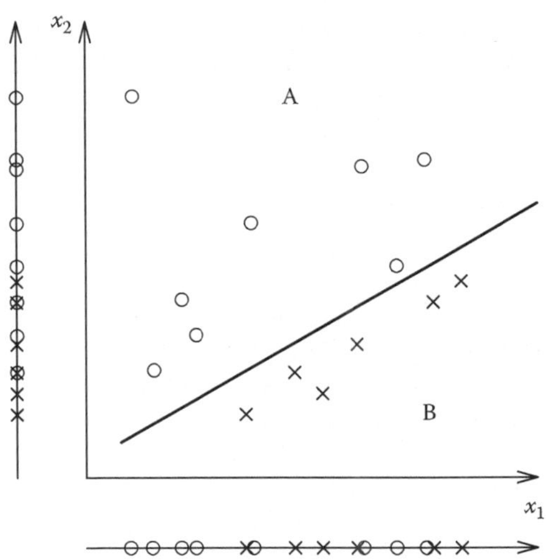

FIGURE 1.2 Artificial data for two sample classes A (denoted by circles, $n_1 = 8$) and B (denoted by crosses, $n_2 = 6$), and two variables x_1 and x_2 ($m = 2$). Each single variable is useless for a separation of the two classes, both together perform well.

1.5.2 NITROGEN CONTENT OF CEREALS COMPUTED FROM NIR DATA

For $n = 15$ cereal samples from barley, maize, rye, triticale, and wheat, the nitrogen contents, y, have been determined by the Kjeldahl method; values are between 0.92 and 2.15 mass% of dry sample. From the same samples near infrared (NIR) reflectance spectra have been measured in the range 1100 to 2298 nm in 2 nm intervals; each spectrum consists of 600 data points. NIR spectroscopy can be performed much easier and faster than wet-chemistry analyses; therefore, a mathematical model that relates NIR data to the nitrogen content may be useful. Instead of the original absorbance data, the first derivative data have been used to derive a regression equation of the form

$$\hat{y} = b_0 + b_1x_1 + b_2x_2 + \cdots + b_mx_m \tag{1.1}$$

with $\hat{y}$ being the modeled (predicted) y, and x_1 to x_m the independent variables (first derivative data at m wavelengths). The regression coefficients b_j and the intercept b_0 have been estimated by the widely used method, partial least-squares regression (PLS, Section 4.7). This method can handle data sets containing more variables than samples and accepts highly correlating variables, as is often the case with chemistry data; furthermore, PLS models can be optimized in some way for best prediction that means low absolute prediction errors $|\hat{y} - y|$ for new cases.

Figure 1.3a shows the experimental nitrogen content, y, plotted versus the nitrogen content predicted from all $m = 600$ NIR data, $\hat{y}$, using the so-called "calibration" mode. In calibration mode, all samples are used for model building and the obtained model is applied to the same data. For Figure 1.3b the "full CROSS VALIDATION (full CV, Section 4.2.5)" mode has been applied, that means one of the samples has been left out and from the remaining $n - 1$ samples a model has been calculated and applied to the left out sample giving a predicted value for this sample. This procedure has been repeated n times with each object left out once (therefore also called "leave-one-out CV"). Note that the prediction errors from CV are considerably greater than from calibration mode, but are more realistic estimations of the prediction performance than results from calibration mode. Two measures are used in Figure 1.3 to characterize the prediction performance. First, r^2 is the squared Pearson correlation coefficient between y and $\hat{y}$, which is for a good model close to 1. The other measure is the standard deviation of the prediction errors used as a criterion for the distribution of the prediction errors. For a good model the errors are small, the distribution is narrow, and the standard deviation is small. In calibration mode this standard deviation is called SEC, the standard error of calibration; in CV mode it is called SEP_{CV}, the standard error of prediction estimated by CV (Section 4.2.3). If the prediction errors are normally distributed—which is often the case—an approximative 99% tolerance interval for prediction errors can be estimated by $\pm 2.5\ SEP_{CV}$.

We may suppose that not all 600 wavelengths are useful for the prediction of nitrogen contents. A variable selection method called genetic algorithm (GA, Section 4.5.6) has been applied resulting in a subset with only five variables (wavelengths). Figure 1.3c and d shows that models with these five variables are better than models

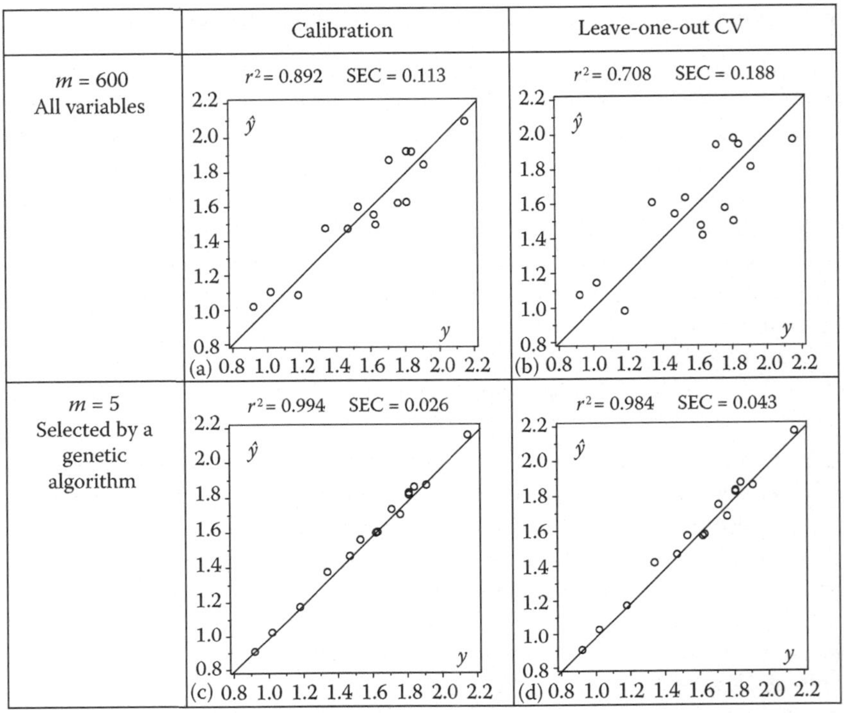

FIGURE 1.3 Modeling the nitrogen content of cereals by NIR data; $n = 15$ samples. m, number of used variables (wavelengths); y, nitrogen content from Kjeldahl analysis; $\hat{y}$, nitrogen content predicted from NIR data (first derivative); r^2, squared Pearson correlation coefficient between y and $\hat{y}$. SEC and SEP_{CV} are the standard deviations of prediction errors for calibration mode and full CV, respectively. Models using a subset of five wavelengths (c, d) give better results than models using all 600 wavelengths (a, b). Prediction errors for CV are larger than for calibration mode but are a more realistic estimation for the prediction performance.

with 600 variables; again the CV prediction errors are larger than the prediction errors in calibration mode.

For this example, the commercial software products *The Unscrambler* (Unscrambler 2004) has been used for PLS and *MobyDigs* (MobyDigs 2004) for GA; results have been obtained within few minutes. Some important aspects of multivariate calibration have been mentioned together with this example; others have been left out, for instance, full CV is not always a good method to estimate the prediction performance.

1.5.3 Elemental Composition of Archaeological Glasses

Janssen et al. (1998) analyzed 180 archaeological glass vessels from the fifteenth to seventeenth century using x-ray methods to determine the concentrations of 13 elements present in the glass. The goal of this study was to learn about glass production and about trade connections between the different renowned producers. The data set consists of four groups, each one corresponding to a different type of glass. The group sizes are very different: group 1 consists of 145 objects (glass samples), group 2 has 15, and groups 3 and 4 have only 10 objects. Each glass sample can be considered to be represented by a point in a 13-dimensional space with the coordinates of a point given by the elemental concentrations. A standard method to project the high-dimensional space on to a plane (for visual inspection) is PRINCIPAL COMPONENT ANALYSIS (PCA, Chapter 3). Actually, a PCA plot of the concentration data visualizes the four groups of glass samples very well as shown in Figure 1.4. Note that PCA does not utilize any information about the group membership of the samples; the clustering is solely based on the concentration data. The projection axes are called the principal components (PC1 and PC2). A quality measure how good the projection reflects the situation in the high-dimensional space is the percent variance preserved by the projection axes. Note that variance is considered here as potential information about group memberships. The first PC reflects 49.2% of the total variance (equal to the sum of the variances of all 13 concentrations); the first two PCs explain 67.5% of the total variance of the multivariate data; these high percentages indicate that the PCA plot is informative.

PCA is a type of EXPLORATORY DATA ANALYSIS without using information about group memberships. Another aim of data analysis can be to estimate how accurately a new glass vessel could be assigned to one of the four groups. For this CLASSIFICATION problem we will use LINEAR DISCRIMINANT ANALYSIS (LDA, Section 5.2) to derive discriminant rules that will allow to predict the group membership of a new object. Since no new object is available, the data at hand can be split into training and test data. LDA is then applied to the training data and the prediction is made for the

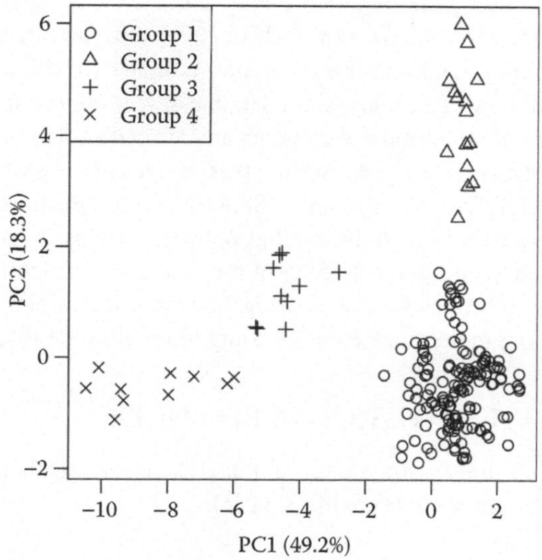

FIGURE 1.4 Projection of the glass vessels data set on the first two PCs. Both PCs together explain 67.5% of the total data variance. The four different groups corresponding to different types of glass are clearly visible.

TABLE 1.1
Classification Results of the Glass Vessels Data Using LDA

	Is Assigned (%)			
Sample	**To Group 1**	**To Group 2**	**To Group 3**	**To Group 4**
From group 1	99.99	0.01	0	0
From group 2	1.27	98.73	0	0
From group 3	0	0	99.97	0.03
From group 4	0	0	11.53	88.47

Note: Relative frequencies of the assignment of each object to one of the four groups are computed by the bootstrap technique.

test data. Since the true group membership is known for the test data, it is possible to count the number of misclassified objects for each group. Assigning objects to training and test data is often done randomly. In this example we build a training data set with the same number of objects as the original data set (180 samples) by drawing random samples with replication from the original data. Using random sampling with replication gives each object the same chance of being drawn again. Of course, the training set contains some samples more than once. One can show that around one third of the objects of the original data will not be used in the training set, and actually these objects will be taken for the test set. Although training and test sets were generated using random assignment, the results of LDA could be too optimistic or too pessimistic—just by chance. Therefore, the whole procedure is repeated 1000 times, resulting in 1000 pairs of training and test sets. This repeating procedure is called BOOTSTRAP (Section 4.2.6). For each of the 1000 pairs of data sets, LDA is applied for the training data and the prediction is made for the test data. For each object we can count how often it is assigned to one of the four groups. Afterwards, the relative frequencies of the group assignments are computed, and the average is taken in each of the four data groups. The resulting percentages of the group assignments are presented in Table 1.1. Except for group 4 (88.47% correct) the misclassification rates are very low. There is only a slight overlap between groups 1 and 2. Objects from group 4 are assigned to group 3 in 11.53% of the cases. The difficulty here is the small number of objects in groups 2–4, and using bootstrap it can happen that the small groups are even more underrepresented, leading to an unstable discriminant rule.

1.6 UNIVARIATE STATISTICS—A REMINDER

For the following sections assume a set of n data $x_1, x_2, \ldots, x_n$ that are considered as the components of a vector $\boldsymbol{x}$ (variable x in $\mathbb{R}$).

1.6.1 EMPIRICAL DISTRIBUTIONS

The distribution of data plays an important role in statistics; chemometrics is not so happy with this concept because the number of available data is often small or the

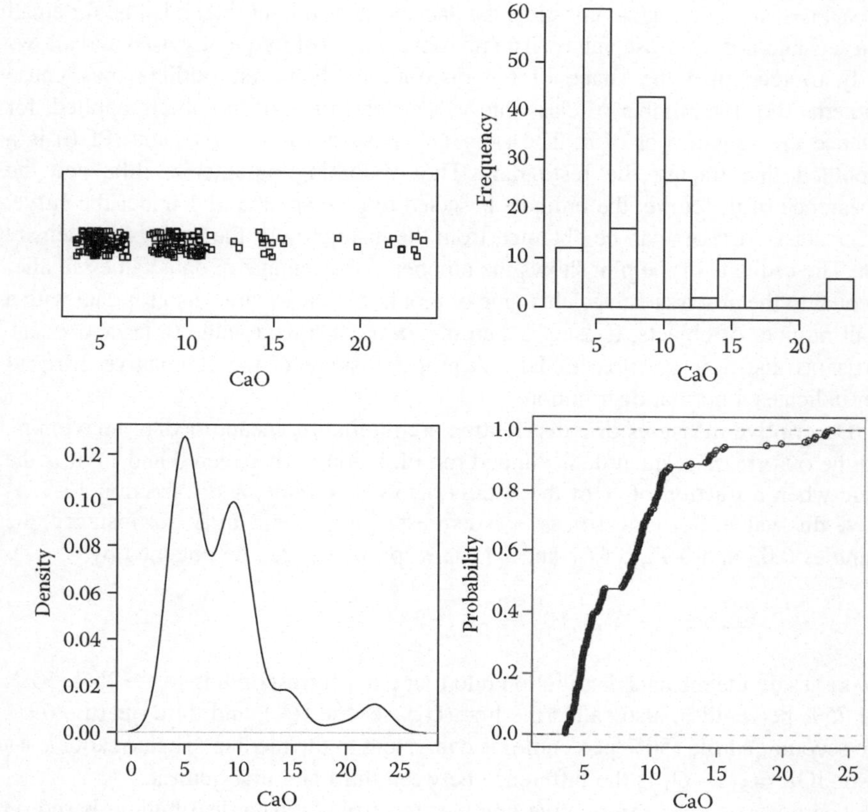

FIGURE 1.5 Graphical representations of the distribution of data: one-dimensional scatter plot (upper left), histogram (upper right), probability density plot (lower left), and ECDF (lower right). Data used are CaO concentrations (%) of 180 archaeological glass vessels.

type of a distribution is unknown. The values of a variable x (say the concentration of a chemical compound in a set with n samples) have an EMPIRICAL DISTRIBUTION; whenever possible it should be visually inspected to obtain a better insight into the data. A number of different plots can be used for this purpose as summarized in Figure 1.5. The corresponding commands in ℝ are as follows (using default parameters); the data set "glass" of the ℝ-package chemometrics contains the CaO contents (%) of $n = 180$ archaeological glass vessels (Janssen et al. 1998):

```
R:  library(chemometrics)                  # includes data
    data(glass)                            # glass data
    CaO <- glass[,"CaO"]                   # data for CaO
    stripchart(CaO, method="jitter")       # 1-dim. scatter plot
    hist(CaO)                              # histogram
    plot(density(CaO))                     # density trace
    boxplot(CaO)                           # boxplot
```

A ONE-DIMENSIONAL SCATTER PLOT plots the data along a straight line; it is fine for small data sets and not too close values. The HISTOGRAM is most frequently used and allows easily to recognize the shape of the distribution; however, outliers may cause problems. For the number of class intervals several rules of thumb are applied, for instance the square root of n. The PROBABILITY DENSITY FUNCTION/TRACE (PDF) is a smoothed line tracing the histogram. The smoothing parameters influence the appearance of the curve; the ordinate is scaled to give an area of 1 under the curve.

CUMULATIVE PLOTS can be obtained from the histogram or the probability density plot. The ordinate of the plot shows the number or percentage of data values smaller or equal to the abscissa value. This type of plot is well applicable also for data with a small number of objects, if using a step of $1/n$ at each data value (EMPIRICAL CUMULATIVE DISTRIBUTION FUNCTION, ECDF). A proper S-shape of the cumulative distribution indicates a normal distribution.

Descriptive measures of a distribution are minimum, mean, median, maximum, and the QUANTILES. A quantile is defined for a fraction $\propto$ (between 0 and 1); it is the value when a fraction of $\propto$ of the data is below this value, and a fraction $1-\propto$ is above this value. For PERCENTILES, α is expressed in percent. In $\mathbb{R}$, for instance, the quantiles 0.05 and 0.95, `q[1]` and `q[2]`, respectively, can be obtained by

```
R:  q <- quantile(x,c(0.05,0.95))
```

QUARTILES divide the data distribution into four parts corresponding to the 25%, 50%, and 75% percentiles, also called the first (Q_1), second (Q_2), and third quartile (Q_3). The second quartile (50% percentile) is equivalent to the median. The INTERQUARTILE RANGE $\text{IQR} = Q_3 - Q_1$ is the difference between third and first quartile.

The BOXPLOT is an informative graphics to display a data distribution, based on median and quartiles. The boxplot can be defined as follows (Frank and Todeschini 1994): The height of the box is given by the first and third quartile, and the mid line shows the median; the width of the box has usually no meaning. One whisker extends from the first quartile to the smallest data value in the interval Q_1 to $Q_1 - 1.5$ IQR and is called the lower whisker. The other whisker extends from the third quartile to the largest data value in the interval Q_3 to $Q_3 + 1.5$ IQR and is called the upper whisker. Outliers—not within the range $[Q_1 - 1.5 \text{ IQR}, Q_3 + 1.5 \text{ IQR}]$—are plotted as individual points. Figure 1.6 (left) shows a boxplot of the variable CaO content of the glass vessels data, together with explanations of its definition. Boxplots are especially powerful for comparing different data distributions. For this reason, boxplots of the single data sets are displayed side by side. Figure 1.6 (right) shows four parallel boxplots referring to the CaO contents of the four groups of glass vessels. It is clearly seen that the data values of groups 3 and 4 are nonoverlapping and different to those of the first two groups. The corresponding commands in $\mathbb{R}$ are

```
R:  library(chemometrics)      # includes data
    data(glass)                # glass data
    data(glass.grp)            # groups for glass data
    CaO <- glass[,"CaO"]       # data for CaO
    boxplot(CaO)               # boxplot
    boxplot(CaO ~ glass.grp)   # separate boxplots for groups
```

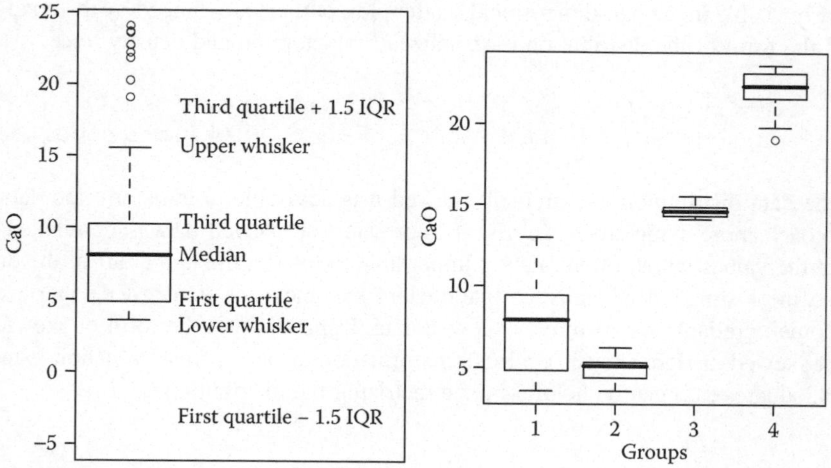

FIGURE 1.6 Boxplot of the CaO data with definitions and outlier boundaries (left) and parallel boxplots of the CaO data for four individual groups of glass vessels (right).

Some of the above plots can be combined in one graphical display, like one-dimensional scatter plot, histogram, probability density plot, and boxplot. Figure 1.7 shows this so-called EDAPLOT (exploratory data analysis plot) (Reimann et al. 2008). It provides deeper insight into the univariate data distribution: The single groups are

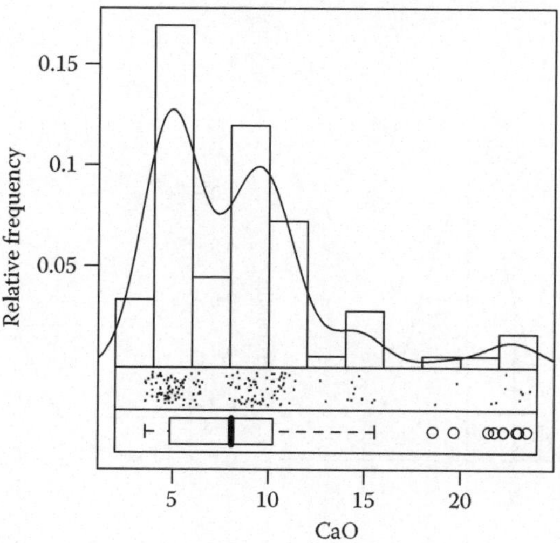

FIGURE 1.7 EDAPLOT of data combines one-dimensional scatter plot, histogram, probability density trace, and boxplot. Data used are CaO concentrations (%) of 180 archaeological glass vessels.

clearly visible in the one-dimensional scatter plot; outliers are flagged by the boxplot; and the form of the distribution is visualized by histogram and density trace.

```
R: library(StatDA)                    # load library StatDA
   edaplot(CaO, H.freq=FALSE)         # EDAPLOT (4 plots combined)
```

If the data distribution is extremely skewed it is advisable to transform the data to approach more symmetry. The visual impression of skewed data is dominated by extreme values which often make it impossible to inspect the main part of the data. Also the estimation of statistical parameters like mean or standard deviation can become unreliable for extremely skewed data. Depending on the form of skewness (left skewed or right skewed), a log-transformation or power transformation (square root, square, etc.) can be helpful in symmetrizing the distribution.

1.6.2 Theoretical Distributions

A few relevant mathematically defined distributions are summarized here together with tools for using them in ℝ.

The normal distribution, $N(\mu, \sigma^2)$, has a mean (expectation) μ and a standard deviation σ (variance σ^2). Figure 1.8 (left) shows the probability density function of the normal distribution $N(\mu, \sigma^2)$, and Figure 1.8 (right) the cumulative distribution function with the typical S-shape. A special case is the standard normal distribution, $N(0, 1)$, with $\mu = 0$ and standard deviation $\sigma = 1$. The normal distribution plays an important role in statistical testing.

Data values x following a normal distribution $N(\mu, \sigma^2)$ can be transformed to a standard normal distribution by the so-called z-transformation

$$z = (x - \mu)/\sigma \tag{1.2}$$

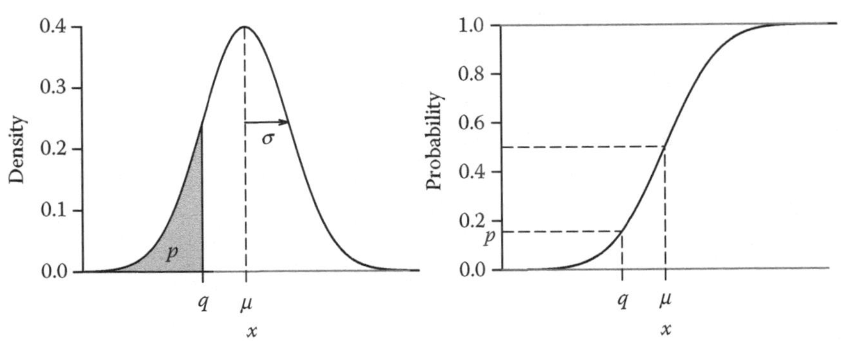

FIGURE 1.8 Probability density function (PDF) (left) and cumulative distribution function (right) of the normal distribution $N(\mu, \sigma^2)$ with mean μ and standard deviation σ. The quantile q defines a probability p.

The probability density, d, at value x is defined by

$$N(0,1):\ d(x) = \frac{1}{\sqrt{2\pi}} \exp\{-x^2/2\} \tag{1.3}$$

$$N(\mu,\sigma^2):\ d(x) = \frac{1}{\sigma\sqrt{2\pi}} \exp\left\{-\frac{(x-\mu)^2}{2\sigma^2}\right\} \tag{1.4}$$

```
R:  d <- dnorm(x,mean=0,sd=1)
    d <- dnorm(x,mean=mu,sd=sigma)
```

The area (probability, integral) under the probability density curve, p, from $-\infty$ to x can be calculated by

```
R:  p <- pnorm(x,mean=mu,sd=sigma)
```

The quantile, q, for a probability p (area from $-\infty$ to q) can be calculated by

```
R:  q <- qnorm(p,mean=mu,sd=sigma)
```

Figure 1.8 explains graphically how probabilities and quantiles are defined for a normal distribution. For instance the 1%-percentile ($p=0.01$) of the standard normal distribution is -2.326, and the 99%-percentile ($p=0.99$) is 2.326; both together define a 98% interval.

For simulation purposes n random numbers from a normal distribution $N(\mu,\sigma^2)$ can be generated as components of a vector $\boldsymbol{x}$ by

```
R:  x <- rnorm(n,mean=mu,sd=sigma)
```

Equivalent functions are provided in ℝ, for instance, for uniformly distributed data ("unif" instead of "norm" in the commands); n uniformly distributed data in the range a to b are generated by

```
R:  x <- runif(n,a,b)
```

A set of n integer numbers in the range `low` to `high` (including these values) is obtained by

```
R:  x <- round(runif(n,low-0.5,high+0.5),0)
```

Other important, mathematically defined distributions are the t-distribution (Figure 1.9, left), the chi-square distribution (Figure 1.9, right), and the F-distribution (Figure 1.10). These distributions are used in various statistical tests (Section 1.6.5); no details are given here, but only information about their use within ℝ. The form of these distributions depends on one or two parameters, called degrees of

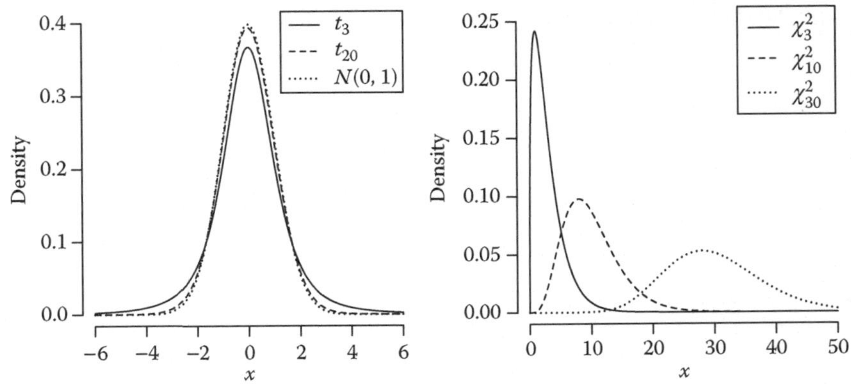

FIGURE 1.9 t-Distributions with 3 and 20 DF, respectively, and standard normal distribution corresponding to a t-distribution with $DF = \infty$ (left). Chi-square distribution with 3, 10, and 30 DF, respectively (right).

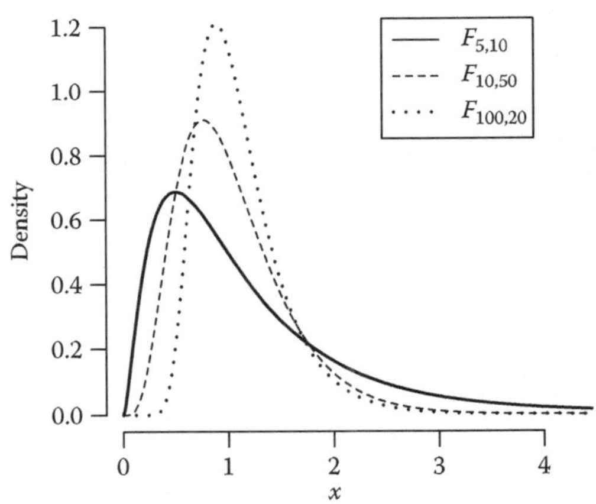

FIGURE 1.10 The form of the F-distribution is determined by two parameters.

freedom (DF) which are related to the number of data used in a statistical test. Quantiles for probabilities p of these distributions can be computed in $\mathbb{R}$ as follows:

t-Distribution:	$\mathbb{R}$: `q <- qt(p,df)`
Chi-square distribution:	$\mathbb{R}$: `q <- qchisq(p,df)`
F-Distribution:	$\mathbb{R}$: `q <- qf(p,df1,df2)`

For computing probabilities, the q in the names of the functions has to be replaced by p, for computing densities it has to be replaced by d, and for generating random numbers it has to be replaced by r (using the appropriate function parameters); see the help files in $\mathbb{R}$ for further information.

1.6.3 Central Value

Section 1.6.2 discussed some theoretical distributions which are defined by more or less complicated mathematical formulae; they aim at modeling real empirical data distributions or are used in statistical tests. There are some reasons to believe that phenomena observed in nature indeed follow such distributions. The normal distribution is the most widely used distribution in statistics, and it is fully determined by the mean value μ and the standard deviation σ. For practical data these two parameters have to be estimated using the data at hand. This section discusses some possibilities to estimate the mean or central value, and the next section mentions different estimators for the standard deviation or spread; the described criteria are listed in Table 1.2. The choice of the estimator depends mainly on the data quality. Do the data really follow the underlying hypothetical distribution? Or are there outliers or extreme values that could influence classical estimators and call for robust counterparts?

The classical and most used estimator for a central value is the ARITHMETIC MEAN $\bar{x}$ (x_mean in ℝ-notation). Throughout this book the term "mean" will be used for the arithmetic mean. For a normal or approximately normal distribution the mean is the best (most precise) central value.

$$\bar{x} = \frac{1}{n}\sum_{i=1}^{n} x_i \qquad \mathbb{R}:\quad \texttt{x\_mean <- mean(x)} \tag{1.5}$$

The GEOMETRIC MEAN $\bar{x}_G$ (x_gmean) could be used with advantage for right skewed distributions (e.g., lognormal distributions). However, the geometric mean is rarely applied and requires $x_i > 0$ for all i.

$$\bar{x}_G = \left(\prod_{i=1}^{n} x_i\right)^{1/n} \qquad \mathbb{R}:\quad \texttt{x\_gmean <- prod(x)\^{}(1/length(x))} \tag{1.6}$$

TABLE 1.2
Basic Statistical Measures of Data

Type of Measure	Measure	Symbol	Equation Number	Remark
Central value	Arithmetic mean	$\bar{x}$	1.5	
	Geometric mean	$\bar{x}_G$	1.6	$x_i > 0$
	Median	$\bar{x}_M$		Robust
	Huber M-estimator	$\bar{x}_{HUBER}$	1.7	Robust
Spread	Standard deviation	s	1.9	
	Coefficient of variation (relative standard deviation)	RSD	1.11	
	(Sample) variance	v, s^2	1.12	
	Interquartile range	IQR		Robust
	Standard deviation from IQR	s_{IQR}	1.8	Robust
	Median absolute deviation	MAD		Robust
	Standard deviation from MAD	s_{MAD}	1.10	Robust

A robust measure for the central value—much less influenced by outliers than the mean—is the MEDIAN $\bar{x}_M$ (x_median). The median divides the data distribution into two equal halves; the number of data higher than the median is equal to the number of data lower than the median. In the case that n is an even number, there exist two central values and the arithmetic mean of them is taken as the median. Because the median is solely based on the ordering of the data values, it is not affected by extreme values.

```
ℝ:  x_median <- median(x)
```

There are several other possibilities for robustly estimating the central value. Well known are M-estimators for location (Huber 1981). The basic idea is to use a function ψ that defines a weighting scheme for the objects. The M-estimator is then the solution of the implicit equation

$$\sum \psi(x_i - \bar{x}_{\text{HUBER}}) = 0 \tag{1.7}$$

and can be computed by

```
ℝ:  library(MASS)
    x_huber <- huber(x)$mu
```

Depending on the choice of the ψ-function, the influence of outlying objects is bounded, resulting in a robust estimator of the central value. Moreover, theoretical properties of the estimator can be computed (Maronna et al. 2006).

1.6.4 SPREAD

The RANGE of the data is defined as the difference between the maximum and minimum value; it is very sensitive to outliers which coincide with either the maximum or minimum or both. The robust counterpart of the range is the INTERQUARTILE RANGE (IQR, x_iqr).

```
ℝ:  x_iqr <- IQR(x)
```

IQR is the difference between the third and the first quartile, and thus is not influenced by up to 25% of the lowest and 25% of the largest data. In the case of a normal distribution the theoretical standard deviation σ can be estimated from IQR by

$$s_{\text{IQR}} = 0.7413 \cdot \text{IQR} \tag{1.8}$$

The classical and most used measure for the variation of data is the STANDARD DEVIATION s (x_sd in ℝ) which should not be mixed up with the theoretical standard deviation σ of the normal distribution.

$$s = \left(\frac{1}{n-1} \sum_{i=1}^{n} (x_i - \bar{x})^2 \right)^{1/2} \qquad \mathbb{R}: \quad \texttt{x\_sd <- sd(x)} \tag{1.9}$$

The standard deviation is very sensitive to outliers; if the data are skewed, not only the mean will be biased, but also s will be even more biased because squared deviations are used. In the case of normal or approximately normal distributions, s is the best measure of spread because it is the most precise estimator for σ; unfortunately in practice the classical standard deviation is often uncritically used instead of robust measures for the spread.

A robust counterpart to the standard deviation s is the MEDIAN ABSOLUTE DEVIATION (MAD, `x_mad`).

$\mathbb{R}$: `x_mad <- mad(x)`

MAD is based on the median $\bar{x}_M$ as central value; the absolute differences $|x_i - \bar{x}_M|$ are calculated and MAD is defined as the median of these differences. In the case of a normal distribution MAD can be used for a robust estimation of the theoretical standard deviation σ by

$$s_{MAD} = 1.483 \cdot \text{MAD} \tag{1.10}$$

The above measures of spread are expressed in the same unit as the data. If data with different units should be compared or the spread should be given in percent of the central value it is better to use a measure which is dimension free. Such a measure is the COEFFICIENT OF VARIATION or RELATIVE STANDARD DEVIATION (RSD, `x_rsd`). It is expressed in percent and defined by

$$\text{RSD} = 100 s/\bar{x} \qquad \mathbb{R}: \quad \texttt{x\_rsd <- 100 * sd(x)/mean(x)} \tag{1.11}$$

A robust coefficient of variation can be calculated by using s_{IQR} or s_{MAD} instead of s, and the median instead of the mean.

In statistics it is useful to work with the concept of squared deviations from a central value; the average of the squared deviations is denoted VARIANCE. Consequently, the unit of the variance is the squared data unit. The classical estimator of the variance is the SAMPLE VARIANCE (v, `x_var`), defined as the squared standard deviation.

$$v = s^2 = \frac{1}{n-1} \sum_{i=1}^{n} (x_i - \bar{x})^2 \qquad \mathbb{R}: \quad \texttt{x\_var <- var(x)} \tag{1.12}$$

Robust estimations of the variance are squared standard deviations as obtained from IQR or MAD, $(s_{IQR})^2$, and $(s_{MAD})^2$, respectively.

1.6.5 Statistical Tests

Statistical hypothesis testing requires the formulation of a so-called NULL HYPOTHESIS H_0 that should be tested, and an alternative hypothesis H_1 which expresses the alternative. In most cases there are several alternatives, but the alternative to test has to be fixed. For example, if two distributions have to be tested for equality of the means, the alternative could be unequal means, or that one mean is smaller/larger than the other one. For simplicity we will only state the null hypothesis in this overview below but not the alternative hypothesis. For the example of testing for equality of the means of two random samples `x1` and `x2` the ℝ command for the two-sample t-test is

```
R:  t.test(x1,x2,alternative="two.sided")
```

For the parameter `alternative` the default is `"two.sided"`, and thus it does not need to be provided. However, one could also take `"less"` or `"greater"` as alternatives. It is recommended to look at the help pages of the functions for the detailed definition of the parameters.

The outcome of a test is the so-called p-value, a probability which directly helps with the decision: If the p-value is larger than (or equal to) a predefined significance level (5%), the test is said to be "not significant," and thus the null hypothesis H_0 cannot be rejected. On the other hand, if the p-value is smaller than the significance level, H_0 has to be rejected, and it is said the test "is significant." The significance level has to be determined in advance; it is often taken as 5%. That means in 5% of the cases we will reject H_0 although it is valid (type 1 error, probability of incorrectly rejecting H_0). On the other hand we make a type 2 error if we accept H_0 although the alternative hypothesis is valid. Both sources of error should be kept as small as possible, but decreasing one error source usually means to increase the other one.

Coming back to the example of the two-sample t-test and ℝ, all results can be saved in an ℝ object, for instance, named `result` from which particular results can be extracted as follows:

```
R:  result <- t.test(x1,x2,alternative="two.sided")
    p <- result$p.value  # p-value of the test
```

All test results that are saved in `result` as list elements can either be seen in the help file or can be listed with

```
R:  str(result)  # gives the detailed structure of the
                 # object result
```

Like other statistical methods, the user has to be careful with the requirements of a statistical test. For many statistical tests the data have to follow a normal distribution. If this data requirement is not fulfilled, the outcome of the test can be biased and misleading. A possible solution to this problem are nonparametric tests that are much less restrictive with respect to the data distribution. There is a rich literature on statistical testing; a basic reference is Lehmann (1959). Nonparametric tests are treated in Hollander and Wolfe (1973) and Conover (1998). A more recent book on testing is Kanji (1999).

Some tests that are widely used in univariate statistics are listed here together with hints for their use within ℝ and the necessary requirements, but without any mathematical treatment. In multivariate statistics these tests are rarely applied to single variables but often to latent variables; for instance, a discriminant variable can be defined via a two-sample *t*-test.

Tests for the Distribution

SHAPIRO–WILK TEST. H_0: the data distribution follows a normal distribution.

ℝ: `shapiro.test(x)`

KOLMOGOROV-SMIRNOV TEST. H_0: the data distribution follows a given hypothetical distribution. For instance, for a hypothetical normal distribution $N(\mu = 10, \sigma^2 = 2^2)$ the test can be performed by

ℝ: `ks.test(x,"pnorm",10,2)`

A test for a hypothetical uniform distribution with the limits `low` and `high` (inclusive) can be performed by

ℝ: `ks.test(x,"punif",low,high)`

Tests for the Central Value

ONE-SAMPLE T-TEST. H_0: the central value of the data distribution is equal to $\mu = m$. Requirements: normal distribution, independent samples.

ℝ: `t.test(x,mu=m)`

WILCOXON SIGNED-RANK TEST (nonparametric). H_0: the data distribution is symmetric around the median. Requirements: continuous, symmetric distribution.

ℝ: `wilcox.test(x)`

Tests for Two Central Values of Independent Samples

TWO-SAMPLE T-TEST. H_0: the central values of the two data distributions are equal. Requirements: normal distribution of both data sets, independent samples.

ℝ: `t.test(x1,x2)`

WILCOXON RANK SUM TEST (=MANN–WHITNEY *U*- TEST) (nonparametric). H_0: the two data distributions are equal; in the sense the medians are equal. Requirements: independent samples, continuous distributions.

ℝ: `wilcox.test(x1,x2)`

Tests for Two Central Values of Dependent Pairs of Samples

PAIRED T-TEST. H_0: the distribution of the differences of the pairs of both data distributions has a central value of zero. Requirements: normal distribution of both data sets, independent sample pairs.

ℝ: `t.test(x1,x2,paired=TRUE)`

WILCOXON TEST (nonparametric). H_0: the median of the differences of the pairs of samples is zero. Requirements: independent samples pairs, continuous and symmetric distributions.

ℝ: `wilcox.test(x1,x2,paired=TRUE)`

Tests for Several Central Values of Independent Samples

(ONE-WAY) ANOVA. H_0: the central values of the data distributions are equal. Requirements: normal distribution of all distributions, independent samples, equal variances.

ℝ: `anova(lm(x~group))`

Here `x` includes the data and `group` is a grouping variable.

If, for example, the data of three groups are stored in vectors `x1`, `x2`, and `x3`, then `x<-c(x1,x2,x3)` combines all data in a vector `x`. The grouping variable `group` can be generated by

```
group <-
c(rep(1,length(x1)),rep(2,length(x2)),rep(3,length(x3)))
```

and will contain the numbers 1, 2, 3 indicating the group memberships.

KRUSKAL-WALLIS TEST (nonparametric). H_0: central values of the data distributions are equal. Requirements: independent samples, continuous distributions.

ℝ: `kruskal.test(x,group)`

Tests for Two Variances of Independent Samples

F-TEST. H_0: the variances of two data distributions are equal. Requirements: normal distribution of both data sets, independent samples.

ℝ: `var.test(x1,x2)`

ANSARI–BRADLEY TEST (nonparametric). H_0: the variances of two data distributions are equal. Requirements: independent samples.

ℝ: `ansari.test(x1,x2)`

Tests for Several Variances of Independent Samples

BARTLETT-TEST. H_0: the variances of the data distributions are equal. Requirements: normal distribution of all data sets, independent samples.

ℝ: `bartlett.test(x ~ group)`

FLIGNER TEST (nonparametric). H_0: the variances of the data distributions are equal. Requirements: independent samples.

ℝ: `fligner.test(x ~ group)`

REFERENCES

Adams, M. J.: *Chemometrics in Analytical Spectroscopy*. The Royal Society of Chemistry, Cambridge, United Kingdom, 1995.

Beebe, K. R., Pell, R. J., Seasholtz, M. B.: *Chemometrics: A Practical Guide*. Wiley, New York, 1998.

Brereton, R. G.: *Chemometrics—Applications of Mathematics and Statistics to Laboratory Systems*. Ellis Horwood, New York, 1990.

Brereton, R. G. (Ed.): *Multivariate Pattern Recognition in Chemometrics, Illustrated by Case Studies*. Elsevier, Amsterdam, the Netherlands, 1992.

Brereton, R. G.: *Chemometrics—Data Analysis for the Laboratory and Chemical Plant*. Wiley, Chichester, United Kingdom, 2006.

Brereton, R. G.: *Applied Chemometrics for Scientists*. Wiley, Chichester, United Kingdom, 2007.

Burgard, D. R., Kuznicki, J. T.: *Chemometrics: Chemical and Sensory Data*. CRC Press, Boca Raton, FL, 1990.

Buydens, L. C. M., Melssen, W. J. (Eds.): *Chemometrics—Exploring and Exploiting Chemical Information*. Catholic University of Nijmegen, Nijmegen, the Netherlands, 1994.

Carlson, R.: *Design and Optimization in Organic Synthesis*. Elsevier, Amsterdam, the Netherlands, 1992.

Chau, F. T., Liang, Y. Z., Gao, J., Shao, X. G.: *Chemometrics—From Basics to Wavelet Transform*. Wiley, Hoboken, NJ, 2004.

Clerc, J. T., Naegeli, P., Seibl, J.: *Chimia* 27, 1973, 639–639. Artificial intelligence.

Conover, W. J.: *Practical Nonparametric Statistics*. Wiley, New York, 1998.

Crawford, L. R., Morrison, J. D.: *Anal. Chem.* 40, 1968, 1469–1474. Computer methods in analytical mass spectrometry. Empirical identification of molecular class.

Crawford, L. R., Morrison, J. D.: *Anal. Chem.* 43, 1971, 1790–1795. Computer methods in analytical mass spectrometry. Development of programs for analysis of low resolution mass spectra.

Dalgaard, P.: *Introductory Statistics with R*. Springer, New York, 2002.

Danzer, K., Hobert, H., Fischbacher, C., Jagemann, K.U.: *Chemometrik—Grundlagen und Anwendungen*. Springer, Berlin, Germany, 2001.

Deming, S. N., Morgan, S. L.: *Experimental Design: A Chemometric Approach*. Elsevier, Amsterdam, the Netherlands, 1987.

Duda, R. O., Hart, P. E.: *Pattern Classification and Scene Analysis*. Wiley, New York, 1973.

Efron, B., Tibshirani, R. J.: *An Introduction to the Bootstrap*. Chapman & Hall, London, United Kingdom, 1993.

Einax, J. W., Zwanziger, H. W., Geiss, S.: *Chemometrics in Environmental Analysis*. VCH-Wiley, Weinheim, Germany, 1997.

Eriksson, L., Antti, H., Holmes, E., Johansson, E., Lundstedt, T., Shockcor, J., and Wold, S.: in Gasteiger, J. (Ed.), *Handbook of Chemoinformatics*, Vol. 3, Wiley-VCH, Weinheim, Germany, 2003, pp. 1134–1166. Partial least squares (PLS) in cheminformatics.

Esbensen, K., Geladi, P.: *J. Chemom.* 4, 1990, 389–412. The start and early history of chemometrics: Selected interviews. Part 2.

Flury, B., Riedwyl, H.: *Multivariate Statistics: A Practical Approach*. Chapman & Hall, Boca Raton, FL, 1988.

Fox, J.: *Applied Regression Analysis, Linear Models, and Related Methods*. Sage Publications, Thousand Oaks, CA, 1997.

Fox, J.: *An R and S-Plus Companion to Applied Regression*. Sage Publications, Thousand Oaks, CA, 2002.

Frank, I. E., Todeschini, R.: *The Data Analysis Handbook*. Elsevier, Amsterdam, the Netherlands, 1994.

Fukunaga, K.: *Introduction to Statistical Pattern Recognition*. Academic Press, New York, 1972.

Gasteiger, J. (Ed.), *Handbook of Chemoinformatics—From Data to Knowledge (4 volumes)*. Wiley-VCH, Weinheim, Germany, 2003.

Gasteiger, J., Engel, T.: *Chemoinformatics—A Textbook*. Wiley-VCH, Weinheim, Germany, 2003.

Geladi, P., Esbensen, K.: *J. Chemom.* 4, 1990, 337–354. The start and early history of chemometrics: Selected interviews. Part 1.

Harper, A. M., Duewer, D. L., Kowalski, B. R., Fasching, J. L.: in Kowalski, B. R. (Ed.), *Chemometrics: Theory and Application*, ACS Symp. Ser. Vol. 52, American Chemical Society, Washington, DC, 1977, pp. 14–52. ARTHUR, an experimental data analysis: the heuristic use of a polyalgorithm.

Hastie, T., Tibshirani, R. J., Friedman, J.: *The Elements of Statistical Learning*. Springer, New York, 2001.

Henrion, R., Henrion, G.: *Multivariate Datenanalyse*. Springer, Berlin, Germany, 1995.

Hoeskuldsson, A.: *Prediction Methods in Science and Technology*. Thor Publishing, Holte, Denmark, 1996.

Hollander, M., Wolfe, D. A.: *Nonparametric Statistical Inference*. Wiley, New York, 1973.

Huber, P. J.: *Robust Statistics*. Wiley, New York, 1981.

Isenhour, T. L., Jurs, P. C.: in Mark, H. B. Jr., Mattson, J. S., Macdonald, H. C. Jr. (Eds.), *Computer Fundamentals for Chemists*, Marcel Dekker, New York, 1973, pp. 285–330. Learning machines.

Jackson, J. E.: *A User's Guide to Principal Components*. Wiley, New York, 2003.

Janssen, K. H. A., De Raedt, I., Schalm, O., Veeckman, J.: *Microchim. Acta* 15(suppl.), 1998, 253–267. Compositions of 15th–17th century archaeological glass vessels excavated in Antwerp.

Johnson, R. A., Wichern, D. W.: *Applied Multivariate Statistical Analysis*. Prentice Hall, Upper Saddle River, NJ, 2002.

Joliffe, I. T.: *Principal Component Analysis*. Springer Verlag, New York, 1986.

Jonker, G. (Ed.): *Chemometrics Tutorials*. Elsevier, Amsterdam, the Netherlands, 1990.

Jurs, P. C., Isenhour, T. L.: *Chemical Applications of Pattern Recognition*. Wiley, New York, 1975.

Jurs, P. C., Kowalski, B. R., Isenhour, T. L.: *Anal. Chem.* 41, 1969a, 21–27. Computerized learning machines applied to chemical problems. Molecular formula determination from low resolution mass spectrometry.

Jurs, P. C., Kowalski, B. R., Isenhour, T. L., Reilley, C. N.: *Anal. Chem.* 41, 1969b, 690–695. Computerized learning machines applied to chemical problems. Convergence rate and predictive ability of adaptive binary pattern classifiers.

Jurs, P. C., Kowalski, B. R., Isenhour, T. L., Reilley, C. N.: *Anal. Chem.* 41, 1969c, 1949–1959. Investigation of combined patterns from diverse analytical data using computerized learning machines.

Kanji, G. K.: *100 Statistical Tests*. Sage Publications, London, United Kingdom, 1999.

Kaufmann, L., Rousseeuw, P. J.: *Finding Groups of Data*. Wiley, New York, 1990.

Kemsley, E. K.: *Discriminant Analysis and Class Modelling of Spectroscopic Data*. Wiley, Chichester, United Kingdom, 1998.

Kessler, W.: *Multivariate Datenanalyse für die Pharma-, Bio- und Prozessanalytik*. Wiley-VCH, Weinheim, Germany, 2007.

Kohonen, T.: *Self-Organizing Maps*. Springer, Berlin, Germany, 1995.

Kowalski, B. R.: *J. Chem. Inf. Comput. Sci.* 15, 1975, 201–203. Chemometrics: Views and propositions.

Kowalski, B. R. (Ed.): *Chemometrics: Theory and Application*. American Chemical Society, Washington, DC, 1977.

Kowalski, B. R. (Ed.): *Chemometrics—Mathematics and Statistics in Chemistry*. D. Reidel, Dordrecht, the Netherlands, 1983.

Kowalski, B. R., Bender, C. F.: *J. Am. Chem. Soc.* 94, 1972, 5632–5639. Pattern recognition. A powerful approach to interpreting chemical data.

Kowalski, B. R., Brown, S., Vandeginste, B. G. M.: *J. Chemom.* 1, 1987, 1–2. Editorial (starting the *Journal of Chemometrics*).

Kowalski, B. R., Jurs, P. C., Isenhour, T. L., Reilley, C. N.: *Anal. Chem.* 41, 1969a, 1945–1959. Computerized learning machines applied to chemical problems. Interpretation of infrared spectrometry data.

Kowalski, B. R., Jurs, P. C., Isenhour, T. L., Reilley, C. N.: *Anal. Chem.* 41, 1969b, 690–700. Computerized learning machines applied to chemical problems. Multicategory pattern classification by least squares.

Kramer, R.: *Chemometric Techniques for Quantitative Analysis*. Marcel Dekker, New York, 1998.

Lachenbruch, P. A.: *Discriminant Analysis*. Hafner Press, New York, 1975.

Lehmann, E. L.: *Testing Statistical Hypotheses*. Wiley, New York, 1959.

Liddell, R. W., Jurs, P. C.: *Anal. Chem.* 46, 1974, 2126–2130. Interpretation of infrared spectra using pattern recognition techniques.

Lindberg, W., Persson, J. A., Wold, S.: *Anal. Chem.* 55, 1983, 643–648. Partial least-squares method for spectrofluorimetric analysis of mixtures of humic acid and ligninsulfonate.

Looney, C. G.: *Pattern Recognition Using Neural Networks*. Oxford University Press, New York, 1997.

Mager, P. P.: *Multivariate Chemometrics in QSAR (Quantitative Structure-Activity Relationships)—A Dialogue*. Research Studies Press, Letchworth, United Kingdom, 1988.

Malinowski, E. R.: *Factor Analysis in Chemistry*. Wiley, New York, 2002.

Malinowski, E. R., Howery, D. G.: *Factor Analysis in Chemistry*. Wiley, New York, 1980.

Maronna, R., Martin, D., Yohai, V.: *Robust Statistics: Theory and Methods*. Wiley, Toronto, ON, Canada, 2006.

Martens, H., Martens, M.: *Multivariate Analysis of Quality—An Introduction*. Wiley, Chichester, United Kingdom, 2000.

Martens, H., Naes, T.: *Multivariate Calibration*. Wiley, Chichester, United Kingdom, 1989.

Massart, D. L., Dijkstra, A., Kaufmann, L.: *Evaluation and Optimization of Laboratory Methods and Analytical Procedures*. Elsevier, Amsterdam, the Netherlands, 1978.

Massart, D. L., Kaufmann, L.: *The Interpretation of Analytical Chemical Data by the Use of Cluster Analysis*. Wiley, New York, 1983.

Massart, D. L., Vandeginste, B. G. M., Buydens, L. C. M., De Jong, S., Smeyers-Verbeke, J.: *Handbook of Chemometrics and Qualimetrics: Part A*. Elsevier, Amsterdam, the Netherlands, 1997.

Massart, D. L., Vandeginste, B. G. M., Deming, S. N., Michotte, Y., Kaufmann, L.: *Chemometrics: A Textbook*. Elsevier, Amsterdam, the Netherlands, 1988.

Meisel, W. S.: *Computer-Oriented Approaches to Pattern Recognition*. Academic Press, New York, 1972.

Miller, J. N., Miller, J. C.: *Statistics and Chemometrics for Analytical Chemistry*. Pearson Education Ltd, Harlow, United Kingdom, 2000.

MobyDigs: Software. Talete srl, www.talete.it, Milan, Italy, 2004.

Naes, T., Isaksson, T., Fearn, T., Davies, T.: *A User-Friendly Guide to Multivariate Calibration and Classification*. NIR Publications, Chichester, United Kingdom, 2004.

Nilsson, N. J.: *Learning Machines*. McGraw Hill, New York, 1965.

Otto, M.: *Chemometrie*. VCH-Wiley, Weinheim, Germany, 1997.

Otto, M.: *Chemometrics—Statistics and Computer Application in Analytical Chemistry*. Wiley-VCH, Weinheim, Germany, 2007.

Pomerantsev, A. L. (Ed.): *Progress in Chemometrics Research*. Nova Science Publishers, New York, 2005.

Preuss, D. R., Jurs, P. C.: *Anal. Chem.* 46, 1974, 520–525. Pattern recognition techniques applied to the interpretation of infrared spectra.

Reimann, C., Filzmoser, P., Garrett, R. G., Dutter, R.: *Statistical Data Analysis Explained. Applied Environmental Statistics with R*. Wiley, Chichester, United Kingdom, 2008.

Ripley, B. D.: *Pattern Recognition and Neural Networks*. Cambridge University Press, Cambridge, 1996.

Rose, J. R.: in Schleyer, P. v. R., Allinger, N. L., Clark, T., Gasteiger, J., Kollman, P. A., Schaefer, I. H. F., Schreiner, P. R. (Eds.), *The Encyclopedia of Computational Chemistry*, Vol. 3, Wiley, Chichester, United Kingdom, 1998, pp. 1082–1097. Inductive learning methods.

Rousseeuw, P. J., Leroy, A. M.: *Robust Regression and Outlier Detection*. Wiley, New York, 1987.

Sharaf, M. A., Illman, D. L., Kowalski, B. R.: *Chemometrics*. Wiley, New York, 1986.

Smilde, A., Bro, R., Geladi, P.: *Multi-Way Analysis with Applications in the Chemical Sciences*. Wiley, Chichester, United Kingdom, 2004.

Tenenhaus, M.: *La Regression PLS. Theorie et Practique*. Editions Technip, Paris, France, 1998.

Ting, K. L. H., Lee, R. C. T., Milne, G. W. A., Guarino, A. M.: *Science* 180, 1973, 417–420. Applications of artificial intelligence: Relationships between mass spectra and pharmacological activity of drugs.

Todeschini, R., Consonni, V.: *Handbook of Molecular Descriptors*. Wiley-VCH, Weinheim, Germany, 2000.

Tou, J. T., Gonzalez, R. C.: *Pattern Recognition Principles*. Addison-Wesley, London, United Kingdom, 1974.

Unscrambler: Software. Camo Process AS, www.camo.no, Oslo, Norway, 2004.

van de Waterbeemd, H. (Ed.): *Chemometric Methods in Molecular Design*. VCH, Weinheim, Germany, 1995.

Vandeginste, B. G. M., Massart, D. L., Buydens, L. C. M., De Jong, S., Smeyers-Verbeke, J.: *Handbook of Chemometrics and Qualimetrics: Part B*. Elsevier, Amsterdam, the Netherlands, 1998.

Varmuza, K.: *Pattern Recognition in Chemistry*. Springer, Berlin, Germany, 1980.

Varmuza, K.: in Gasteiger, J. (Ed.), *Handbook of Chemoinformatics*, Vol. 3, Wiley-VCH, Weinheim, Germany, 2003, pp. 1098–1133. Multivariate data analysis in chemistry.

Venables, W. N., Ripley, B. D.: *Modern Applied Statistics with S*. Springer, New York, 2003.

von Homeyer, A.: in Gasteiger, J. (Ed.), *Handbook of Chemoinformatics*, Vol. 3, Wiley-VCH, Weinheim, Germany, 2003, pp. 1239–1280. Evolutionary algorithms and their applications.

Wangen, L. E., Frew, B. M., Isenhour, T. L., Jurs, P. C.: *Appl. Spectros*. 25, 1971, 203–207. Investigation of the fourier transform for analyzing spectroscopic data by computerized learning machines.

Watanabe, S. (Ed.): *Methodologies of Pattern Recognition*. Academic Press, New York, 1969.

Willett, P.: *Similarity and Clustering in Chemical Information Systems*. Research Studies Press, Letchworth, United Kingdom, 1987.

Wold, S.: *Kemisk Tidskrift* 3, 1972, 34–37. Spline functions, a new tool in data analysis.

Wold, S.: *Svensk Naturvetenskap* 201, 1974, 206. Chemometrics-chemistry and applied mathematics.

Wold, S., Ruhe, A., Wold, H., Dunn, W. J. I.: *SIAM J. Sci. Stat. Comput*. 5, 1984, 735–743. The collinearity problem in linear regression. The partial least squares approach to generalized inverses.

Wold, S., Sjöström, M.: in Kowalski, B. R. (Ed.), *Chemometrics: Theory and Application*, ACS Symp. Ser. 52, Vol. 52, American Chemical Society, Washington, DC, 1977, pp. 243–282. SIMCA: A method for analyzing chemical data in terms of similarity and analogy.

Wolff, D. D., Parsons, M. L.: *Pattern Recognition Approach to Data Interpretation*. Plenum Press, New York, 1983.

Zupan, J.: *Algorithms for Chemists*. Wiley, Chichester, United Kingdom, 1989.

Zupan, J.: in Gasteiger, J. (Ed.), *Handbook of Chemoinformatics*, Vol. 3, Wiley-VCH, Weinheim, Germany, 2003, pp. 1167–1215. Neural networks.

Zupan, J., Gasteiger, J.: *Neural Networks in Chemistry and Drug Design*. Wiley-VCH, Weinheim, Germany, 1999.

2 Multivariate Data

2.1 DEFINITIONS

A simple form of multivariate data is a rectangular table (matrix, spreadsheet) consisting of n rows, m columns, and each cell containing a numerical value. Each row corresponds to an OBJECT, for instance a sample; each column corresponds to a particular FEATURE of the objects (VARIABLE, for instance a measurement on the objects). We call these data the matrix $\boldsymbol{X}$, with element x_{ij} in row i and column j. A column vector $\boldsymbol{x}_j$ contains the values of variable j for all objects; a row vector, $\boldsymbol{x}_i^{\mathrm{T}}$, is a transposed vector and contains all features for object i (Table 2.1).

Geometrically, each object can be considered as a point in an m-dimensional space (variable space, feature space) with the coordinates given by the m variables. Human imagination and drawing of scatter plot of course allow only two or three dimensions. However, many methods in multivariate data analysis can be understood easily by a geometric representation of a two-dimensional $\boldsymbol{X}$-matrix (Figures 2.1 and 2.2), and this approach will be used as far as possible throughout this book. An important concept is to consider the distance between object points (data points) as a measure of similarity of the objects. Based on the distances, clusters of objects and outliers can be detected. In the scatter plot in Figure 2.2, we may detect two clusters (groups) of objects and one outlier not belonging to any of the two clusters.

Matrix data are frequent in science and technology. Typical EXAMPLES IN CHEMISTRY comprise: samples are characterized by the concentrations of chemical compounds or elements (determined for instance by chromatography or atom absorption spectroscopy), or samples are characterized by spectroscopic data (the columns are often infrared absorptions). Typical matrix size is 20–1000 objects and 5–500 variables. With only an $\boldsymbol{X}$-matrix available and not using other information about the objects, the aim of data analysis is often a cluster analysis; this means a search for groups of similar objects (forming clusters) and a search for outlier objects, but also a search for similar (correlating) variables. The most important method for this purpose is principal component analysis (PCA), which allows a visual inspection of the clustering of objects (or variables); other important (nonlinear) methods are hierarchical cluster analysis (dendrograms) and Kohonen maps. This type of data evaluation is addressed by the terms EXPLORATORY DATA ANALYSIS or UNSUPERVISED LEARNING.

Additionally to the x-data, a PROPERTY y may be known for each object (Figure 2.3). The property can be a continuous number, such as the concentration of a compound, or a chemical/physical/biological property, but may also be a discrete number that encodes a class membership of the objects. The properties are usually the interesting facts of the objects, but often they cannot be determined directly or only with high cost; on the other hand, the x-data are often easily available. Methods from

TABLE 2.1
Nomenclature in Chemometrics and Statistics

This Book	Synonyms
Data set (multivariate)	Data matrix, random sample (of observations)
Object	Sample, observation, case, compound
Variable	Feature, measurement, descriptor, parameter
Data value	Matrix element

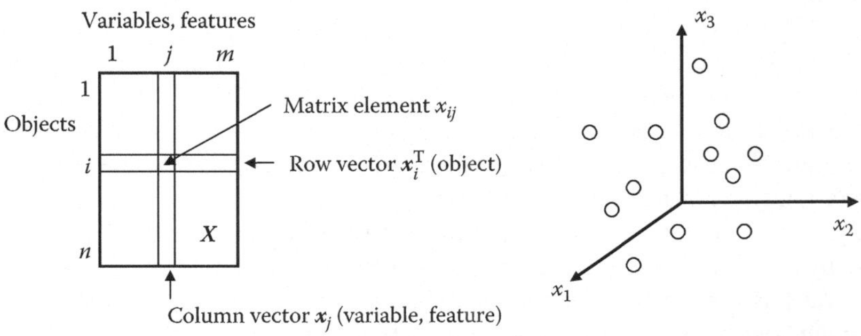

FIGURE 2.1 Simple multivariate data: matrix X with n rows (objects) and m columns (variables, features). An example (right) with $m = 3$ variables shows each object as a point in a three-dimensional coordinate system.

x_1	x_2
1.0	1.0
1.5	3.0
2.0	1.5
3.0	2.5
4.5	5.5
5.0	6.5
5.5	4.5
6.0	6.0
6.5	2.0
7.0	5.0

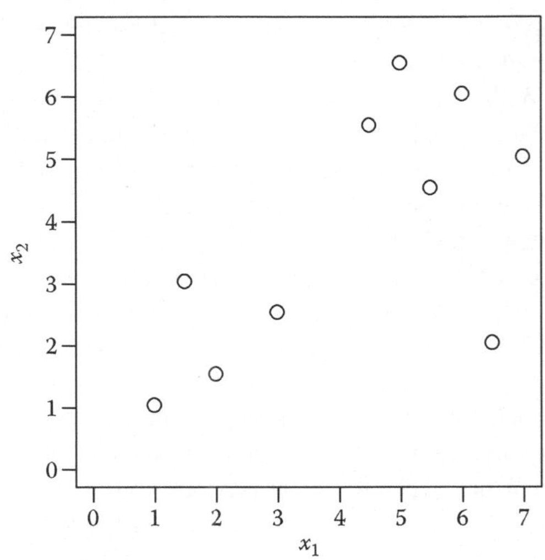

FIGURE 2.2 Data matrix and scatter plot: Geometrical representation of objects and variables defined by a two-dimensional X-matrix with 10 objects.

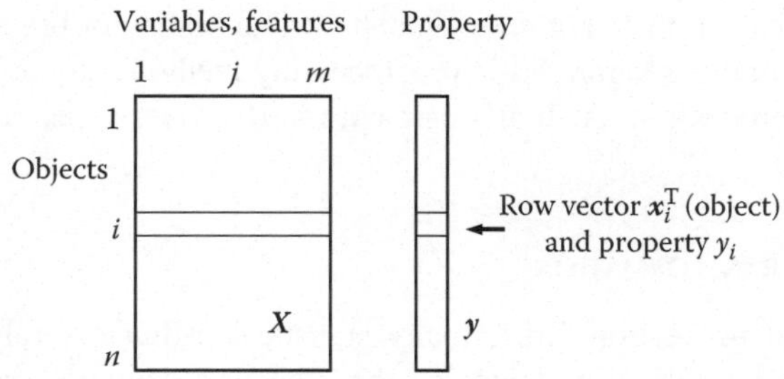

FIGURE 2.3 Variable (feature) matrix ***X*** and a property vector ***y***. The property may be a continuous number (a physical, chemical, biological, or technological property), as well as a discrete number or categorical variable defining a class membership of the objects.

MULTIVARIATE CALIBRATION are applied to develop mathematical models that allow to predict a continuous y from variables $x_1, \ldots, x_m$. Typical examples in chemistry are the quantitative analysis of compounds in complex mixtures (without isolation of the interesting compounds and often by using near infrared data), or a prediction of a chemical/physical/biological property from chemical structures (that can be characterized by numerical molecular descriptors). MULTIVARIATE CLASSIFICATION methods are for instance applied to determine the origin of samples (based on chemical–analytical data) or for the recognition of a chemical structure class of organic compounds (based on spectroscopic data). Widely used methods in chemometrics are partial least-squares (PLS) regression, and principal component regression (PCR) for calibration, as well as PLS, linear discriminant analysis (LDA), and *k*-nearest neighbor classification (*k*-NN) for classification. Application of artificial neural networks (ANN) is a nonlinear approach for these problems. Because modeling and prediction of *y*-data is a defined aim of data analysis, this type of data evaluation is called SUPERVISED LEARNING.

If more than one property is relevant, then we have an ***X***-matrix and a corresponding ***Y***-matrix. If the properties are highly correlated, a combined treatment of all properties is advisable, otherwise each property can be handled separately as described above. Mostly used for a joined evaluation of ***X*** and ***Y*** is PLS (then sometimes called PLS2); a nonlinear method is a Kohonen counter propagation network.

More complex than vectors or matrices (***X***, ***X*** and ***y***, ***X*** and ***Y***) are three-way data or MULTIWAY DATA (Smilde et al. 2004). Univariate data can be considered as one-way data (one measurement per sample, a vector of numbers); two-way data are obtained for instance by measuring a spectrum for each sample (matrix, two-dimensional array, "classical" multivariate data analysis); three-way data are obtained by measuring a spectrum under several conditions for each sample (a matrix for each sample, three-dimensional array). This concept can be generalized to multiway data.

2.2 BASIC PREPROCESSING

The used variables may originate from the same source (for instance data from a single spectroscopic method), but may also have very different origins and magnitudes. At least in the latter case, an appropriate scaling or preprocessing of the data is required. In this section, basic preprocessing methods are described; application-specific preprocessing, such as treatment of IR or mass spectra, is included in Chapter 7.

2.2.1 Data Transformation

As already noted in Section 1.6.1, many statistical estimators rely on symmetry of the data distribution. For example, the standard deviation can be severely increased if the data distribution is much skewed. It is thus often highly recommended to first transform the data to approach a better symmetry. Unfortunately, this has to be done for each variable separately, because it is not sure if one and the same transformation will be useful for symmetrizing different variables. For right-skewed data, the LOG-TRANSFORMATION is often useful (that means taking the logarithm of the data values). More flexible is the POWER TRANSFORMATION which uses a power p to transform values x into x^p. The value of p has to be optimized for each variable; any real number is reasonable for p, except $p = 0$ where a log-transformation has to be taken. A slightly modified version of the power transformation is the BOX–COX TRANSFORMATION, defined as

$$x_{\text{BOX-COX}} = \begin{cases} (x^p - 1)/p & \text{for } p \neq 0 \\ \log(x) & \text{for } p = 0 \end{cases} \tag{2.1}$$

The optimal parameter p can be found by maximum-likelihood estimation, but even the optimal p will not guarantee that the Box–Cox transformed values are symmetric. Note that all these transformations are only defined for positive data values. In case of negative values, a constant has to be added to make them positive. Within R, the Box–Cox transformation can be performed to data of a vector $\boldsymbol{x}$ as follows:

```
R:  library(geoR)                    # load library geoR
    p <- boxcox.fit(x)$lambda        # determine optimal p
    library(car)                     # load library car
    x_boxcox <- box.cox(x,p)         # Box-Cox transformation
```

Especially for data which are proportions in the range of 0–1, the LOGIT TRANSFORMATION can be useful to approach a normal distribution. It is defined for a data value x as

$$x_{\text{LOGIT}} = \frac{1}{2}\log\left(\frac{x}{1-x}\right) \tag{2.2}$$

```
R:  x_logit <- 0.5*log(x/(1-x))
```

Figure 2.4 shows the effect of the logit transformation on a uniformly distributed variable x. The left figure is the density function of the uniform distribution in

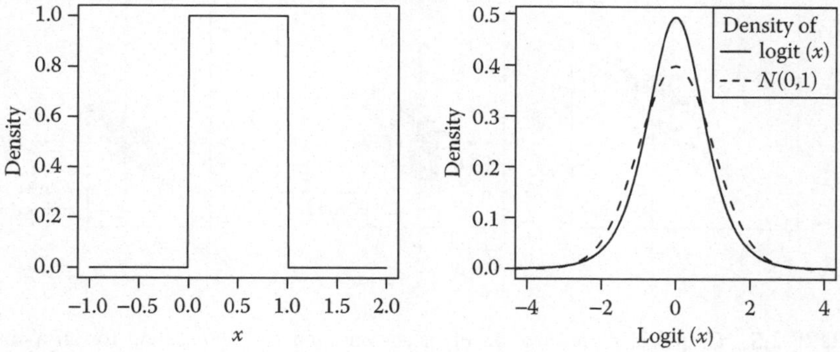

FIGURE 2.4 Probability density function of the uniform distribution (left), and the logit-transformed values as solid line and the standard normal distribution as dashed line (right).

the interval 0–1, and the right figure shows the probability density of the logit-transformed values (solid line), overlaid with the probability density of the standard normal distribution (dashed line).

2.2.2 Centering and Scaling

Centering and scaling is usually applied after data transformation. It also refers to column-wise manipulations of a matrix $\boldsymbol{X}$ with the purpose that all columns have mean zero (centering) and the same variance (scaling). Depending on the task, only centering but not scaling is applied. Let $\bar{x}_j$ be the mean and s_j the standard deviation of a variable j. After MEAN-CENTERING, the variable has a mean of zero; the data are shifted by $\bar{x}_j$ and the center of the data becomes the new origin—consequently the information about the origin is lost; the distances between the data points remain unchanged. Mean-centering simplifies many methods in multivariate data analysis. Notation in $\mathbb{R}$ is given for a matrix $\boldsymbol{X}$.

$$x_{ij}\text{ (mean-centered)} = x_{ij}\text{ (original)} - \bar{x}_j \tag{2.3}$$

$\mathbb{R}$: `X_cent <- scale(X,center=TRUE,scale=FALSE)`

VARIANCE SCALING standardizes each variable j by its standard deviation s_j; usually, it is combined with mean-centering and is then called AUTOSCALING (or z-transformation).

$$x_{ij}\text{ (autoscaled)} = \frac{x_{ij}\text{ (original)} - \bar{x}_j}{s_j} \tag{2.4}$$

$\mathbb{R}$: `X_auto <- scale(X,center=TRUE,scale=TRUE)`

Autoscaled data have a mean of zero and a variance (or standard deviation) of one, thereby giving all variables an equal statistical weight. Autoscaling shifts the centroid

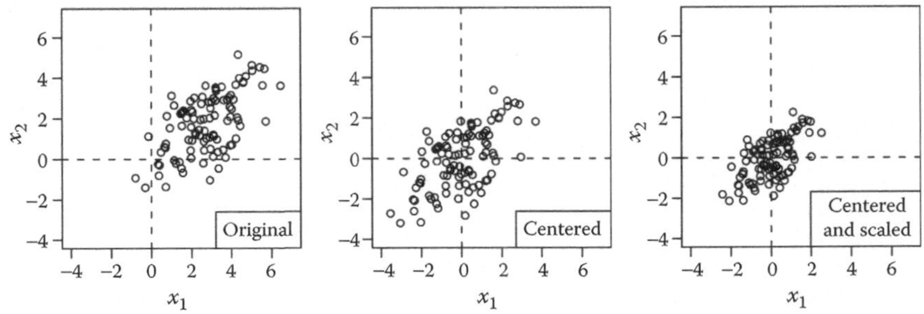

FIGURE 2.5 Graphical representation of mean-centering and autoscaling for an X-matrix with two variables.

of the data points to the origin and changes the scaling of the axes, consequently the relative distances between the data points are changed (except the original variances of all variables are equal, Figure 2.5). Autoscaling is the most used preprocessing in chemometrics; a disadvantage is a blow-up of variables with small values (for instance originating from less precise measurements near the detection limit).

If the data set contains variables from different origins, a BLOCK SCALING is advisable (Eriksson et al. 2006). For example, cereal samples may be characterized by $m_1 = 500$ IR absorbances and $m_2 = 3$ mass percentage of the elements C, H, and N. With simple autoscaling, the IR data would highly dominate any multivariate data analysis. If both variable subsets should be given equal weight, then for instance the variables in each set can be scaled to give equal sums of their variances (for instance 1). A soft block-scaling would result in a sum of variances in each block, which is proportional to the square root of the number of variables in the block. Other weighting schemes can be based on user-defined importances of the variable blocks.

If the data include outliers, it is advisable to use robust versions of centering and scaling. The simplest possibility is to replace the arithmetic means of the columns by the column medians, and the standard deviations of the columns by the median absolute deviations (MAD), see Sections 1.6.3 and 1.6.4, as shown in the following ℝ-code for a matrix X.

```
R: X_auto_robust <-
   scale(X,center=apply(X,2,median),scale=apply(X,2,mad))
```

DOUBLE CENTERING centers both columns and rows; it is the basic transformation for correspondence factor analysis—a method conceptually similar to PCA, and often applied to contingency tables (Greenacre 1992).

2.2.3 NORMALIZATION

Normalization refers to row-wise transformations of a matrix X. Usual transformations are normalization to a CONSTANT SUM of the variables (for instance for concentration data), a CONSTANT MAXIMUM value of the variables (as common for mass

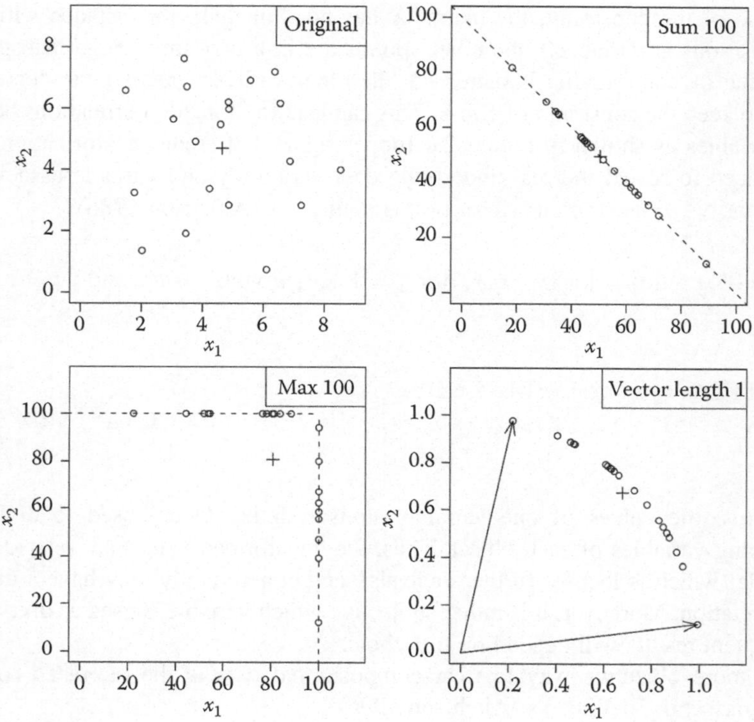

FIGURE 2.6 Graphical representation of normalization to a constant sum of the variables, to a constant maximum of the variables, and to a constant vector length. $\boldsymbol{X}$-matrix with two variables; center marked by +.

spectra), or normalization to a CONSTANT VECTOR LENGTH (constant sum of squared variables). The graphical representation of these transformations in Figure 2.6 makes clear that the dimensionality of the data is reduced and that the distances of the data points are modified. Such data are called to be CLOSED and may give artifacts in some methods of data analysis; for appropriate further data transformation, see Section 2.2.4.

```
ℝ:  X_sum100 <- X/apply(X,1,sum)*100            # sum 100
    X_max100 <- X/apply(X,1,max)*100            # maximum 100
    X_length1 <- X/sqrt(apply(X^2,1,sum))       # length 1
```

2.2.4 Transformations for Compositional Data

While there are transformations that normalize a data matrix $\boldsymbol{X}$ to row sums equal to 1 (see Section 2.2.3), some data sets are originally provided in this form that the values of an object sum up to for instance 100%, like relative concentrations of chemical compounds in a mixture. Such data are called COMPOSITIONAL DATA or CLOSED DATA.

Because of the constraint, the matrix $\boldsymbol{X}$ has not full rank; for instance with three compositions (x-variables), the object points are in a two-dimensional subspace. If the value of one variable is increased, the other variable values must decrease in order to keep the constraint of 100%. This can lead to "forced" correlations between the variables as shown in Figure 2.6 (upper right). Different transformations were introduced to reveal the real underlying data variability and variable associations. The ADDITIVE LOGRATIO TRANSFORMATION is defined as (Aitchison 1986)

$$x_{ij} \text{ (add-log ratio)} = \log (x_{ij}/x_{ik}) \quad \text{for } j = 1, \ldots, m \text{ with } j \neq k \quad \text{and} \quad i = 1, \ldots, n \tag{2.5}$$

```
R: library(chemometrics)
   X_alr <- alr(X,divisor=3)  # column 3 of matrix X is
                              # used as divisor variable
```

In words, the values of one variable (divisor, index k) are used to divide the remaining variables of each object. Thus, the transformed values are related to one variable (which is lost for further analysis), and consequently they have a different interpretation. Moreover, it is rather subjective which variable is used as divisor, and subsequent results will depend on this choice.

A more objective way to treat compositional data is the CENTERED LOGRATIO TRANSFORMATION, defined as (Aitchison 1986)

$$x_{ij} \text{ (cent-logratio)} = \log (x_{ij}/\bar{x}_{G,i}) \quad \text{for } j = 1, \ldots, m \quad \text{and} \quad i = 1, \ldots, n \tag{2.6}$$

with $\bar{x}_{G,i}$ being the geometric mean (see Section 1.6.3) of all m variables in object i. One can show that the transformed data again have not full rank because the rows sum up to 0.

The ISOMETRIC LOGRATIO TRANSFORMATION (Egozcue et al. 2003) repairs this reduced rank problem by taking an orthonormal basis system with one dimension less. Mathematical details of these methods are out of the scope of the book; however, use within $\mathbb{R}$ is easy.

```
R: library(chemometrics)
   X_ilr <- ilr(X)  # isomeric logratio transformation
   X_clr <- clr(X)  # centered logratio transformation
```

2.3 COVARIANCE AND CORRELATION

2.3.1 OVERVIEW

Correlation analysis estimates the extent of the relationship between two or more variables, often only linear relationships are considered. The variables $x_1, \ldots, x_m$ may be measurements for instance of concentrations on n compounds. The correlations between all pairs of variables can then be summarized in a CORRELATION MATRIX

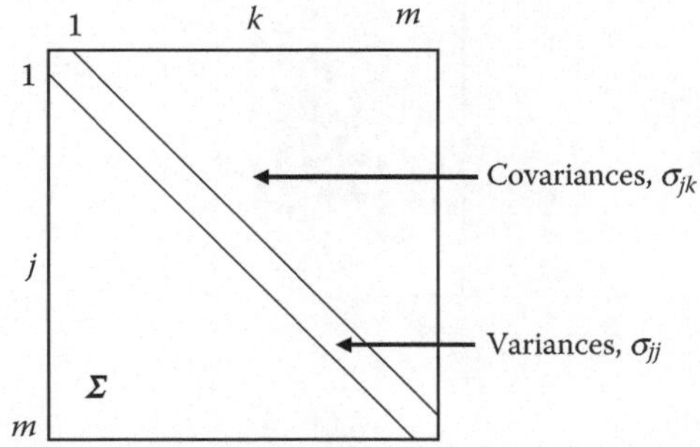

FIGURE 2.7 Covariance matrix $\boldsymbol{\Sigma}$ for m variables.

of dimension $m \times m$. The main diagonal includes only correlations of 1 (correlation of a variable with itself). Due to the definition of correlations, all other values must be in the interval $[-1, +1]$ because the correlation is a standardized measure of variable relation.

The basis for calculating the correlation between two variables x_j and x_k is the COVARIANCE σ_{jk}. The covariance also measures the relation but is not standardized to the interval $[-1, +1]$. Thus, only the sign of the covariance allows an interpretation of the type of relation, but not the value itself. The covariances σ_{jk} can be summarized in the COVARIANCE MATRIX $\boldsymbol{\Sigma}$ (dimension $m \times m$), which is a quadratic, symmetric matrix. The cases $j = k$ (main diagonal) are "covariances" between one and the same variable, which are in fact the variances σ_{jj} of the variables x_j for $j = 1, \ldots, m$ (note that in Chapter 1 variances were denoted as σ^2); therefore, this matrix is also called variance–covariance matrix (Figure 2.7). Matrix $\boldsymbol{\Sigma}$ refers to a data population of infinite size, and should not be confused with estimations of it as described in Section 2.3.2, for instance the sample covariance matrix $\boldsymbol{C}$.

Using the square root of the variances, the covariances can be standardized to correlations by

$$\text{correlation } (x_j, x_k) = \frac{\sigma_{jk}}{\sqrt{\sigma_{jj}\sigma_{kk}}} \tag{2.7}$$

for all variable pairs. This demonstrates that the covariance matrix is indeed the basis for the correlation matrix. The covariance matrix contains information about the distribution of the object points in the variable space, and is a central item for many methods in multivariate data analysis. In Figure 2.8, three different distributions of two-dimensional data are shown as scatter plots and with the covariance matrices. For these examples, $n = 200$ bivariate normally distributed points (`X`) have been simulated, with a mean vector `mean` and a covariance matrix `sigma`; in the ℝ-code below, values for `sigma` are from $\boldsymbol{\Sigma}_1$ in Figure 2.8.

```
ℝ:  library(mvtnorm)
    sigma <- matrix(c(1,0.8,0.8,1),ncol=2)  # sigma1 in Fig. 2.8
    X <- rmvnorm(200,mean=c(0,0),sigma=sigma)
```

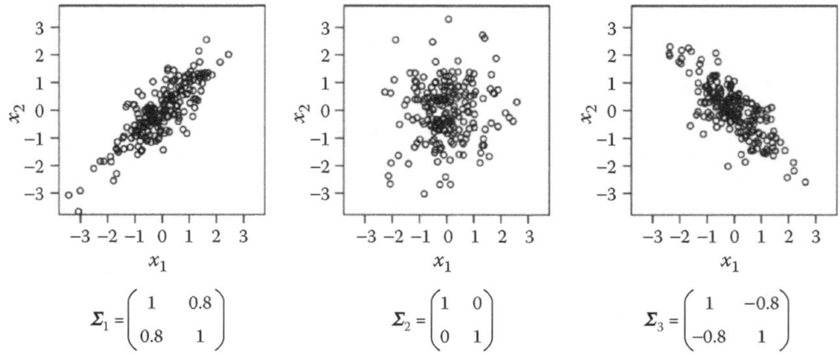

FIGURE 2.8 Covariance matrices Σ for different distributions of two-dimensional data. In the main diagonal of Σ are the variances of x_1 and x_2, respectively. For each covariance matrix 200 bivariate normally distributed points have been simulated.

Highly correlating (COLLINEAR) variables make the covariance matrix singular, and consequently the inverse cannot be calculated. This has important consequences on the applicability of several methods. Data from chemistry often contain collinear variables, for instance the concentrations of "similar" elements, or IR absorbances at neighboring wavelengths. Therefore, chemometrics prefers methods that do not need the inverse of the covariance matrix, as for instance PCA, and PLS regression. The covariance matrix becomes singular if

- At least one variable is proportional to another
- At least one variable is a linear combination of other variables
- At least one variable is constant
- Number of variables is larger than the number of objects

2.3.2 Estimating Covariance and Correlation

In Sections 1.6.3 and 1.6.4, different possibilities were mentioned for estimating the central value and the spread, respectively, of the underlying data distribution. Also in the context of covariance and correlation, we assume an underlying distribution, but now this distribution is no longer univariate but multivariate, for instance a multivariate normal distribution. The covariance matrix Σ mentioned above expresses the covariance structure of the underlying—unknown—distribution. Now, we can measure n observations (objects) on all m variables, and we assume that these are random samples from the underlying population. The observations are represented as rows in the data matrix $X(n \times m)$ with n objects and m variables. The task is then to estimate the covariance matrix from the observed data X. Naturally, there exist several possibilities for estimating Σ (Table 2.2). The choice should depend on the distribution and quality of the data at hand. If the data follow a multivariate normal distribution, the classical covariance measure (which is the basis for the Pearson correlation) is the best choice. If the data distribution is skewed, one could either transform them to more symmetry and apply the classical methods, or alternatively

TABLE 2.2
Measures of Relationship between Two Variables x_j and x_k

Measure	Symbol	Equation No.	Remark
Covariance (sample)	c_{jk}	2.8	$-\infty, \ldots, +\infty$
Pearson's correlation	r_{jk}	2.12	$-1, \ldots, +1$
Spearman's rank correlation	ρ_{jk}		$-1, \ldots, +1$, robust
Kendall's tau correlation	τ_{jk}		$-1, \ldots, +1$, robust
MCD	γ_{jk}		$-1, \ldots, +1$, robust

use nonparametric measures (Spearman's or Kendall's correlation). If the data include outliers (univariate outliers or multivariate outliers being not visible in a single variable), a robust measure of covariance (correlation) should be preferred (Table 2.2).

The classical measure of covariance between two variables x_j and x_k is the SAMPLE COVARIANCE, c_{jk}, defined by

$$c_{jk} = \frac{1}{n-1} \sum_{i=1}^{n} (x_{ij} - \bar{x}_j)(x_{ik} - \bar{x}_k) \tag{2.8}$$

```
ℝ:  c_jk <- cov(xj,xk)
```

For mean-centered variables and written in vector notation

$$c_{jk} = \frac{1}{n-1} \sum_{i=1}^{n} (x_{ij} x_{ik}) = \frac{1}{n-1} \mathbf{x}_j^{\mathrm{T}} \mathbf{x}_k \tag{2.9}$$

Basically, each variable j can be characterized by its ARITHMETIC MEAN, $\bar{x}_j$, VARIANCE v_j, and STANDARD DEVIATION s_j (Figure 2.9). The means $\bar{x}_1$ to $\bar{x}_m$ form the MEAN VECTOR $\bar{\mathbf{x}}$;

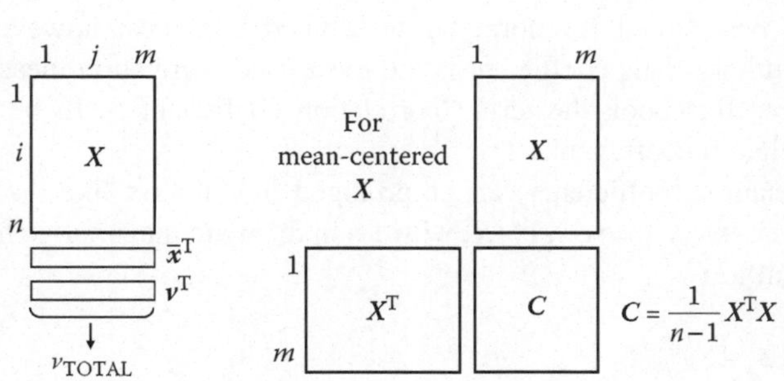

FIGURE 2.9 Basic statistics of multivariate data and covariance matrix. $\bar{\mathbf{x}}^{\mathrm{T}}$, transposed mean vector; $\mathbf{v}^{\mathrm{T}}$, transposed variance vector; v_{TOTAL}, total variance (sum of variances $v_1, \ldots, v_m$). $\mathbf{C}$ is the sample covariance matrix calculated from mean-centered $\mathbf{X}$.

the components are the coordinates of the centroid (center) of all objects in the variable space. The sum of the variances of all variables is called the TOTAL VARIANCE, v_{TOTAL}, of the data set; variances of single variables can be given in percent of total variance. Variance is considered as potential information about the objects. Consider that a constant variable for instance cannot contribute to separate objects into classes or to a model for the prediction of a property; on the other hand, a variable with a high variance is not necessarily informative but may be only noise.

Based on the definition of the covariance c_{jk} in Equation 2.9, the sample covariance matrix $\boldsymbol{C}$ can be calculated for mean-centered $\boldsymbol{X}$ by

$$\boldsymbol{C} = \frac{1}{n-1}\boldsymbol{X}^{\mathrm{T}}\boldsymbol{X} \tag{2.10}$$

In general, $\boldsymbol{C}$ is given by

$$\boldsymbol{C} = \frac{1}{n-1}(\boldsymbol{X} - \mathbf{1}\bar{\boldsymbol{x}})^{\mathrm{T}}(\boldsymbol{X} - \mathbf{1}\bar{\boldsymbol{x}}) \tag{2.11}$$

with $\mathbf{1}$ being a vector of ones of length n, and is calculated in $\mathbb{R}$ (mean-centering of $\boldsymbol{X}$ not necessary) by

$\mathbb{R}$: `C <- cov(X)`

The classical correlation coefficient is the PEARSON CORRELATION COEFFICIENT, (r_{jk}, `r`) which is according to Equation 2.7 the sample covariance, standardized by the standard deviations s_j and s_k of the variables.

$$r_{jk} = \frac{c_{jk}}{s_j s_k} \qquad \mathbb{R}: \ \texttt{r <-cor(xj,xk)} \tag{2.12}$$

The range of r_{jk} is -1 to $+1$; a value of $+1$ indicates a perfect linear relationship, a value of -1 indicates a perfect inverse linear relationship; absolute values of approximately <0.3 indicate a poor or no linear relationship. The Pearson correlation coefficient is best suited for normally distributed variables; however, it is very sensitive to outliers. This coefficient is the most used correlation measure; as usual also throughout this book the term "correlation coefficient" will be used for the Pearson correlation coefficient.

The correlation coefficients can be arranged in a matrix like the covariances. The resulting CORRELATION MATRIX ($\boldsymbol{R}$, with 1's in the main diagonal) is for autoscaled x-data identical to $\boldsymbol{C}$.

$\mathbb{R}$: `R <- cor(X)`

A robust, nonparametric (distribution free) measure for the correlation of variables is the SPEARMAN RANK CORRELATION (symbol ρ_{jk}, in $\mathbb{R}$ `r_Spearman`). It is not limited to linear relationships, but measures the continuously increasing or decreasing association. The data x_j and x_k are first separately replaced by their rank numbers (ranks). Next the Pearson correlation coefficient is calculated using the ranks. The matrix `R_Spearman` with Spearman's correlation values for a data matrix $\boldsymbol{X}$ can be calculated in $\mathbb{R}$ by

```
R:  R_Spearman <- cor(X,method="spearman")
```

The range of ρ is -1 to $+1$; the value is invariant to variable transformations that keep the sequence of the values; that means that for instance a logarithmic transformation has no influence. Because the Spearman correlation is based on ranks, it is relatively robust against outliers.

KENDALL'S TAU CORRELATION (τ_{jk}, `r_Kendall`) also measures the extent of monotonically increasing or decreasing relationships between the variables. It is also a nonparametric measure of association. It is computationally more intensive than the Spearman rank correlation because all slopes of pairs of data points have to be computed. Then Kendall's tau correlation is defined as the average of the signs of all pairwise slopes. The range of τ is -1 to $+1$; the method is relatively robust against outliers; for many applications ρ and τ give similar answers.

```
R:  R_Kendall <- cor(X,method="kendall")
```

A more robust correlation measure, γ_{jk}, can be derived from a robust covariance estimator such as the MINIMUM COVARIANCE DETERMINANT (MCD) estimator. The MCD estimator searches for a subset of h observations having the smallest determinant of their classical sample covariance matrix. The robust LOCATION ESTIMATOR—a robust alternative to the mean vector—is then defined as the arithmetic mean of these h observations, and the robust covariance estimator is given by the sample covariance matrix of the h observations, multiplied by a factor. The choice of h determines the robustness of the estimators: taking about half of the observations for h results in the most robust version (because the other half of the observations could be outliers). Increasing h leads to less robustness but higher efficiency (precision of the estimators). The value $0.75n$ for h is a good compromise between robustness and efficiency.

```
R:  library(robustbase)
    C_MCD <- covMcd(X,alpha=0.75)$cov  # robust estimation of
                                       # covariance matrix of X
```

The corresponding robust correlation matrix `R_MCD`, containing the elements γ_{jk}, is obtained from the robust covariance matrix by dividing by robust standard deviations; in $\mathbb{R}$ by

```
R:  R_MCD <- covMcd(X,alpha=0.75,cor=TRUE)$cor
```

There exist other estimators for robust covariance or correlation, like S-estimators (Maronna et al. 2006). In general, there are restrictions for robust estimations of the covariance matrix: not more than about 50–100 variables, and n/m (the ratio of sample size to number of variables) at least 2. This is a limitation for applications in chemometrics (Table 2.2).

2.4 DISTANCES AND SIMILARITIES

A fundamental idea in multivariate data analysis is to regard the distance between objects in the variable space as a measure of the similarity of the objects. Distance and similarity are inverse: a large distance means a low similarity. Two objects are considered to belong to the same category or to have similar properties if their distance is small. The distance between objects depends on the selected distance definition, the used variables, and on the scaling of the variables. Distance measurements in high-dimensional space are extensions of distance measures in two dimensions (Table 2.3).

Let two objects be defined by the vectors $\boldsymbol{x}_A$ (with components/variables $x_{A1}, x_{A2}, \ldots, x_{Am}$) and $\boldsymbol{x}_B$ (with components/variables $x_{B1}, x_{B2}, \ldots, x_{Bm}$). Most used is the EUCLIDEAN DISTANCE, d(Euclid)—equivalent to distance in daily life. Applying Pythagoras' rule gives (Figure 2.10)

$$d\,(\text{Euclid}) = \left(\sum_{j=1}^{m}(x_{Bj}-x_{Aj})^2\right)^{1/2} = \left(\sum_{j=1}^{m}\Delta x_j^2\right)^{1/2}$$
$$= \left[(\boldsymbol{x}_B-\boldsymbol{x}_A)^{\mathrm{T}}(\boldsymbol{x}_B-\boldsymbol{x}_A)\right]^{1/2} \qquad (2.13)$$

```
ℝ:  d_Euclid <- dist(X,method="euclidean")
```

Here `X` denotes the data matrix, and the result `d_Euclid` is a distance matrix. Even for two vectors, the result is a 2×2 matrix with 0's in the main diagonal, and the

TABLE 2.3
Distance and Similarity Measures

Measure	Symbol	Equation No.	Remark
Euclidean distance	d (Euclid)	2.13	
City block (Manhattan) distance	d (city)	2.14	
Minkowski distance	d (Minkowski)	2.15	
Correlation coefficient (cos α), similarity	r, cos α	2.16	$-1, \ldots, +1$
Similarity of vectors	s_{AB}	2.17	$0, \ldots, +1$
Mahalanobis distance	d (Mahalanobis)	2.18 and 2.19	

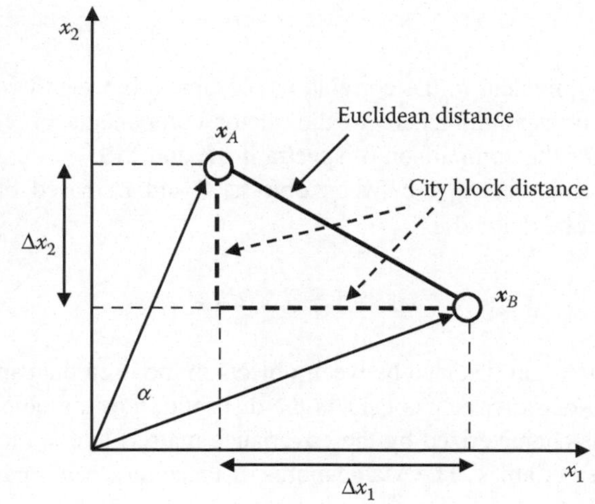

FIGURE 2.10 Euclidean distance and city block distance (Manhattan distance) between objects represented by vectors or points $\boldsymbol{x}_A$ and $\boldsymbol{x}_B$. The cosine of the angle between the object vectors is a similarity measure and corresponds to the correlation coefficient of the vector components.

distance in the two elements of the other diagonal. The Euclidean distance is based on squared differences $\Delta x_j = x_{Bj} - x_{Aj}$ of the variables. The CITY BLOCK DISTANCE (or MANHATTAN DISTANCE) is the sum of the absolute differences of the variables.

$$d\ (\text{city}) = \sum_{j=1}^{m} |x_{Bj} - x_{Aj}| \tag{2.14}$$

```
R: d_city <- dist (X,method="manhattan")
```

In general, the MINKOWSKI DISTANCE is defined by

$$d\ (\text{Minkowski}) = \left(\sum_{j=1}^{m} (x_{Bj} - x_{Aj})^p \right)^{1/p} \tag{2.15}$$

```
R: d_Minkowski <- dist(X,method="minkowski",p=p)
```

The COSINE OF THE ANGLE α between the object vectors is a similarity measure; it is independent from the vector lengths and therefore only considers relative values of the variables.

$$\cos\alpha = \frac{\boldsymbol{x}_A^{\mathrm{T}}\boldsymbol{x}_B}{\sqrt{\left(\boldsymbol{x}_A^{\mathrm{T}}\boldsymbol{x}_A\right)\left(\boldsymbol{x}_B^{\mathrm{T}}\boldsymbol{x}_B\right)}} = \frac{\boldsymbol{x}_A^{\mathrm{T}}\boldsymbol{x}_B}{\|\boldsymbol{x}_A\| \cdot \|\boldsymbol{x}_B\|} \tag{2.16}$$

```
R:  cos_alpha <- t(xa)%*%xb/(sqrt(sum(xa^2))*sqrt(sum(xb^2)))
```

This measure is equivalent to the correlation coefficient between two sets of mean-centered data—corresponding here to the vector components of $\boldsymbol{x}_A$ and $\boldsymbol{x}_B$. It is frequently used for the comparison of spectra in IR and MS.

A similarity measure, s_{AB}, between objects A and B, based on any distance measure, d_{AB}, can be defined as

$$s_{AB} = 1 - d_{AB}/d_{\max} \tag{2.17}$$

with $d_{\max}$ the maximum distance between objects in the used data set.

The MAHALANOBIS DISTANCE considers the distribution of the object points in the variable space (as characterized by the covariance matrix) and is independent from the scaling of the variables. The Mahalanobis distance between a pair of objects $\boldsymbol{x}_A$ and $\boldsymbol{x}_B$ is defined as

$$d\ (\text{Mahalanobis}) = \left[(\boldsymbol{x}_B - \boldsymbol{x}_A)^{\mathrm{T}} \cdot \boldsymbol{C}^{-1} \cdot (\boldsymbol{x}_B - \boldsymbol{x}_A)\right]^{0.5} \tag{2.18}$$

It is a distance measure that accounts for the covariance structure, here estimated by the sample covariance matrix $\boldsymbol{C}$. Clearly, one could also take a robust covariance estimator. The Mahalanobis distance can also be computed from each observation to the data center, and the formula changes to

$$d(\boldsymbol{x}_i) = \left[(\boldsymbol{x}_i - \bar{\boldsymbol{x}})^{\mathrm{T}} \boldsymbol{C}^{-1} (\boldsymbol{x}_i - \bar{\boldsymbol{x}})\right]^{0.5} \quad \text{for } i = 1, \ldots, n \tag{2.19}$$

Here, $\boldsymbol{x}_i$ is an object vector, and the center is estimated by the arithmetic mean vector $\bar{\boldsymbol{x}}$, alternatively robust central values can be used. In $\mathbb{R}$ a vector `d_Mahalanobis` of length n containing the Mahalanobis distances from n objects in $\boldsymbol{X}$ to the center `x_mean` can be calculated by

```
R:  x_mean <- apply(X,2,mean)
    d_Mahalanobis <- sqrt(mahalanobis(X,x_mean,cov(X)))
```

Points with a constant Euclidean distance from a reference point (like the center) are located on a hypersphere (in two dimensions on a circle); points with a constant Mahalanobis distance to the center are located on a hyperellipsoid (in two dimensions on an ellipse) that envelops the cluster of object points (Figure 2.11). That means the Mahalanobis distance depends on the direction. Mahalanobis distances are used in classification methods, by measuring the distances of an unknown object to prototypes (centers, centroids) of object classes (Chapter 5). Problematic with the Mahalanobis distance is the need of the inverse of the covariance matrix which cannot be calculated with highly correlating variables. A similar approach without this drawback is the classification method SIMCA based on PCA (Section 5.3.1, Brereton 2006; Eriksson et al. 2006).

2.5 MULTIVARIATE OUTLIER IDENTIFICATION

The concept of the Mahalanobis distance can also be used in the context of detecting multivariate outliers. In Figure 2.11 (right), we have shown ellipses corresponding to certain Mahalanobis distances from the origin of the two-dimensional data. Obviously, data points further away from the center receive a larger Mahalanobis distance. A multivariate outlier can be defined as data value being exceptionally far away from the center with respect to the underlying covariance structure. Thus, multivariate outliers have an exceptionally high Mahalanobis distance. The question remains what means "exceptionally high?"

If it can be assumed that the multivariate data follow a multivariate normal distribution with a certain mean and covariance matrix, then it can be shown that the squared Mahalanobis distance approximately follows a chi-square distribution χ^2_m with m degrees of freedom (m is the number of variables). One can then take a quantile of the chi-square distribution, like the 97.5% quantile $\chi^2_{m,0.975}$ as cutoff value.

```
ℝ:  cutoff <- qchisq(0.975,m)
```

If an object has a larger squared Mahalanobis distance than the cutoff it is exceptionally high and can therefore be considered as potential multivariate outlier.

For identifying outliers, it is crucial how center and covariance are estimated from the data. Since the classical estimators arithmetic mean vector $\bar{\boldsymbol{x}}$ and sample covariance matrix $\boldsymbol{C}$ are very sensitive to outliers, they are not useful for the purpose of outlier detection by taking Equation 2.19 for the Mahalanobis distances. Instead, robust estimators have to be taken for the Mahalanobis distance, like the center and

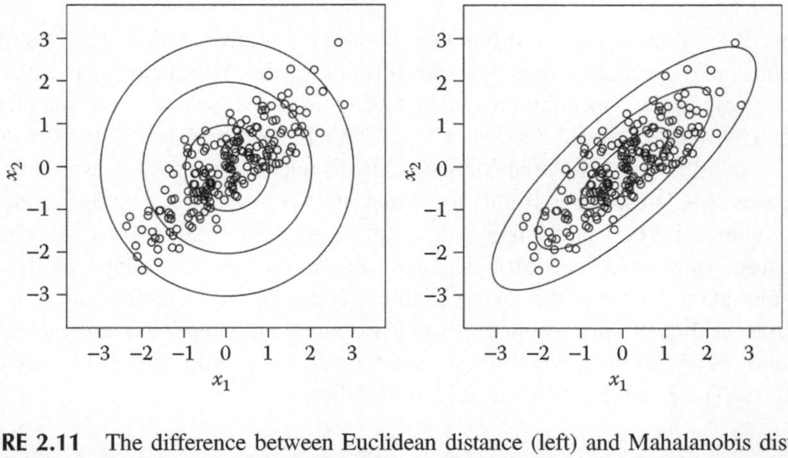

FIGURE 2.11 The difference between Euclidean distance (left) and Mahalanobis distance (right) is shown. The three lines (circles and ellipses) correspond to distances of 1, 2, and 3, from the origin, respectively. The Mahalanobis distance also accounts for the covariance structure (correlation of the variables) of the data.

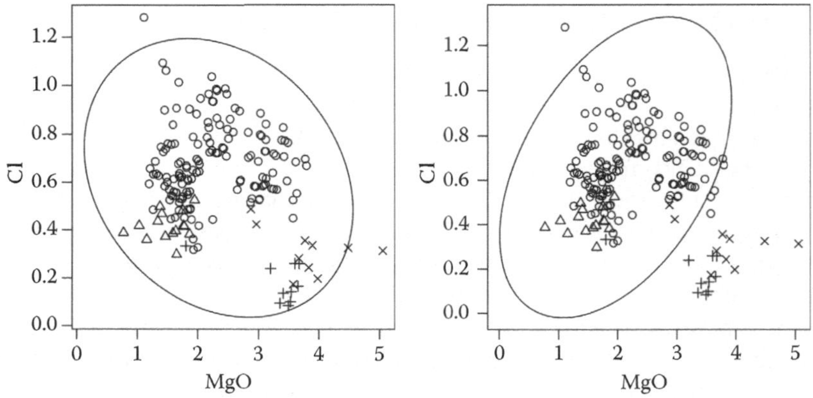

FIGURE 2.12 Concentrations of MgO and Cl in glass vessels samples (Janssen et al. 1998). The Mahalanobis distances are computed using classical (left) and robust (right) estimates for center and covariance. The ellipses correspond to the value of $\chi^2_{2,0.975} = 7.38$ of the resulting squared Mahalanobis distance. Using the robust estimates an outlier group in the lower-right corner of the plot can be identified.

covariance matrix coming from the MCD estimator (see Section 2.3.2). The ℝ code for computing robust Mahalanobis distances for all objects of a data matrix $\boldsymbol{X}$ to the center is as follows:

```
ℝ: library(robustbase)
   X_mcd <- covMcd(X)  # robust center and covariance by MCD
   d_Mahalanobis_robust <-
   sqrt(mahalanobis(X,center=X_mcd$center,cov=X_mcd$cov))
```

Figure 2.12 illustrates the difference between classical and robust estimation of center and covariance matrix in the formula of the Mahalanobis distance. The data shown are the concentrations of MgO and Cl in the glass vessels data (Janssen et al. 1998) as used in Section 1.5.3. The ellipses correspond to a value of $\chi^2_{2,0.975} = 7.38$ for the squared Mahalanobis distance with classical estimates (left) and robust MCD estimates (right) for center and covariance. An outlier group with high values of MgO and small values of Cl has big influence on the classical estimates. They even lead to a negative correlation (negative slope of the main ellipsoid axis) between the two variables. Note that the outlier group is only identified with the robust estimates. Another outlier in the upper left corner is visible. The outliers are multivariate outliers since they do not clearly stick out in any of the variables. The ℝ code for Figure 2.12 is as follows:

```
ℝ: library(chemometrics)
   par(mfrow=c(1,2))          # arranging 2 plots side by side
   data(glass)                # glass data
```

```
X <- glass[,c("MgO","Cl")]           # select MgO and Cl
data(glass.grp)              # groups for glass data
drawMahal(X,center=apply(X,2,mean),covariance=cov(X),
          quantile=0.975,pch=glass.grp)
                # generates left plot in Fig. 2.12
library(robustbase)
X_mcd <- covMcd(X)
drawMahal(X,center=X_mcd$center, covariance=X_mcd$cov,
          quantile=0.975,pch=glass.grp)
                # generates right plot in Fig. 2.12
```

Outlier identification according to Figure 2.12 is only possible for two-dimensional data. For higher-dimensional data, it is no longer possible to show an ellipse as cutoff value. However, in this case, it is still possible to compute the Mahalanobis distances which are always univariate, and to show them in a plot. This has been done for the above data in Figure 2.13. The left plot shows the Mahalanobis distance using classical estimates for center and covariance, the right plot is the robustified version using the MCD estimator. The horizontal line indicates the cutoff value $\sqrt{\chi^2_{2,0.975}} = 2.72$. While the left plot would only identify three outliers, in the right plot a whole group and some additional outliers are visible (compare Figure 2.12, right). The $\mathbb{R}$ code for Figure 2.13 is as follows.

```
R:  library(chemometrics)
    res <- Moutlier(X,quantile=0.975,pch=glass.grp)
              # generates the plot in Fig. 2.13 and computes
              # classical and robust Mahalanobis distances
```

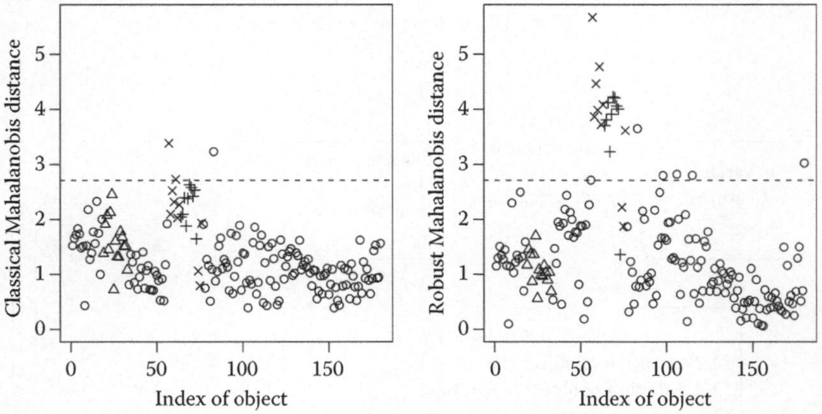

FIGURE 2.13 Concentrations of MgO and Cl in glass vessels samples (Janssen et al. 1998). The plots show the Mahalanobis distances versus the object number; the distances are computed using classical (left) and robust (right) estimates for location and covariance. The horizontal lines correspond to the cutoff value $\sqrt{\chi^2_{2,0.975}} = 2.72$. Using the robust estimates, several outliers are identified.

The Mahalanobis distance used for multivariate outlier detection relies on the estimation of a covariance matrix (see Section 2.3.2), in this case preferably a robust covariance matrix. However, robust covariance estimators like the MCD estimator need more objects than variables, and thus for many applications with $m > n$ this approach is not possible. For this situation, other multivariate outlier detection techniques can be used like a method based on robustified principal components (Filzmoser et al. 2008). The $\mathbb{R}$ code to apply this method on a data set $\boldsymbol{X}$ is as follows:

```
R:  library(mvoutlier)
    out <- pcout(X,makeplot=TRUE)
                        # generates a plot indicating outliers
    out$wfinal01        # n values 0/1 (outliers/ non-outliers)
```

The above method is very fast to compute and can deal with large data sets, ranging into thousands of objects and/or variables.

2.6 LINEAR LATENT VARIABLES

2.6.1 OVERVIEW

An essential concept in multivariate data analysis is the mathematical combination of several variables into a new variable that has a certain desired property (Figure 2.14). In chemometrics such a new variable is often called a LATENT VARIABLE, other names are COMPONENT or FACTOR. A latent variable can be defined as a formal combination (a mathematical function or a more general algorithm) of the variables; a latent variable "summarizes" the variables in an appropriate way to obtain a certain property. The value of a latent variable is called SCORE. Most often LINEAR LATENT VARIABLES are used given by

$$u = b_1x_1 + b_2x_2 + \cdots + b_mx_m \tag{2.20}$$

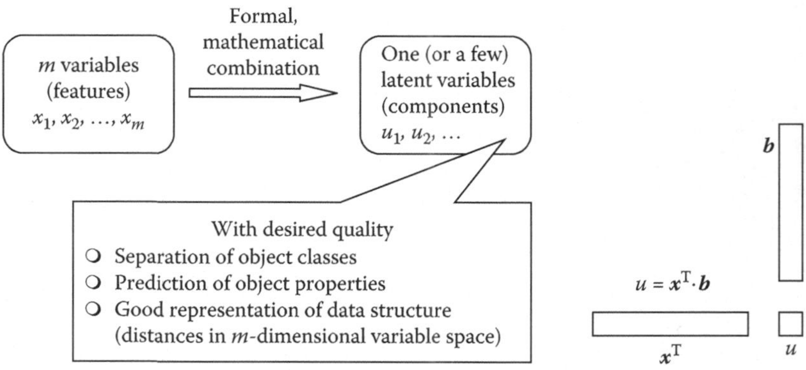

FIGURE 2.14 General concept of a latent variable (left), and calculation of the score u of a linear latent variable from the transposed variable vector $\boldsymbol{x}^{\mathrm{T}}$ and the loading vector $\boldsymbol{b}$ as a scalar product (right).

where
u SCORES, value of linear latent variable
b_j LOADINGS, coefficients describing the influence of the variables on the score
x_j VARIABLES, features

The variables form an object vector $\boldsymbol{x} = (x_1, \ldots, x_m)^T$; the loadings form a loading vector $\boldsymbol{b} = (b_1, \ldots, b_m)^T$, so that the score can be calculated by the scalar product of $\boldsymbol{x}^T$ and $\boldsymbol{b}$ (Figure 2.14)

$$u = \boldsymbol{x}^T \cdot \boldsymbol{b} \tag{2.21}$$

The loading vector is usually scaled to a length of 1, that means $\boldsymbol{b}^T \cdot \boldsymbol{b} = 1$; it defines a direction in the variable space.

Depending on the aim of data analysis different mathematical criteria are applied for the definition of latent variables:

- For principal component analysis (PCA), the criterion is maximum variance of the scores, providing an optimal representation of the Euclidean distances between the objects.
- In multivariate classification, the latent variable is a discriminant variable possessing optimum capability to separate two object classes.
- In multivariate calibration, the latent variable has maximum correlation coefficient or covariance with a y-property, and can therefore be used to predict this property.

2.6.2 Projection and Mapping

Calculation of scores as described by Equations 2.20 and 2.21 can be geometrically considered as an orthogonal PROJECTION (a linear MAPPING) of a vector $\boldsymbol{x}$ on to a straight line defined by the loading vector $\boldsymbol{b}$ (Figure 2.15). For n objects, a score vector $\boldsymbol{u}$ is obtained containing the scores for the objects (the values of the linear latent variable for all objects).

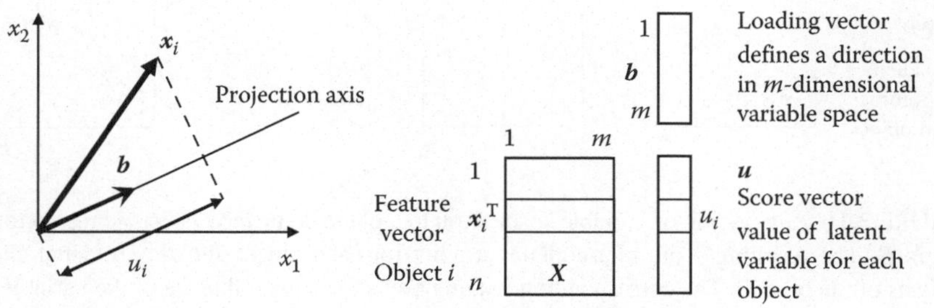

FIGURE 2.15 Rectangular projection of a variable vector $\boldsymbol{x}_i$ on to an axis defined by loading vector $\boldsymbol{b}$ resulting in the score u_i.

$$\boldsymbol{u} = \boldsymbol{X} \cdot \boldsymbol{b} \tag{2.22}$$

If several linear latent variables are calculated, the corresponding loading vectors are collected in a loading matrix $\boldsymbol{B}$, and the scores form a score matrix $\boldsymbol{U}$ (Figure 2.16).

$$\boldsymbol{U} = \boldsymbol{X} \cdot \boldsymbol{B} \tag{2.23}$$

In PCA, for instance, each pair j, k of loading vectors is orthogonal (all scalar products $\boldsymbol{b}_j^{\mathrm{T}} \cdot \boldsymbol{b}_k$ are zero); in this case, matrix $\boldsymbol{B}$ is called to be orthonormal and the projection corresponds to a rotation of the original coordinate system.

Any pair of latent variables defines a projection of the m-dimensional variable space on to a plane given by the loading vectors (Figure 2.16). The projection coordinates are given by the scores; the SCORE PLOT is a scatter plot using two score vectors and contains a point for each object. Depending on the aim of data analysis (and the applied method), the scatter plot may indicate clusters containing similar objects, or may display a good separation of object classes. Note that the projection coordinates u_1 and u_2 are different linear combinations of the same variables. Usually, an appropriate score plot gives more information about the data structure than any variable–variable plot.

In a LOADING PLOT, the loadings of two latent variables are used as coordinates giving a scatter plot with a point for each variable. This plot indicates the similarities of the variables and their influence on the scores. Variables near the origin have small loadings and have—on average—less influence on the scores (in Figure 2.16: variable 4). Variables located for instance in the upper right corner (variables 2 and 5) have in general high values for objects located in the score plot in the same area. Such a simple visual interpretation of corresponding score and loading plots helps to

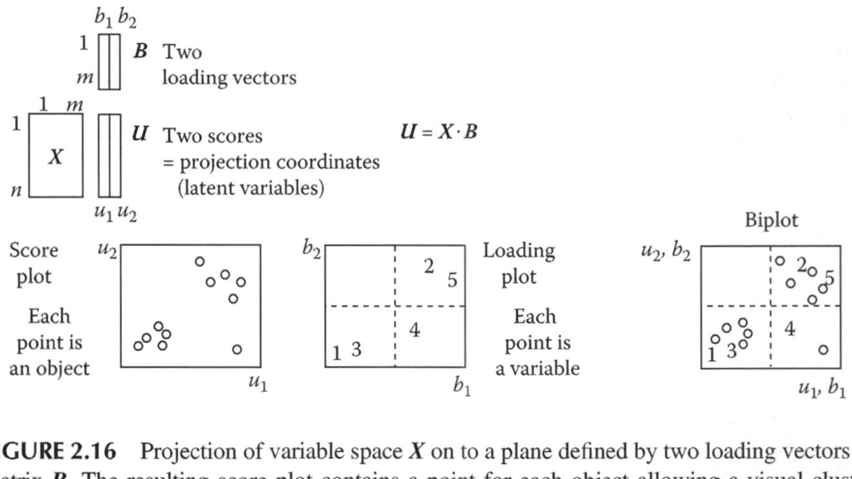

FIGURE 2.16 Projection of variable space $\boldsymbol{X}$ on to a plane defined by two loading vectors in matrix $\boldsymbol{B}$. The resulting score plot contains a point for each object allowing a visual cluster analysis of the objects. The corresponding loading plot shows the influence of the variables on the locations of the object points in the score plot; variables near the origin have less influence; variables forming a cluster have high correlations. The biplot is a combination of a score plot and a corresponding loading plot.

find the characteristics of object clusters. Loading plots are often drawn with arrows from the origin to the points that represent the variables. The length of an arrow is proportional to the contribution of the variable to the two components; the angle between two arrows is an approximate measure of the correlation between the two variables.

A BIPLOT combines a score plot and a loading plot. It contains points for the objects and points (or arrows) for the variables, and can be an instructive display of PCA results; however, the number of objects and the number of variables should not be too high. An appropriate scaling of the scores and the loadings is necessary, and mean-centering of the variables is useful. The biplot gives information about clustering of objects and of variables (therefore "bi"), and about relationships between objects and variables.

The described projection method with scores and loadings holds for all linear methods, such as PCA, LDA, and PLS. These methods are capable to compress many variables to a few ones and allow an insight into the data structure by two-dimensional scatter plots. Additional score plots (and corresponding loading plots) provide views from different, often orthogonal, directions.

Also nonlinear methods can be applied to represent the high-dimensional variable space in a smaller dimensional space (eventually in a two-dimensional plane); in general such data transformation is called a MAPPING. Widely used in chemometrics are Kohonen maps (Section 3.8.3) as well as latent variables based on artificial neural networks (Section 4.8.3.4). These methods may be necessary if linear methods fail, however, are more delicate to use properly and are less strictly defined than linear methods.

2.6.3 EXAMPLE

The used artificial two-dimensional data set $\boldsymbol{X}$ contains 10 objects with the first 5 objects belonging to class 1, and the 5 others to class 2 (Table 2.4). Figure 2.17 shows the $x_1 - x_2$-variable plot. Variances of the variables x_1 and x_2 are 6.29 and 3.03, respectively; the total variance v_{TOTAL} is the sum 9.32. Among the infinite possible directions in the two-dimensional variable space, two are selected for a projection. The 45° straight line from the lower left to the upper right corner is approximately the direction with the maximum spread of the object points, and is defined by a loading vector $\boldsymbol{b}_{45} = [(1/2)^{0.5}, (1/2)^{0.5}] = [0.7071, 0.7071]$. Note that the loading vector is normalized to a length of one. Projection on to this direction preserves well the distances (considered as inverse similarities) in the two-dimensional variable space; the scores have a variance of 6.97 which is 74.8% of v_{TOTAL}. The optimum projection direction for this purpose could be found by PCA, and gives a slightly different vector $\boldsymbol{b}_{\text{PCA}} = [0.8878, 0.4603]$, with the maximum possible variance of the scores of 7.49 (80.4% of v_{TOTAL}).

Another projection axis considered is the 135° straight line from the lower right to the upper left corner (Figure 2.18); it is a direction for a good separation of class 1 and class 2, and is defined by a loading vector $\boldsymbol{b}_{135} = [-0.7071, 0.7071]$. Actually, the classes are well separated by the scores obtained from $\boldsymbol{b}_{135}$. The optimum discriminant direction could be found by LDA, and gives a slightly different vector $\boldsymbol{b}_{\text{LDA}} = [-0.4672, 0.8841]$. Projection on to $\boldsymbol{b}_{\text{LDA}}$ (resulting in LDA scores) gives

TABLE 2.4
Two-Dimensional Artificial Data for 10 Objects from Two Classes

No.	x_1	x_2	y	Score for b_{45}	Score for b_{PCA}	Score for b_{135}	Score for b_{LDA}
1	0.80	3.50	1	3.04	2.32	1.91	2.72
2	3.00	4.00	1	4.95	4.50	0.71	2.13
3	4.20	4.80	1	6.36	5.94	0.42	2.28
4	6.00	6.00	1	8.49	8.09	0.00	2.50
5	6.70	7.10	1	9.76	9.22	0.28	3.15
6	1.50	1.00	2	1.77	1.79	−0.35	0.18
7	4.00	2.50	2	4.60	4.70	−1.06	0.34
8	5.50	3.00	2	6.01	6.26	−1.77	0.08
9	7.30	3.50	2	7.64	8.09	−2.69	−0.32
10	8.50	4.50	2	9.19	9.62	−2.83	0.01
s	2.51	1.74		2.64	2.74	1.53	1.35
v	6.29	3.03		6.97	7.49	2.35	1.83
v%	67.50	32.50		74.82	80.36	25.18	19.64
$\bar{x}$	4.75	3.99		6.18	6.05	−0.54	1.31
$\bar{x}$ class 1	4.14	5.08		6.52	6.01	0.66	2.56
$\bar{x}$ class 2	5.36	2.90		5.84	6.09	−1.74	0.06
s class 1	2.37	1.47		2.69	2.76	0.74	0.40
s class 2	2.76	1.29		2.86	3.04	1.06	0.24
t	−1.20	2.93		0.64	−0.07	4.01	6.97

Note: y, denotes the class of the objects. For projections to four different directions $\boldsymbol{b}_{45}$, $\boldsymbol{b}_{PCA}$, $\boldsymbol{b}_{135}$, and $\boldsymbol{b}_{LDA}$, the scores are given. s, standard deviation; v, variance; v%, variance in percent of total variance; $\bar{x}$, mean; t, t-test value for comparison of class means.

a better separation of the classes; the result from a t-test is 6.97, the maximum possible value.

The ℝ code for this example is as follows (methods PCA and LDA are described in Chapter 3 and Section 5.2, respectively):

```
ℝ:
X <- as.matrix(read.table("x_10_2.txt"))  # read x-data from
                                          # ASCII text file
y <- as.matrix(read.table("y_10.txt"))    # read y-data from
                                          # ASCII text file
plot(X,pch=y)                             # scatter plot x1, x2
                                          # with y for classes
b45 <- c(0.7071, 0.7071)                  # loading vector b45
u_b45 <- X%*%b45                          # scores for b45
```

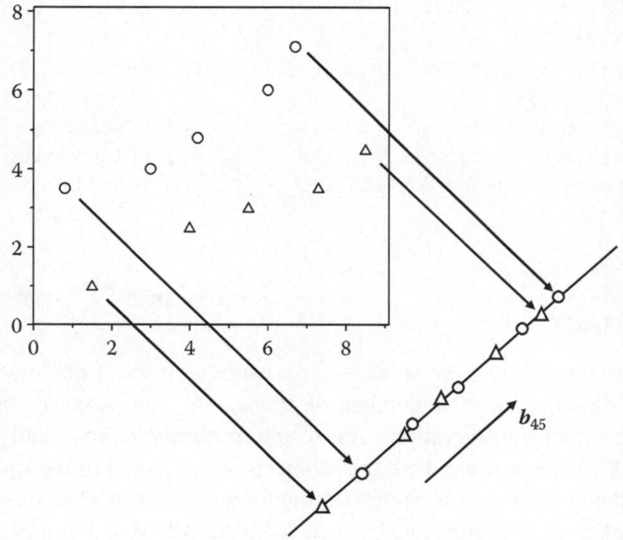

FIGURE 2.17 Projection of the object points from a two-dimensional variable space on to a direction $\boldsymbol{b}_{45}$ giving a latent variable with a high variance of the scores, and therefore a good preservation of the distances in the two-dimensional space.

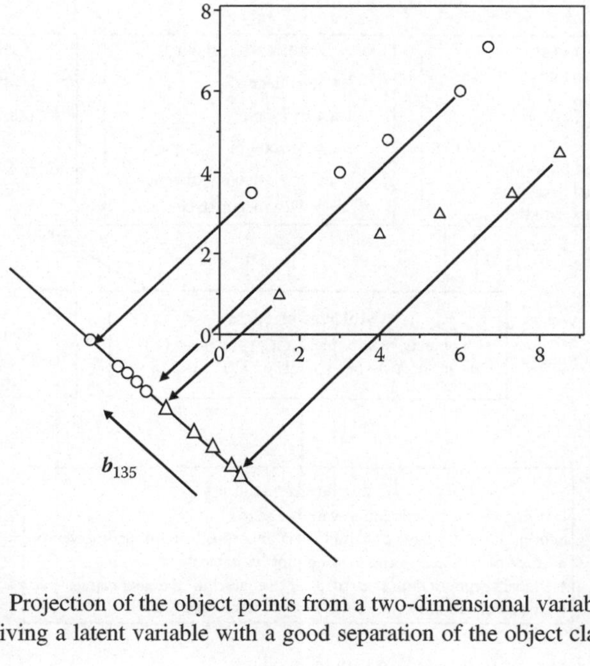

FIGURE 2.18 Projection of the object points from a two-dimensional variable space on to a direction $\boldsymbol{b}_{135}$ giving a latent variable with a good separation of the object classes.

```
b135 <- c(-0.7071, 0.7071)                   # loading vector b135
u_b135 <- X%*%b135                           # scores for b135
Xc <- scale(X,center=TRUE,scale=FALSE)       # mean-centering X
b_PCA <- svd(Xc)$v[,1]                       # PCA loading vector
u_PCA <- X%*% b_PCA                          # PCA scores
b <- solve(t(Xc)%*%Xc)%*%t(Xc)%*%y           # LDA loading vector
b_LDA <- b/sqrt(drop(t(b)%*%b))              # length 1
u_LDA <- X%*% b_LDA                          # LDA scores
```

2.7 SUMMARY

For UNIVARIATE data, only ONE VARIABLE is measured at a set of OBJECTS (samples) or is measured on one object a number of times. For MULTIVARIATE data, SEVERAL VARIABLES are under consideration. The resulting numbers are usually stored in a DATA MATRIX $\boldsymbol{X}$ of size $n \times m$ where the n objects are arranged in the rows and the m variables in the columns. In a geometric interpretation, each object can be considered as a point in an m-dimensional VARIABLE SPACE. Additionally, a PROPERTY of the objects can be stored in a vector $\boldsymbol{y}$ $(n \times 1)$ or several properties in a matrix $\boldsymbol{Y}$ $(n \times q)$ (Figure 2.19).

DATA TRANSFORMATIONS can be applied to change the distributions of the values of the variables, for instance to bring them closer to a normal distribution. Usually, the data are MEAN-CENTERED (column-wise), often they are AUTOSCALED (means of all

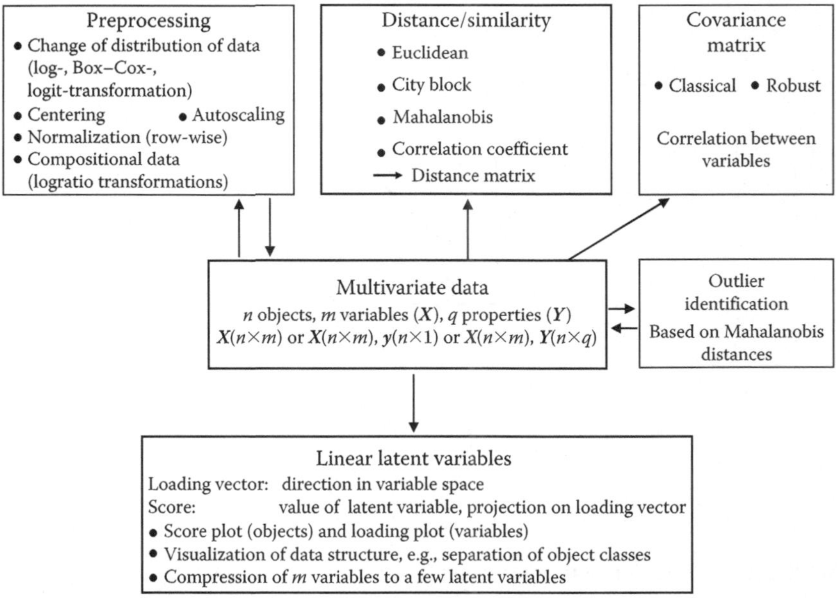

FIGURE 2.19 Basic principles of multivariate data.

variables become zero, and variances become 1) to achieve an equal statistical influence of the variables on the result. So-called COMPOSITIONAL DATA (with constant row sums) may require special transformations.

The DISTANCE between object points is considered as an inverse SIMILARITY of the objects. This similarity depends on the variables used and on the distance measure applied. The distances between the objects can be collected in a DISTANCE MATRIX. Most used is the EUCLIDEAN DISTANCE, which is the commonly used distance, extended to more than two or three dimensions. Other distance measures (CITY BLOCK DISTANCE, CORRELATION COEFFICIENT) can be applied; of special importance is the MAHALANOBIS DISTANCE which considers the spatial distribution of the object points (the correlation between the variables). Based on the Mahalanobis distance, MULTIVARIATE OUTLIERS can be identified. The Mahalanobis distance is based on the COVARIANCE MATRIX of $\boldsymbol{X}$; this matrix plays a central role in multivariate data analysis and should be estimated by appropriate methods—mostly robust methods are adequate.

Another fundamental concept of multivariate data analysis is the use of LATENT VARIABLES (COMPONENTS, FACTORS). A latent variable is computed by a "useful" mathematical combination of the variables, most often a linear relation. "Useful" means optimal for a certain aim of data analysis, for instance preserving the distances of the object points in the high-dimensional variable space or suitable for separating objects into given classes or suitable for modeling a property. A linear latent variable is defined by a LOADING VECTOR which corresponds to a certain direction in the variable space. The SCORES are the values of the latent variable; they are obtained by projecting the object points to the direction of a loading vector. Thus for each object, a new value—the score—is obtained that summarizes the values of the variables in an appropriate way.

PROJECTION of the variable space on to a plane (defined by two loading vectors) is a powerful approach to visualize the distribution of the objects in the variable space, which means detection of clusters and eventually outliers. Another aim of projection can be an optimal separation of given classes of objects. The SCORE PLOT shows the result of a projection to a plane; it is a scatter plot with a point for each object. The corresponding LOADING PLOT (with a point for each variable) indicates the relevance of the variables for certain object clusters.

EXPLORATORY DATA ANALYSIS has the aim to learn about the data distribution (clusters, groups of similar objects). In multivariate data analysis, an $\boldsymbol{X}$-matrix (objects/samples characterized by a set of variables/measurements) is considered. Most used method for this purpose is PCA, which uses latent variables with maximum variance of the scores (Chapter 3). Another approach is cluster analysis (Chapter 6).

MULTIVARIATE CALIBRATION has the aim to develop mathematical models (latent variables) for an optimal prediction of a property y from the variables $x_1, \ldots, x_m$. Most used method in chemometrics is partial least squares regression, PLS (Section 4.7). An important application is for instance the development of quantitative structure–property/activity relationships (QSPR/QSAR).

MULTIVARIATE CLASSIFICATION has the aim to assign objects correctly to given classes (for instance different origins of samples). One approach is to use a latent variable with a good discrimination power (LDA and related methods, Chapter 5) (Chapter 5). Several other classification methods are available, for instance *k*-NN classification (Section 5.3.3) which is based on distance measurements between object points and is related to spectra similarity searches in molecular spectroscopy.

REFERENCES

Aitchison, J.: *The Statistical Analysis of Compositional Data*. Chapman & Hall, London, United Kingdom, 1986.

Brereton, R. G.: *Chemometrics—Data Analysis for the Laboratory and Chemical Plant*. Wiley, Chichester, United Kingdom, 2006.

Egozcue, J. J., Pawlowsky-Glahn, V., Mateu-Figueros, G., Barcelo-Vidal, C.: *Math. Geol.* 35, 2003, 279–300. Isometric logratio transformation for compositional data analysis.

Eriksson, L., Johansson, E., Kettaneh-Wold, N., Trygg, J., Wikström, C., Wold, S.: *Multi- and Megavariate Data Analysis*. Umetrics AB, Umea, Sweden, 2006.

Filzmoser, P., Maronna, R., Werner, M.: *Comput. Stat. Data Anal.* 52, 2008, 1694–1711. Outlier identification in high dimensions.

Greenacre, M.: *Stat. Methods Med. Res.* 1, 1992, 97–117. Correspondence analysis in medical research.

Janssen, K. H. A., De Raedt, I., Schalm, O., Veeckman, J.: *Microchim. Acta* 15(suppl.), 1998, 253–267. Compositions of 15th–17th century archaeological glass vessels excavated in Antwerp.

Maronna, R., Martin, D., Yohai, V.: *Robust Statistics: Theory and Methods*. Wiley, Toronto, ON, Canada, 2006.

Smilde, A., Bro, R., Geladi, P.: *Multi-Way Analysis with Applications in the Chemical Sciences*. Wiley, Chichester, United Kingdom, 2004.

3 Principal Component Analysis

3.1 CONCEPTS

Principal component analysis (PCA) can be considered as "the mother of all methods in multivariate data analysis." The aim of PCA is dimension reduction and PCA is the most frequently applied method for computing linear latent variables (components). PCA can be seen as a method to compute a new coordinate system formed by the latent variables, which is orthogonal, and where only the most informative dimensions are used. Latent variables from PCA optimally represent the distances between the objects in the high-dimensional variable space—remember, the distance of objects is considered as an inverse similarity of the objects. PCA considers all variables and accommodates the total data structure; it is a method for exploratory data analysis (unsupervised learning) and can be applied to practical any $\boldsymbol{X}$-matrix; no y-data (properties) are considered and therefore not necessary.

Dimension reduction by PCA is mainly used for

- Visualization of multivariate data by scatter plots
- Transformation of highly correlating x-variables into a smaller set of uncorrelated latent variables that can be used by other methods
- Separation of relevant information (described by a few latent variables) from noise
- Combination of several variables that characterize a chemical-technological process into a single or a few "characteristic" variables

PCA is successful for data sets with correlating variables as it is often the case with data from chemistry. Constant variables or highly correlating variables cause no problems for PCA; however, outliers may have a severe influence on the result, and also scaling is important.

The direction in a variable space that best preserves the relative distances between the objects is a latent variable which has MAXIMUM VARIANCE of the scores (these are the projected data values on the latent variable). This direction is called by definition the FIRST PRINCIPAL COMPONENT (PC1). It is defined by a loading vector

$$\boldsymbol{p}_1 = (p_1, p_2, \ldots, p_m) \tag{3.1}$$

In chemometrics, the letter p is widely used for loadings in PCA (and partial least-squares [PLS]). It is common in chemometrics to normalize the lengths of loading vectors to 1; that means $\boldsymbol{p}_1^{\mathrm{T}}\boldsymbol{p}_1 = 1$; m is the number of variables. The corresponding scores (projection coordinates of the objects, in chemometrics widely denoted by letter t) are linear combinations of the loadings and the variables (see Section 2.6 about latent variables). For instance, for object i, defined by a vector $\boldsymbol{x}_i$ with elements x_{i1} to x_{im}, the score t_{i1} of PC1 is

$$t_{i1} = x_{i1} p_1 + x_{i2} p_2 + \cdots + x_{im} p_m = \boldsymbol{x}_i^{\mathrm{T}} \cdot \boldsymbol{p}_1 \tag{3.2}$$

The last part of Equation 3.2 expresses this orthogonal projection of the data on the latent variable. For all n objects arranged as rows in the matrix $\boldsymbol{X}$ the score vector, $\boldsymbol{t}_1$, of PC1 is obtained by

$$\boldsymbol{t}_1 = \boldsymbol{X} \cdot \boldsymbol{p}_1 \tag{3.3}$$

Data for a demo example with 10 objects and two mean-centered variables x_1 and x_2 are given in Table 3.1; the feature scatter plot in Figure 3.1. The loading vector for PC1, $\boldsymbol{p}_1$, has the components 0.839 and 0.544 (in Section 3.6 we describe methods to calculate such values). Note that a vector in the opposite direction (−0.839, −0.544) would be equivalent. The scores $\boldsymbol{t}_1$ of PC1 cover more than 85% of the total variance.

The SECOND PRINCIPAL COMPONENT (PC2) is defined as an orthogonal direction to PC1 and again possessing the maximum possible variance of the scores. For two-dimensional data, only one direction, orthogonal to PC1, is possible for PC2. In general further PCs can be computed up to the number of variables. Subsequent PCs are orthogonal to all previous PCs, and their direction has to cover the maximum

TABLE 3.1
Demo Example for PCA with 10 Objects and Two Mean-Centered Variables x_1 and x_2

i	x_1	x_2	t_1	t_2
1	−5.0	−3.5	−6.10	−0.21
2	−4.0	0.5	−3.08	2.60
3	−3.0	−3.0	−4.15	−0.88
4	−2.0	−2.0	−2.77	−0.59
5	−0.5	−0.5	−0.69	−0.15
6	1.0	−1.5	0.02	−1.80
7	1.5	3.5	3.16	2.12
8	2.5	4.0	4.27	1.99
9	4.0	0.5	3.63	−1.76
10	5.5	2.0	5.70	−1.32
$\bar{x}$	0.00	0.00	0.00	0.00
v	12.22	6.72	16.22	2.72
v%	64.52	35.48	85.64	14.36

Note: i, Object number; t_1 and t_2 are the PCA scores of PC1 and PC2, respectively; $\bar{x}$, mean; v, variance; v%, variance in percent of total variance.

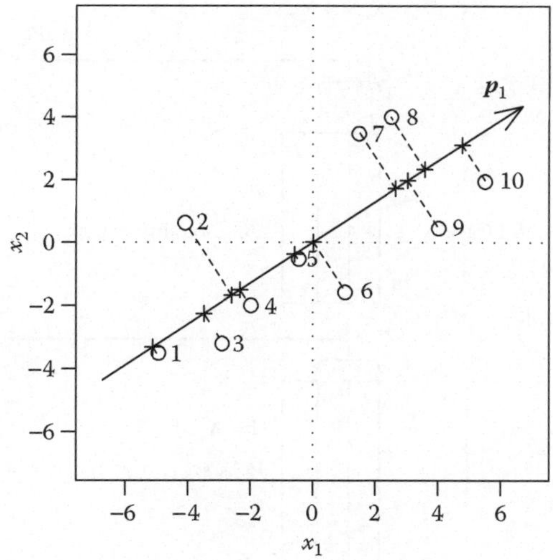

FIGURE 3.1 Scatter plot of demo data from Table 3.1. The first principal component (PC1) is defined by a loading vector $\boldsymbol{p}_1 = [0.839, 0.544]$. The scores are the orthogonal projections of the data on the loading vector.

possible variance of the data projected on this direction. Because the loading vectors of all PCs are orthogonal to each other—as the axes in the original x-coordinate system—this data transformation is a rotation of the coordinate system. For orthogonal vectors, the scalar product is zero, so for all pairs of PCA loading vectors we have

$$\boldsymbol{p}_j^{\mathrm{T}} \cdot \boldsymbol{p}_k = 0 \quad j, k = 1, \ldots, m \tag{3.4}$$

For many practical data sets, the variances of PCs with higher numbers become very small or zero; usually, the first two to three PCs, containing the main amount of variance (potential information), are used for scatter plots. Furthermore, the maximum number of PCs is limited by the minimum of n and m.

All loading vectors are collected as columns in the LOADING MATRIX, $\boldsymbol{P}$, and all score vectors in the SCORE MATRIX, $\boldsymbol{T}$ (Figure 3.2).

$$\boldsymbol{T} = \boldsymbol{X} \cdot \boldsymbol{P} \tag{3.5}$$

The PCA scores have a very powerful mathematical property. They are orthogonal to each other, and since the scores are usually centered, any two score vectors are uncorrelated, resulting in a zero correlation coefficient. No other rotation of the coordinate system except PCA has this property.

$$\boldsymbol{t}_j^{\mathrm{T}} \cdot \boldsymbol{t}_k = 0 \quad j, k = 1, \ldots, m \tag{3.6}$$

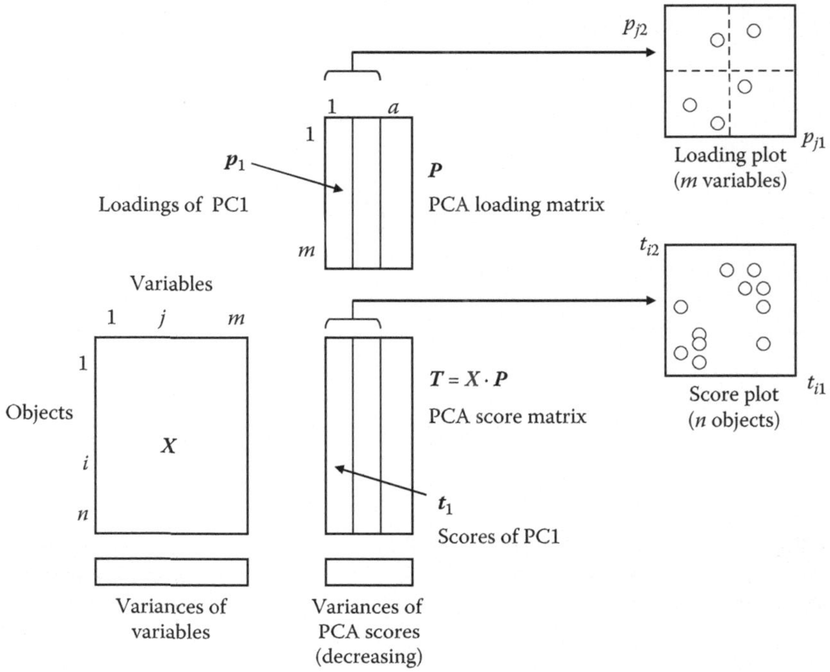

FIGURE 3.2 Matrix scheme for PCA. Since the aim of PCA is dimension reduction and the variables are often highly correlated $a < \min(n, m)$ PCs are used.

The $\boldsymbol{X}$-matrix can be reconstructed from the PCA scores, $\boldsymbol{T}$. Usually, only a few PCs are used (the maximum number is the minimum of n and m), corresponding to the main structure of the data. This results in an approximated $\boldsymbol{X}$-matrix with reduced noise (Figure 3.3). If all possible PCs would be used, the error (residual) matrix $\boldsymbol{E}$ would be zero.

$$\boldsymbol{X}_{\text{appr}} = \boldsymbol{T} \cdot \boldsymbol{P}^{\text{T}} \qquad \boldsymbol{X} = \boldsymbol{T} \cdot \boldsymbol{P}^{\text{T}} + \boldsymbol{E} \qquad \boldsymbol{E} = \boldsymbol{X} - \boldsymbol{X}_{\text{appr}} \tag{3.7}$$

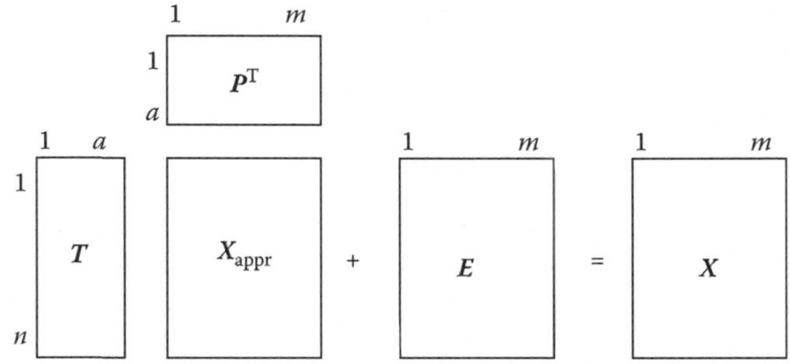

FIGURE 3.3 Approximate reconstruction, $\boldsymbol{X}_{\text{appr}}$, of the $\boldsymbol{X}$-matrix from PCA scores $\boldsymbol{T}$ and the loading matrix $\boldsymbol{P}$ using a components; $\boldsymbol{E}$ is the error (residual) matrix, see Equation 3.7.

3.2 NUMBER OF PCA COMPONENTS

The principal aim of PCA is dimension reduction; that means to explain as much variability (usually variance) as possible with as few PCs as possible. Figure 3.4 shows different data distributions with three variables. If the correlations between the variables are small, no dimension reduction is possible without a severe loss of variance (potential information); the INTRINSIC DIMENSIONALITY of the data in Figure 3.4 (left) is 3; the number of necessary PCA components is 3; in this case, the variances retained by PC1, PC2, and PC3 would be similar and around 33% of the total variance; consequently PCA is not useful. If the object points are more or less located in a two-dimensional plane (Figure 3.4, middle), the intrinsic dimensionality is 2; PC1 and PC2 together are able to represent the data structure well; the sum of variances preserved by PC1 and PC2 would be near 100%; the variance of PC3 would be very small. If the object points are distributed along a straight line (Figure 3.4, right), the number of variables is still 3, but the intrinsic dimensionality is 1; the data structure can be well represented by only one latent variable, best by PC1; it will preserve almost 100% of the total variance.

The variances of the PCA scores—preferably given in PERCENT OF THE TOTAL VARIANCE (equivalent to the sum of the variable variances)—are important indicators. If in a score plot, using the first two PCs t_2 versus t_1, more than about 70% of the total variance is preserved, the scatter plot gives a good picture of the high-dimensional data structure. If more than 90% of the total variance is preserved, the two-dimensional representation is excellent, and most distances between object points will reflect well the distances in the high-dimensional variable space. If the sum of variances preserved by PC1 and PC2 is small, additional score plots, for instance, t_3 versus t_1 and/or t_4 versus t_3, provide additional views at the data structure and help to recognize the data structure. Note that interesting information is not necessarily displayed in the PC2 versus PC1 plot.

If the PCA scores are used in subsequent methods as uncorrelated new variables, the optimum number of PCs can be estimated by several techniques. The strategies applied use different criteria and usually give different solutions. Basics are the variances of the PCA scores, for instance, plotted versus the PC number (Figure 3.5, left). According to the definition, the PC1 must have the largest variance, and the variances decrease with increasing PC number. For many data sets, the plot shows a steep descent after a few components because most of the variance is covered by the first components. In the example, one may conclude that the data structure is mainly influenced by two driving factors, represented by PC1 and PC2. The other

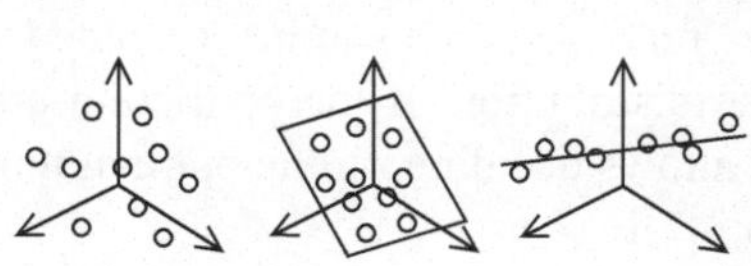

Number of variables	3	3	3
Number of relevant components = intrinsic dimensionality	3	2	1

FIGURE 3.4 Different distributions of object points in a three-dimensional variable space.

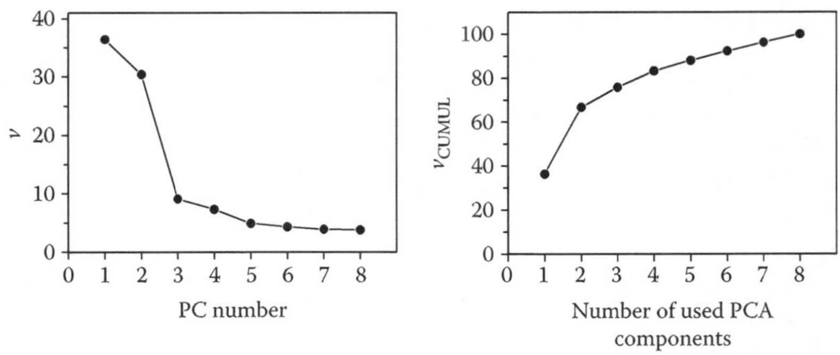

FIGURE 3.5 Scree plot for an artificial data set with eight variables. v, variance of PCA scores (percent of total variance); v_{CUMUL}, cumulative variance of PCA scores.

PCA components with small variances may only reflect noise in the data. Such a plot looks like the profile of a mountain: after a steep slope a more flat region appears that is built by fallen, deposited stones (called scree). Therefore, this plot is often named SCREE PLOT; so to say, it is investigated from the top until the debris is reached. However, the decrease of the variances has not always a clear cutoff, and selection of the optimum number of components may be somewhat subjective. Instead of variances, some authors plot the eigenvalues; this comes from PCA calculations by computing the eigenvectors of the covariance matrix of $\boldsymbol{X}$; note, these eigenvalues are identical with the score variances.

The cumulative variance, v_{CUMUL}, of the PCA scores shows how much of the total variance is preserved by a set of PCA components (Figure 3.5, right). As a rule of thumb, the number of considered PCA components should explain at least 80%, eventually 90% of the total variance.

For autoscaled variables, each variable has a variance of 1, and the total variance is m, the number of variables. For such data, a rule of thumb uses only PCA components with a variance >1, the mean variance of the scores. The number of PCA components with variances larger than 0 is equal to the rank of the covariance matrix of the data.

Cross validation and bootstrap techniques can be applied for a statistically based estimation of the optimum number of PCA components. The idea is to randomly split the data into training and test data. PCA is then applied to the training data and the observations from the test data are reconstructed using 1 to m PCs. The prediction error to the "real" test data can be computed. Repeating this procedure many times indicates the distribution of the prediction errors when using 1 to m components, which then allows deciding on the optimal number of components. For more details see Section 3.7.1.

3.3 CENTERING AND SCALING

The PCA results will change if the origin of the data matrix is changed. Figure 3.6 shows this effect for a two-dimensional data set. In the left picture the original data

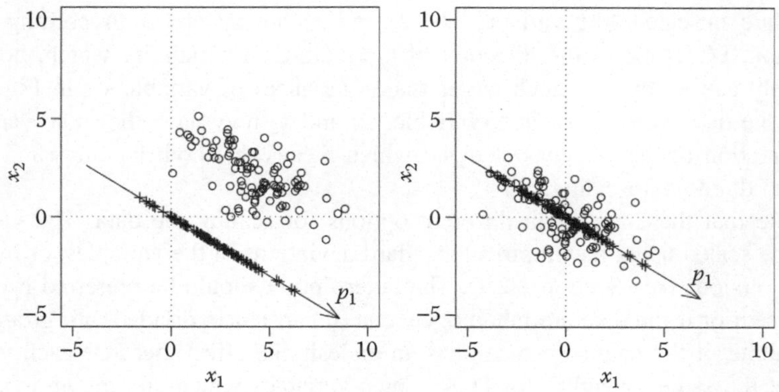

FIGURE 3.6 Effect of mean-centering on PCA. In the left plot the data are not centered at the origin; therefore, the scores are also not centered. The right plot shows centered data which also result in centered scores.

were taken for PCA. The direction of the PC1 is along the largest variability of the data cloud. The scores of the PC1 are the orthogonal projections of the data on this direction, and they are shown with the symbol "+". Obviously, the scores are not centered at zero but they are shifted by a positive constant. This of course, does not change the variance of the scores but it will have consequences later on when the data are reproduced by a smaller number of PCs. Figure 3.6 (right) shows the PCA results on the mean-centered data. In this case also the scores are centered at zero.

For PCA, it is generally recommended to use mean-centered data. Note that there are different possibilities for mean-centering. One could subtract arithmetic column-means from each data column, but also more robust mean-centering methods can be applied (see Section 2.2.2).

Another important aspect of data preparation for PCA is scaling. The PCA results will change if the original (mean-centered) data are taken or if the data were, for instance, autoscaled first. Figure 3.7 (left) shows mean-centered data

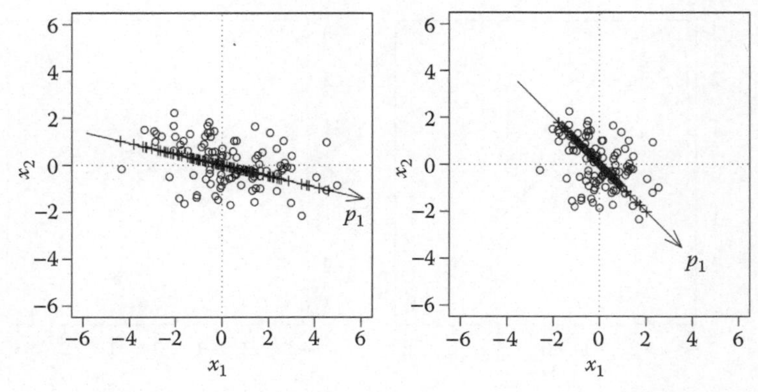

FIGURE 3.7 Effect of autoscaling on PCA. In the left plot the data are not scaled but only centered, in the right plot the data are autoscaled. The results of PCA change.

which are unscaled (the variances of x_1 and x_2 are different). According to the definition, PC1 follows the direction of the main data variability which, however, is mainly caused by the much larger scale (variance) of variable x_1. In Figure 3.7 (right) the data were autoscaled (variables x_1 and x_2 now have the same variance). The direction of the PC1 has changed which means that both loadings and scores changed due to autoscaling.

Note that there are again different options for scaling the data. The variables could be scaled using the empirical standard deviations of the variables, or by using robust versions (see Section 2.2.2). The latter option should be preferred if outliers are present or if the data are inhomogeneous (for instance, divided into groups).

Scaling of the data sometimes has an undesirable effect, because each variable will get the same "weight" for PCA. Thus, variables which are known to include essentially noise will become as important as variables which reflect the true signal. PCA cannot distinguish between important and unimportant information (variance), it will try to express as much of the data variability as possible—also the variability caused by noise. In such cases, the data should not be scaled in order to keep the original importance of the variables.

3.4 OUTLIERS AND DATA DISTRIBUTION

PCA is sensitive with respect to outliers. Outliers are unduly increasing classical measures of variance (that means nonrobust measures), and since the PCs are following directions of maximum variance, they will be attracted by outliers. Figure 3.8 (left) shows this effect for classical PCA. In Figure 3.8 (right), a robust version of PCA was taken (the method is described in Section 3.5). The PCs are defined as DIRECTIONS MAXIMIZING A ROBUST MEASURE OF VARIANCE (see Section 2.3) which is not inflated by the outlier group. As a result, the PCs explain the variability of the nonoutliers which refer to the reliable data information.

The goal of dimension reduction can be best met with PCA if the data distribution is elliptically symmetric around the center. It will not work well as a dimension reduction tool for highly skewed data. Figure 3.9 (left) shows skewed autoscaled

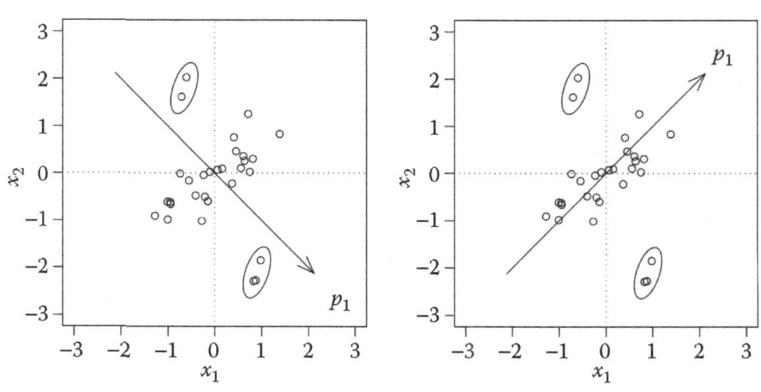

FIGURE 3.8 Classical (left) and robust (right) PCA for data with outliers. Groups of outliers are marked by ellipses; these outliers would not be detected if the variables are used separately.

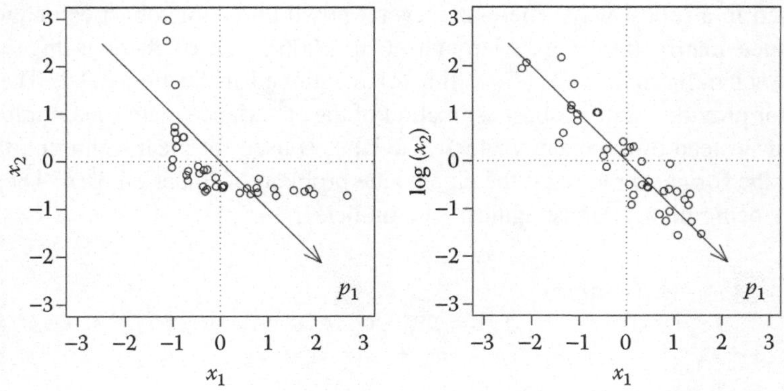

FIGURE 3.9 PCA for skewed autoscaled data: In the left plot PC1 explains 79% of the total variance but fails in explaining the data structure. In the right plot x_2 was log-transformed and then autoscaled; PC1 now explains 95% of the total variance and well follows the data structure.

data. Obviously, PC1 fails to explain the main data variability (although 79% are explained by PC1) because two orthogonal directions would be needed to explain the curvature. In Figure 3.9 (right) variable x_2 was first log-transformed and then autoscaled. Now PC1 is much more informative (95% of the total variance is explained).

3.5 ROBUST PCA

As already noted in Section 3.4, outliers can be influential on PCA. They are able to artificially increase the variance in an otherwise uninformative direction which will be determined as PCA direction. Especially for the goal of dimension reduction this is an undesired feature, and it will mainly appear with classical estimation of the PCs. Robust estimation will determine the PCA directions in such a way that a robust measure of variance is maximized instead of the classical variance. Essential features of robust PCA can be summarized as follows:

- Resulting directions (loading vectors) are orthogonal as in classical PCA.
- Robust variance measure is maximized instead of the classical variance.
- Pearson's correlation coefficient of different robust PCA scores is usually not zero.
- Score plots from robust PCA visualize the main data structure better than classical PCA which may be unduly influenced by outliers.
- Outlier identification is best done with a diagnostic plot based on robust PCA (Section 3.7.3); classical PCA indicates only extreme outliers.

There are essentially two different procedures for robust PCA, a method based on robust estimation of the covariance, and a method based on projection pursuit. For the covariance-based procedure the population covariance matrix $\boldsymbol{\Sigma}$ has to be estimated in a robust way. There are several possibilities for robust estimation of a covariance matrix (see, e.g., Maronna et al. 2006), one of them is the MINIMUM COVARIANCE DETERMINANT (MCD) estimator mentioned in Section 2.3.2. The MCD estimator provides both a robust estimation of the covariance matrix and multivariate location (instead of the mean vector). The latter is used for mean-centering the data matrix, the former for solving the eigenvalue problem of Equation 3.14. The results are PCs being highly robust against data outliers.

```
R:  library(robustbase)
    C_MCD <- covMcd(X,cor=TRUE)   # data robustly autoscaled
    X.rpc <- princomp(X,covmat=C_MCD,cor=TRUE)
    P <- X.rpc$loadings
    T <- X.rpc$scores
```

Diagnostics can and should be done with robustly estimated PCs (Section 3.7.3). The reason is that both score and orthogonal distance are aimed at measuring outlyingness within and from the PCA space, but outliers themselves could spoil PCA if the PCs are not estimated in a robust way.

```
R:  library(chemometrics)
    res <- pcaDiagplot(X,X.rpc,a=2)   # 2 components
```

A limitation of the above robust PCA method is that most robust covariance estimators (also the MCD estimator) need at least twice as many observations than variables. In chemometrics, however, we often have to deal with problems where the number of variables is much larger than the number of observations. In this case, robust PCA can be achieved by the PROJECTION PURSUIT approach. The idea is to go back to the initial definition of PCA where a direction is found by maximizing a measure of variance with the constraint of orthogonality to previously determined directions. Similarly to the nonlinear iterative partial least-squares (NIPALS) algorithm (Section 3.6.4), the PCs are computed one after the other, and they are determined as directions maximizing a robust measure of variance (spread), like the median absolute deviation (MAD) (see Section 1.6.4). For the PROJECTION PURSUIT algorithm, the direction with maximum robust variance of the projected data is "pursued," which practically means to use an efficient algorithm because it is impossible to scan all possible projection directions (Croux et al. 2007). Again, the resulting PCA method is highly robust, and it works in situations where the number of variables is much larger than the number of observations and vice versa.

```
R:  library(pcaPP)
    X.rpc <- PCAgrid(X,k=2,scale=mad)
                        # X is scaled to variance 1 using MAD
                        # 2 principal components are calculated
    P <- X.rpc$loadings                  # PCA loadings
    T <- X.rpc$scores                    # PCA scores
```

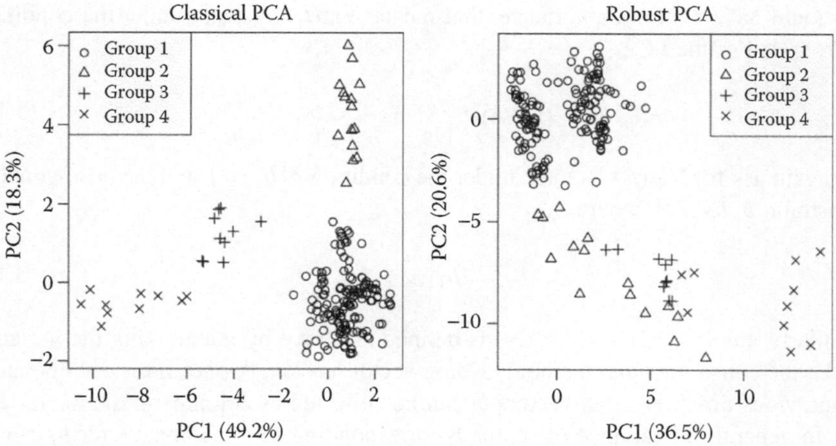

FIGURE 3.10 Plot of the first two PC scores for classical PCA (left) and robust MCD-based PCA (right). The data used are the glass vessels data from Section 1.5.3.

As an example for robust PCA, we use the glass vessels data set from Section 1.5.3 (Janssen et al. 1998). Since much more objects (180) than variables (13) are available, the MCD-based approach for robust PCA is chosen. Figure 3.10 compares the classical (left, see also Figure 1.4) and robust (right) PCA. The scores of the first two PCs are plotted, and the plot symbols refer to the information of the glass type. The classical PCs are mainly attracted by outliers forming different groups. In the robust PCA more details of the main data variability can be seen because the outliers did not determine the directions. It is visible that the objects of group 1 form at least two subgroups. It would be of interest now to compare the origin of the glass vessels of the different subgroups. The percentage of explained variance for robust PCA is considerably lower than for classical PCA. However, for classical PCA the variance measure was artificially increased by the outliers.

3.6 ALGORITHMS FOR PCA

3.6.1 Mathematics of PCA

For readers more interested in the mathematics of PCA an overview is given and the widely applied algorithms are described. For the user of PCA, knowledge of these mathematical details is not necessary.

PCA can be formulated as a mathematical maximization problem with constraints. The PC1 is a linear combination of the variables

$$t_1 = x_1 b_{11} + \cdots + x_m b_{m1} \tag{3.8}$$

with unknown coefficients (loading vector)

$$\boldsymbol{b}_1 = (b_{11}, \ldots, b_{m1})^{\mathrm{T}} \tag{3.9}$$

t_1 should have maximum variance, that means $\mathrm{Var}(t_1) \rightarrow \max$, under the condition $\boldsymbol{b}_1^{\mathrm{T}}\boldsymbol{b}_1 = 1$. For the PC2

$$t_2 = x_1 b_{12} + \cdots + x_m b_{m2} \tag{3.10}$$

we again ask for $\mathrm{Var}(t_2) \rightarrow \max$, under the condition $\boldsymbol{b}_2^{\mathrm{T}}\boldsymbol{b}_2 = 1$ and the orthogonality constraint $\boldsymbol{b}_1^{\mathrm{T}}\boldsymbol{b}_2 = 0$, where

$$\boldsymbol{b}_2 = (b_{12}, \ldots, b_{m2})^{\mathrm{T}} \tag{3.11}$$

Similarly, the kth PC ($3 \leq k \leq m$) is defined as above by maximizing the variance under the constraints that the new loading vector has length one and is orthogonal to all previous directions. All vectors $\boldsymbol{b}_j$ can be collected as columns in the matrix $\boldsymbol{B}$.

In general the variance of scores t_j corresponding to a loading vector $\boldsymbol{b}_j$ can be written as

$$\mathrm{Var}(t_j) = \mathrm{Var}(x_1 b_{1j} + \cdots + x_m b_{mj}) = \boldsymbol{b}_j^{\mathrm{T}}\mathrm{Cov}(x_1, \ldots, x_m)\boldsymbol{b}_j = \boldsymbol{b}_j^{\mathrm{T}}\boldsymbol{\Sigma}\boldsymbol{b}_j \tag{3.12}$$

for $j = 1, \ldots, m$, under the constraints $\boldsymbol{B}^{\mathrm{T}}\boldsymbol{B} = \boldsymbol{I}$. Here, $\boldsymbol{\Sigma}$ is the population covariance matrix (see Section 2.3.1). A maximization problem under constraints can be written as Lagrangian expression

$$\varphi_j = \boldsymbol{b}_j^{\mathrm{T}}\boldsymbol{\Sigma}\boldsymbol{b}_j - \lambda_j(\boldsymbol{b}_j^{\mathrm{T}}\boldsymbol{b}_j - 1) \quad \text{for } j = 1, \ldots, m \tag{3.13}$$

with the Lagrange parameters λ_j. The solution is found by calculating the derivative of this expression with respect to the unknown parameter vectors $\boldsymbol{b}_j$, and setting the result equal to zero.

This gives the equations

$$\boldsymbol{\Sigma}\boldsymbol{b}_j = \lambda_j \boldsymbol{b}_j \quad \text{for } j = 1, \ldots, m \tag{3.14}$$

which is known as EIGENVALUE PROBLEM. This means that the solution for the unknown parameters is found by taking for $\boldsymbol{b}_j$ the EIGENVECTORS of $\boldsymbol{\Sigma}$ and for λ_j the corresponding EIGENVALUES. According to Equation 3.12, the variances for the PCs are equal to the eigenvalues,

$$\mathrm{Var}(t_j) = \boldsymbol{b}_j^{\mathrm{T}}\boldsymbol{\Sigma}\boldsymbol{b}_j = \boldsymbol{b}_j^{\mathrm{T}}\lambda_j\boldsymbol{b}_j = \lambda_j \tag{3.15}$$

and since eigenvectors and their corresponding eigenvalues are arranged in decreasing order, also the variances of the PCs decrease with higher order. Thus, the matrix $\boldsymbol{B}$ with the coefficients for the linear combinations is the eigenvector matrix, also called the loading matrix.

The above formulation and solution of the problem suggests two different procedures for an algorithm to compute the PCs. One possibility, known as JACOBI

ROTATION, is to compute all eigenvectors and eigenvalues of the covariance matrix. This option first requires an estimation of the population covariance matrix $\boldsymbol{\Sigma}$ from the data. Usually the sample covariance matrix $\boldsymbol{C}$ is taken for this purpose, but also robust estimates can be considered (see Section 2.3.2). Since typical data sets in chemometrics have more variables than observations, the covariance-based approach can result in numerical problems because the covariance matrix does not have full rank. As a solution, a different algorithm known as SINGULAR VALUE DECOMPOSITION (SVD) can be taken. The other possibility to compute PCs is a sequential procedure where the PCs are computed one after the other by maximizing the variance and using orthogonality constraints. An algorithm working in this manner and popular in chemometrics is known as NIPALS algorithm.

All algorithms for computing the PCA solution are iterative and for large data sets time consuming. A detailed description and an evaluation of different algorithms from the chemometrics point of view are contained in Vandeginste et al. (1998). Three frequently used algorithms are described in the next sections. In the following, we assume mean-centered multivariate data $\boldsymbol{X}$ with n objects and m variables. The estimated loadings form the vectors $\boldsymbol{p}_j$ and they are collected as columns in the matrix $\boldsymbol{P}$, and the estimated variances of the PCA scores will be denoted by v_j. These are the counterparts to the loading vectors $\boldsymbol{b}_j$ arranged in the eigenvector matrix $\boldsymbol{B}$ and the eigenvalues λ_j of the population covariance matrix $\boldsymbol{\Sigma}$.

3.6.2 Jacobi Rotation

As described above, the PCA loading vectors are taken as the eigenvectors of the estimated covariance matrix of $\boldsymbol{X}$. It is very common to use the sample covariance matrix $\boldsymbol{C}$ as an estimation of the underlying population covariance matrix, but also other estimates being more robust in presence of outliers can be taken (see Section 2.3.2). The eigenvector with the largest eigenvalue determines PC1, the second largest is for PC2, and so on, and the eigenvalues correspond to the variance of the scores of the PCs. The covariance matrix is a symmetric matrix. A proven method for calculation of the eigenvectors and eigenvalues of a real symmetric matrix is Jacobi rotation. The method is described as accurate and stable but rather slow. After computing the eigenvector matrix $\boldsymbol{P}$, the matrix of PC scores $\boldsymbol{T}$ is obtained by multiplying with the mean-centered data matrix $\boldsymbol{X}$,

$$\boldsymbol{T} = \boldsymbol{X} \cdot \boldsymbol{P} \tag{3.16}$$

ℝ:
```
X_jacobi <- princomp(X,cor=TRUE)
P <- X_jacobi$loadings          # column 1 of P is loading
                                # vector of PC1
T <- X_jacobi$scores            # column 1 of T is score
                                # vector of PC1
```

Remark to ℝ-function `princomp`: not applicable if $n < m$.

3.6.3 Singular Value Decomposition

A widely used mathematical tool for PCA is SVD which is a standard method implemented in many mathematical software packages (see also Appendix A.2.7). According to SVD, any matrix $\boldsymbol{X}$ (size $n \times m$) can be decomposed into a product of three matrices (Figure 3.11).

$$\boldsymbol{X} = \boldsymbol{T}_0 \cdot \boldsymbol{S} \cdot \boldsymbol{P}^{\mathrm{T}} \tag{3.17}$$

For mean-centered $\boldsymbol{X}$ the matrix $\boldsymbol{T}_0$ has size $n \times m$ and contains the PCA scores normalized to a length of 1. $\boldsymbol{S}$ is a diagonal matrix of size $m \times m$ containing the so-called singular values in its diagonal which are equal to the standard deviations $\sqrt{\lambda_j}$ of the scores. $\boldsymbol{P}^{\mathrm{T}}$ is the transposed PCA loading matrix with size $m \times m$. The PCA scores, $\boldsymbol{T}$, as defined above are calculated by

$$\boldsymbol{T} = \boldsymbol{T}_0 \cdot \boldsymbol{S} \tag{3.18}$$

```
ℝ:  X_svd <- svd(X)                      # SVD for mean-centered X
    P <- X_svd$v                            # column 1 of P is
                                            # loading vector of PC1
    T <- X_svd$u %*% diag(X_svd$d)          # column 1 of T is score
                                            # vector of PC1
```

For data with many variables and a small number of objects (n is much smaller than m), the above SVD decomposition is very time consuming. Therefore, a more efficient algorithm avoids the computation of the eigenvectors of the $m \times m$ matrix $\boldsymbol{X}^{\mathrm{T}} \cdot \boldsymbol{X}$. It can be shown that $\boldsymbol{T}_0$ is the matrix of eigenvectors of $\boldsymbol{X} \cdot \boldsymbol{X}^{\mathrm{T}}$, and $\boldsymbol{P}$ is the matrix of eigenvectors of $\boldsymbol{X}^{\mathrm{T}} \cdot \boldsymbol{X}$, both normalized to length 1 (see also Appendix A.2.7).

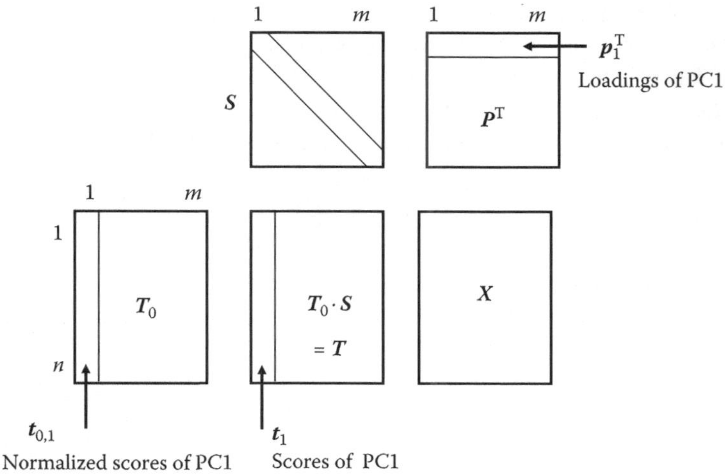

FIGURE 3.11 Matrix scheme for SVD applied to PCA.

Both matrices have the same eigenvalues, namely the squares of the diagonal elements of $\boldsymbol{S}$. Thus the eigenvectors $\boldsymbol{T}_0$ are computed from $\boldsymbol{X} \cdot \boldsymbol{X}^{\mathrm{T}}$ with the eigenvalues (variances of scores) $\boldsymbol{S}^2$, and the PCA scores are $\boldsymbol{T} = \boldsymbol{T}_0 \cdot \boldsymbol{S}$. Using the relation (Equation 3.17), and writing $\boldsymbol{S}^{-2}$ for the inverse of $\boldsymbol{S}^2$, the loadings are computed by

$$\boldsymbol{P} = \boldsymbol{X}^{\mathrm{T}} \cdot \boldsymbol{T}_0 \cdot \boldsymbol{S}^{-1} = \boldsymbol{X}^{\mathrm{T}} \cdot \boldsymbol{T} \cdot \boldsymbol{S}^{-2} \qquad (3.19)$$

```
R:  X_eigen <- eigen(X %*% t(X))   # eigenvectors from XX^T
    T <- X_eigen$vectors %*% diag(sqrt(X_eigen$values))
                 # T: scores
                 # X_eigen$vectors: eigenvectors
                 # diag: makes diagonal matrix
                 # X_eigen$values: eigenvalues
    P <- t(X) %*% T %*% diag(1/X_eigen$values)   # P: loadings
```

Note that since SVD is based on eigenvector decompositions of cross-product matrices, this algorithm gives equivalent results as the Jacobi rotation when the sample covariance matrix $\boldsymbol{C}$ is used. This means that SVD will not allow a robust PCA solution; however, for Jacobi rotation a robust estimation of the covariance matrix can be used.

3.6.4 NIPALS

The nonlinear iterative partial least-squares (NIPALS) algorithm, also called power method, has been popular especially in the early time of PCA applications in chemistry; an extended version is used in PLS regression. The algorithm is efficient if only a few PCA components are required because the components are calculated step-by-step.

NIPALS starts with an initial score vector $\boldsymbol{u}$, that can be arbitrarily chosen from one of the variables, also the variable with the highest variance has been proposed (Figure 3.12a). Next a first approximation, $\boldsymbol{b}$, of the corresponding

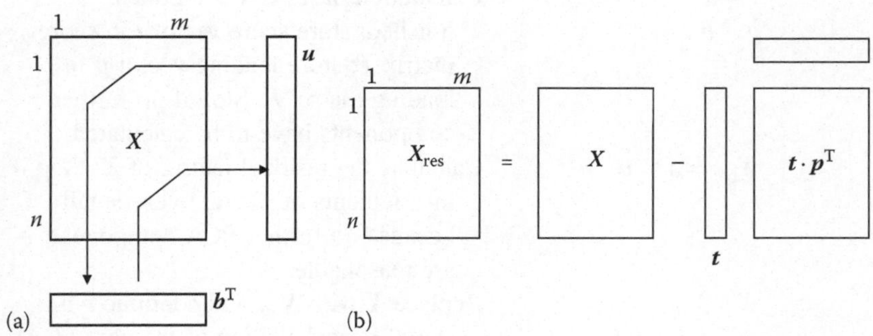

FIGURE 3.12 NIPALS algorithm for PCA. Left scheme (a) shows the iterative procedure for calculation of a PC. The right scheme (b) describes the peeling process (deflation) for elimination of the information of a PC.

loading vector is calculated by $\boldsymbol{X}^{\mathrm{T}} \cdot \boldsymbol{u}$, and $\boldsymbol{b}$ is normalized to length 1. $\boldsymbol{X} \cdot \boldsymbol{b}$ gives a new approximation of the score vector. This cycle is repeated until convergence of $\boldsymbol{b}$ and $\boldsymbol{u}$ is reached; the final values are the PCA loadings $\boldsymbol{p}$ and PCA scores $\boldsymbol{t}$ of the PC1.

After a PC has been calculated the information of this component is "peeled off" from the currently used $\boldsymbol{X}$-matrix (Figure 3.12b). This process is called DEFLATION, and is a projection of the object points on to a subspace which is orthogonal to $\boldsymbol{p}$, the previously calculated loading vector. The obtained $\boldsymbol{X}$-residual matrix $\boldsymbol{X}_{\mathrm{res}}$ is then used as a new $\boldsymbol{X}$-matrix for the calculation of the next PC. The process is stopped after the desired number of PCs is calculated or no further PCA components can be calculated because the elements in $\boldsymbol{X}_{\mathrm{res}}$ are very small.

In mathematical-algorithmic notation NIPALS can be described as follows:

(1)	$\boldsymbol{X}(n \times m)$	$\boldsymbol{X}$ is a mean-centered matrix.
(2)	$\boldsymbol{u} = \boldsymbol{x}_j$	Start with an initial score vector, for instance, with an arbitrarily chosen variable j, or the variable with highest variance.
(3)	$\boldsymbol{b} = \boldsymbol{X}^{\mathrm{T}} \cdot \boldsymbol{u}$ $\boldsymbol{b} = \boldsymbol{b}/\lVert\boldsymbol{b}\rVert$	Calculate a first approximation of the loading vector and normalize it to length 1.
(4)	$\boldsymbol{u}^* = \boldsymbol{X} \cdot \boldsymbol{b}$	Calculate an improved score vector.
(5)	$\boldsymbol{u}_\Delta = \boldsymbol{u}^* - \boldsymbol{u}$ $\Delta u = \boldsymbol{u}_\Delta^{\mathrm{T}} \cdot \boldsymbol{u}$	Calculate the summed squared differences, Δu, between the previous scores and the improved scores. If $\Delta u < \varepsilon$ then convergence is reached; a PCA component has been calculated; continue with step 7, otherwise with step 6 (ε is, for instance, set to 10^{-6}).
(6)	$\boldsymbol{u} = \boldsymbol{u}^*$	Replace the previous score vector by the improved score vector, and continue with step 3.
(7)	$\boldsymbol{t} = \boldsymbol{u}^*$ $\boldsymbol{p} = \boldsymbol{b}$	Calculation of a PCA component is finished; store score vector $\boldsymbol{t}$ in score matrix $\boldsymbol{T}$; store loading vector $\boldsymbol{p}$ in loading matrix $\boldsymbol{P}$. Stop if no further components have to be calculated.
(8)	$\boldsymbol{X}_{\mathrm{res}} = \boldsymbol{X} - \boldsymbol{u} \cdot \boldsymbol{b}^{\mathrm{T}}$	Calculate the residual matrix of $\boldsymbol{X}$. Stop if the elements of $\boldsymbol{X}_{\mathrm{res}}$ are very small because no further PCA components are reasonable.
(9)	$\boldsymbol{X} = \boldsymbol{X}_{\mathrm{res}}$	Replace $\boldsymbol{X}$ with $\boldsymbol{X}_{\mathrm{res}}$ and continue with step 2 for calculation of the next PCA component.

In ℝ a function has been provided for an easy application of the NIPALS algorithm; for the calculation of, for instance, two PCs of a mean-centered matrix X, the ℝ-code is as follows:

```
ℝ:  library(chemometrics)
    X_nipals <- nipals(X,a=2)
    T <- X_nipals$T                # scores
    P <- X_nipals$P                # loadings
```

The NIPALS algorithm is efficient if only a few PCA components are required. Because the deflation procedure increases the uncertainty of following components, the algorithm is not recommended for the computation of many components (Seasholtz et al. 1990). The algorithm fails if convergence is reached already after one cycle; in this case another initial value of the score vector has to be tried (Miyashita et al. 1990).

3.7 EVALUATION AND DIAGNOSTICS

There are several important issues for PCA, like the explained variances of each PC which determine the number of components to select. Moreover, it is of interest if outliers have influenced the PCA calculation, and how well the objects are presented in the PCA space. These and several other questions will be treated below.

3.7.1 Cross Validation for Determination of the Number of Principal Components

Cross validation or bootstrap techniques can be applied for a statistically based estimation of the optimum number of PCA components. The idea is to randomly split the data into training data X_{TRAIN} and test data X_{TEST} either by cross validation or by the bootstrap technique. For cross validation, the data can be randomly subdivided into a number of segments, where the training data consist of the data in all but one segment and the test data are formed by the objects in the omitted segment. PCA is then applied to the training data and the X-training data can be reconstructed by (Figure 3.3)

$$X_{\mathrm{TRAIN}} = {}_a\boldsymbol{T}_{\mathrm{TRAIN}} \cdot {}_a\boldsymbol{P}_{\mathrm{TRAIN}}^{\mathrm{T}} + {}_a\boldsymbol{E}_{\mathrm{TRAIN}} \tag{3.20}$$

using a PCs, with ${}_a\boldsymbol{T}_{\mathrm{TRAIN}}$ the first a PCA scores, and ${}_a\boldsymbol{P}_{\mathrm{TRAIN}}$ the first a PCA loading vectors. The observations from the test data are then projected onto the PCA space (obtained from the training data) and the error matrix of the PCA approximation is computed for the test data.

$${}_a\boldsymbol{T}_{\mathrm{TEST}} = \boldsymbol{X}_{\mathrm{TEST}} \cdot {}_a\boldsymbol{P}_{\mathrm{TRAIN}} \tag{3.21}$$

$${}_a\boldsymbol{E}_{\mathrm{TEST}} = \boldsymbol{X}_{\mathrm{TEST}} - {}_a\boldsymbol{T}_{\mathrm{TEST}} \cdot {}_a\boldsymbol{P}_{\mathrm{TRAIN}}^{\mathrm{T}} \tag{3.22}$$

Finally, a measure of lack of fit using a PCs can be defined using the sum of the squared errors (SSE) from the test set, ${}_a\mathrm{SSE}_{\mathrm{TEST}} = \|{}_a\boldsymbol{E}_{\mathrm{TEST}}\|^2$ (prediction sum of squares). Here, $\|\cdot\|^2$ stands for the sum of squared matrix elements. This measure can be related to the overall sum of squares of the data from the test set, $\mathrm{SS}_{\mathrm{TEST}} = \|\boldsymbol{X}_{\mathrm{TEST}}\|^2$. The quotient of both measures is between 0 and 1. Subtraction from 1 gives a measure of the quality of fit or EXPLAINED VARIANCE for a fixed number of a PCs:

$$
{}_aQ^2_{\mathrm{TEST}} = 1 - \frac{{}_a\mathrm{SSE}_{\mathrm{TEST}}}{\mathrm{SS}_{\mathrm{TEST}}} = 1 - \frac{\|{}_a\boldsymbol{E}_{\mathrm{TEST}}\|^2}{\|\boldsymbol{X}_{\mathrm{TEST}}\|^2} \tag{3.23}
$$

A single split of the data into training and test set may give misleading results—especially for small and inhomogeneous data sets. Therefore, for cross validation, each segment once plays the role of the test data and the corresponding complement is the training data. The quality of fit measure is then averaged. Preferably, training and test data are repeatedly drawn from the complete data set, and the above procedure is applied for each pair of training and test data set. This gives many (averaged) values for ${}_aQ^2_{\mathrm{TEST}}$ which can be displayed, for instance, by a boxplot. The whole procedure is repeated with varying a, the number of PCA components. Looking at the distribution of ${}_aQ^2_{\mathrm{TEST}}$ allows an estimation of the optimal number of PCs.

Figure 3.13 shows the result of this procedure for groups 3 and 4 from the glass vessels data from Section 1.5.3 ($n=20$, $m=13$, (Janssen et al. 1998)). The cross validation procedure with four segments is repeated 100 times, resulting in 100

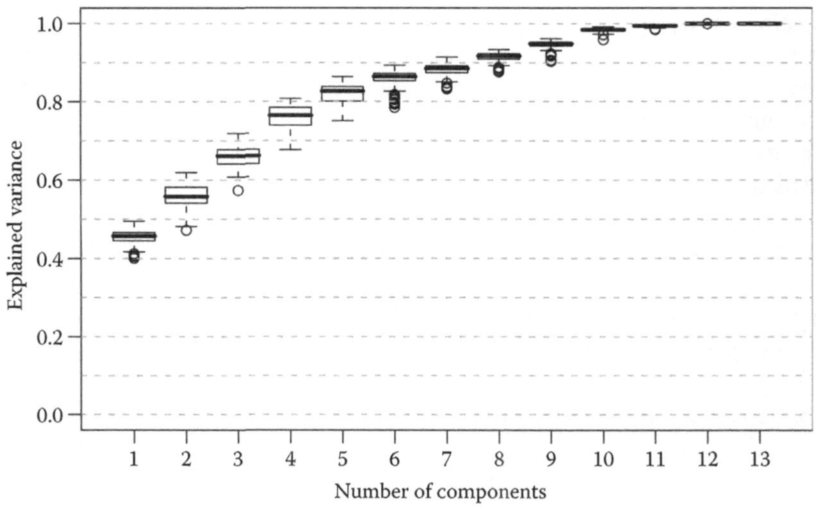

FIGURE 3.13 Determination of the number of PCs with cross validation. The data used are groups 3 and 4 from the glass vessels data from Section 1.5.3.

averaged measures of the explained variance ${}_aQ^2_{\text{TEST}}$ for each number of a PCs. For fixed a, the 100 values are summarized in a boxplot. Using four PCs results in a proportion of explained variance of about 75%.

```
R:  library(chemometrics)
    res <- pcaCV(X,segments=4,repl=100)   # a plot is generated
                # (see Fig. 3.13) and the resulting values of
                # the explained variances are returned
```

3.7.2 Explained Variance for Each Variable

The PCA model gives a representation of the centered (and scaled) data matrix

$$\boldsymbol{X} = {}_a\boldsymbol{T} \cdot {}_a\boldsymbol{P}^{\text{T}} + {}_a\boldsymbol{E} \tag{3.24}$$

where both ${}_a\boldsymbol{T}$ and ${}_a\boldsymbol{P}$ are matrices with a columns, referring to the number of PCs. The more PCs are used the better the approximation of the data matrix, and the smaller the elements of the error matrix ${}_a\boldsymbol{E}$. Now it can be of interest how well each variable is explained by the PCA model using a PCs. Similar to Section 3.7.1 an error measure for each variable can be derived by taking the sum of squared elements of the columns of ${}_a\boldsymbol{E}$. This is

$$\sum_{i=1}^{n} {}_ae_{ij}^2 = \sum_{i=1}^{n} \left(x_{ij} - {}_a\boldsymbol{t}_i^{\text{T}} \cdot {}_a\boldsymbol{p}_j\right)^2 \tag{3.25}$$

where

${}_ae_{ij}$ are the elements of ${}_a\boldsymbol{E}$
${}_a\boldsymbol{t}_i$ and ${}_a\boldsymbol{p}_j$ are the ith and jth row of ${}_a\boldsymbol{T}$ and ${}_a\boldsymbol{P}$, respectively

This measure can be related to the sum of squared elements of the columns of $\boldsymbol{X}$ to obtain a proportion of unexplained variance for each variable. Subtraction from 1 results in a measure ${}_aQ_j^2$ of explained variance for each variable using a PCs

$${}_aQ_j^2 = 1 - \frac{\sum_i {}_ae_{ij}^2}{\sum_i x_{ij}^2} \quad \text{for } j = 1, \ldots, m \tag{3.26}$$

which is a value between 0 and 1. Usually it is desirable that for a fixed number of PCs each variable is explained as good as possible. However, it can happen that single variables obtain a very low value of explained variance while others have a reasonably high value. If this effect is not wanted, the number of PCs has to be increased.

As an example, we consider the glass vessels data and apply PCA using the classical estimators. The left plot in Figure 3.14 shows the values ${}_1Q_j^2$ using one PC to fit the data. The quality of fit is very low for SO_3, K_2O, and PbO. In the right plot two PCs are used and the measures ${}_2Q_j^2$ are shown in barplots. Except for SO_3 the explained variances increased essentially.

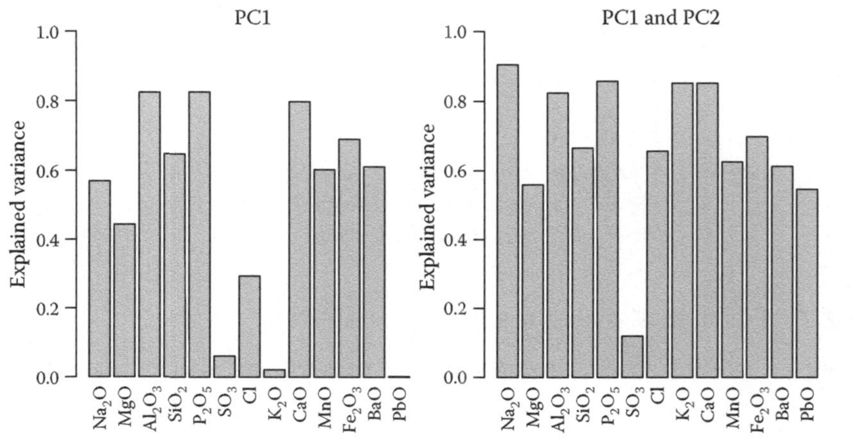

FIGURE 3.14 Explained variance for each variable using one (left) and two (right) PCs. The data used are the glass vessels data from Section 1.5.3.

```
ℝ:  library(chemometrics)
    res <- pcaVarexpl(X,a=2)  # a plot for a=2 components
                 # is generated and the resulting values of
                 # the explained variances are returned
                 # (Fig. 3.14, right)
```

3.7.3 Diagnostic Plots

PCA is sensitive with respect to outlying observations. Outliers could spoil the classical estimation of the PCs, and thus robust PCA versions are preferable (Section 3.5). Basically, there are two different types of outliers that need to be distinguished: LEVERAGE POINTS and ORTHOGONAL OUTLIERS. Figure 3.15 shows three-dimensional data where the first two PCs are used to approximate the data. Most of the objects (regular observations) are lying in the PCA space, and their distribution is visualized by ellipses. However, there are some objects being further away. Object 1 has a large ORTHOGONAL DISTANCE to the PCA space (orthogonal outlier) which is not visible when inspecting the projected data in the PCA space. Such an orthogonal outlier can have an effect on classical PCA. Object 2 has a large orthogonal distance and a large SCORE DISTANCE which means that its projection in the PCA space is far away from the center. Objects of this type are called BAD LEVERAGE POINTS because they can lever the estimation of the PCA space. Finally, object 3 is called a GOOD LEVERAGE POINT because it has a large score distance but a small orthogonal distance. This type of outliers even stabilizes the estimation of the PCA space.

For diagnostics it will be interesting to compute score distance and orthogonal distance for each object and to plot them together with critical boundaries; this will allow distinguishing regular observations from outliers. The score distance SD_i of object i is computed by

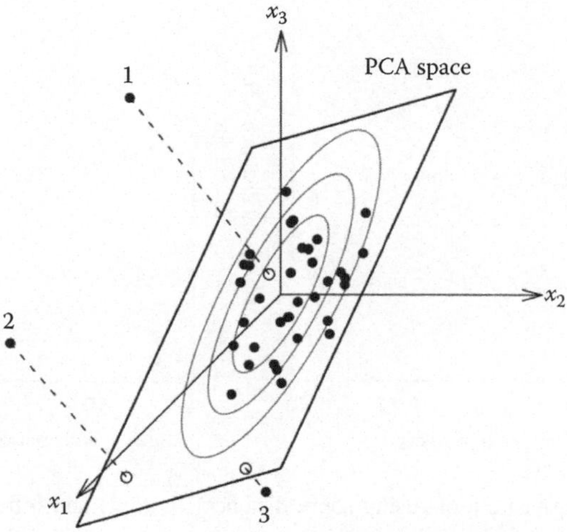

FIGURE 3.15 Visualization of the different types of outliers that can be influential to classical PCA.

$$\mathrm{SD}_i = \left[\sum_{k=1}^{a} \frac{t_{ik}^2}{v_k} \right]^{1/2} \tag{3.27}$$

where

a is the number of PCs forming the PCA space
t_{ik} are the elements of the score matrix $\boldsymbol{T}$
v_k is the variance of the kth PC

If the data majority is multivariate normally distributed, the squared score distances can be approximated by a chi-square distribution, χ_a^2, with a degrees of freedom. Thus, a cutoff value for the score distance is the 97.5% quantile $\sqrt{\chi^2_{a,0.975}}$. Values of SD_i larger than this cutoff value are leverage points (good or bad leverage points, depending on the orthogonal distance).

The orthogonal distance OD_i of object i is computed by

$$\mathrm{OD}_i = ||\boldsymbol{x}_i - \boldsymbol{P} \cdot \boldsymbol{t}_i^{\mathrm{T}}|| \tag{3.28}$$

where

$\boldsymbol{x}_i$ is the ith object of the centered data matrix
$\boldsymbol{P}$ is the loading matrix using a PCs
$\boldsymbol{t}_i^{\mathrm{T}}$ is the transposed score vector of object i for a PCs

For the cutoff value, Hubert et al. (2005) take as approximate distribution of $\mathrm{OD}^{2/3}$ the normal distribution, where center and spread of the $\mathrm{OD}^{2/3}$-values are robustly estimated, e.g., by the median and the MAD. The cutoff value is then

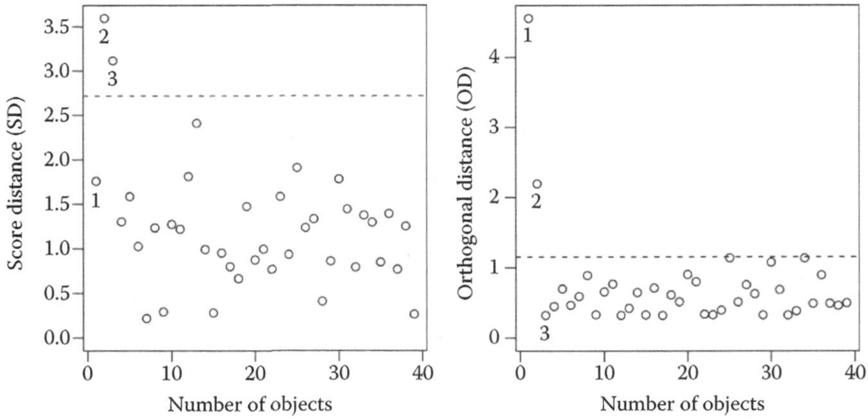

FIGURE 3.16 Diagnostic plots using score distance SD (left) and orthogonal distance OD (right). The lines indicate critical values separating regular observations from outliers (97.5%). Object 1 is an orthogonal outlier not situated in the PCA space; object 2 is a bad leverage outlier (high score distance and high orthogonal distance); object 3 is a good leverage outlier (see Figure 3.15).

$(\text{median}(\text{OD}^{2/3}) + \text{MAD}(\text{OD}^{2/3}) \cdot z_{0.975})^{3/2}$, with $z_{0.975}$ being the 97.5% quantile (value 1.96) of the standard normal distribution.

Figure 3.16 shows the diagnostic plots using the data from Figure 3.15. The left plot shows the score distances; the right plot shows the orthogonal distances. The cutoff values are presented as horizontal lines. Object 1 is identified as outlier in the orthogonal direction to the PCA space. Objects 2 and 3 are leverage points, the first is a bad leverage point having high orthogonal distance, and the second a good leverage point. The cutoff value for SD is $2.72 = \sqrt{\chi^2_{2,0.975}}$. The cutoff value for OD is $1.15 = (0.644 + 0.233 \cdot 1.96)^{3/2}$ (using the median 0.644 of the OD-values, and 0.233 for the MAD $(\text{OD}^{2/3})$).

In $\mathbb{R}$, such a diagnostic plot can be obtained as follows:

```
R:  library(chemometrics)
    X_pca <- princomp(X,cor=TRUE)
    res <- pcaDiagplot(X,X_pca,a=2)  # generates diagnostic
                       # plots using 2 PCA components (Fig. 3.16)
    res$SDist          # score distances
    res$ODist          # orthogonal distances
```

In literature the above diagnostic measures are known under different names. Instead of the score distance from Equation 3.27 which measures the deviation of each observation within the PCA space, often the Hotelling T^2-test is considered. Using this test a confidence boundary can be constructed and objects falling outside this boundary can be considered as outliers in the PCA space. It can be shown that this concept is analogous to the concept of the score distance. Moreover, the score distances are in fact Mahalanobis distances within the PCA space. This is easily seen because for centered data the squared Mahalanobis distance is (compare Equation 2.19)

$$d^2(\boldsymbol{x}_i) = \boldsymbol{x}_i^{\mathrm{T}} \cdot \boldsymbol{C}^{-1} \cdot \boldsymbol{x}_i = \boldsymbol{t}_i^{\mathrm{T}} \cdot \boldsymbol{P}^{\mathrm{T}} \cdot \boldsymbol{C}^{-1} \cdot \boldsymbol{P} \cdot \boldsymbol{t}_i = \boldsymbol{t}_i^{\mathrm{T}} \cdot \boldsymbol{S}^{-2} \cdot \boldsymbol{t}_i = \Sigma_k (t_{ik}^2 / v_k) \quad \text{for } i = 1, \ldots, n \tag{3.29}$$

which corresponds to an equation defining an ellipsoid. The last equality in Equation 3.29 equals the definition of the score distance in Equation 3.27. For $a = 2$ extracted PCs, it is also possible to visualize the critical value $\sqrt{\chi^2_{a,0.975}}$ in the PCA space, because it corresponds to an ellipse that covers 97.5% of the data points (in case of normal distribution) and thus forms a confidence ellipse, see Figure 3.15. For $a > 2$ components, a visualization as in Figure 3.16 is preferable. To summarize, the concept of the score distance is equivalent to the concept of multivariate outlier identification from Section 2.5, but here we are not searching for outliers in the whole space but only in the space of the PCs.

Note that the approximation by the chi-square distribution is only possible for multivariate normally distributed data which somehow is in conflict if outliers are present that should be identified with this measure. We recommend that robust PCA is used whenever diagnostics is done because robust methods tolerate deviations from multivariate normal distribution.

The orthogonal distance can be seen as a measure of LACK OF FIT. Since the orthogonal distance measures the distance in the orthogonal direction of the PCA space, this measure expresses how well the PCs cover the information of an object. If the object is far away, the PCA space would need to be larger (i.e., more components) in order to explain the information of this object. On the other hand, distant objects can lever the PCA space (if it is estimated in a classical way) and thus "attract" PCs. This will cause that such objects are well explained by the spoiled PCA space, but other more homogeneous objects are not. In this way, nonoutliers will appear as objects with large orthogonal distance and outliers that levered the estimation with small orthogonal distance. This effect is usually not wanted and therefore robust PCA should be preferred.

3.8 COMPLEMENTARY METHODS FOR EXPLORATORY DATA ANALYSIS

In chemometrics, PCA is the most used method for exploratory analysis of multivariate data. The following sections describe briefly other methods for exploratory data analysis, most of them are complementary methods to PCA, and are nonlinear. Nonlinearities can be introduced into a linear method by nonlinear transformations of variables (for instance, squares or logarithms), or by adding new variables that are nonlinear functions of original variables (for instance, cross-products). On the other hand, real nonlinear methods exist for exploratory data analysis such as Kohonen mapping, cluster analysis with a dendrogram, and Sammon's nonlinear mapping (NLM). If a linear projection of the variable space is not very informative because of a complicated data structure, such methods may allow a better insight into the data structure than PCA. Surprisingly, many examples of exploratory data analysis in chemometrics can be handled successfully by linear methods like PCA. Advantages of PCA are the clear definition of the method and the interpretability of the results via score and loading plots. Nonlinear approaches have their merits if linear methods fail (often only demonstrated with artificial data), but they suffer from less strictly defined methods, many adjustable parameters, too high expectations, and aspects of fashion. Moreover, for data with few objects and many variables, linear methods are often the only possibility to avoid overfitting.

3.8.1 Factor Analysis

In contrast to PCA which can be considered as a method for basis rotation, factor analysis is based on a statistical model with certain model assumptions. Like PCA, factor analysis also results in dimension reduction, but while the PCs are just derived by optimizing a statistical criterion (spread, variance), the factors are aimed at having a real meaning and an interpretation. Only a very brief introduction is given here; a classical book about factor analysis in chemistry is from Malinowski (2002); many other books on factor analysis are available (Basilevsky 1994; Harman 1976; Johnson and Wichern 2002).

The interpretation of the resulting factors is based on the loading matrix $\boldsymbol{P}_{\mathrm{FA}}$ which are the coefficients for the linear combinations similar to PCA

$$\boldsymbol{X} = \boldsymbol{T}_{\mathrm{FA}} \cdot \boldsymbol{P}_{\mathrm{FA}}^{\mathrm{T}} + \boldsymbol{E} \tag{3.30}$$

with the score matrix $\boldsymbol{T}_{\mathrm{FA}}$ (factors). There is also an error matrix $\boldsymbol{E}$ because the number of factors (components) is reduced to $1 \leq a < m$. Moreover, for each variable a certain proportion of variance (UNIQUENESS) is considered that should not be explained by the joint factors but be left unexplained. Thus the factors only account for the common variability of the variables but not for the variables' unique variability. This should enhance the interpretability of the factors which can be further improved by a rotation. Essentially, after ROTATION the loading matrix should consist of either high (absolute) values or values near zero because such a pattern will allow to see the influence of the variables to the factors (big positive or negative influence, or no influence). A well-known method for orthogonal rotation is VARIMAX, for oblique rotation (the final axes are not orthogonal) there is PROMAX or OBLIMIN.

Factor analysis with the extraction of two factors and varimax rotation can be carried out in ℝ as described below. The factor scores are estimated with a regression method. The resulting score and loading plots can be used as in PCA.

```
ℝ: X_fa <-
   factanal(X,factors=2,rotation="varimax",scores="regression")
   P_fa <- X_fa$loadings                        # loadings
   T_fa<-X_fa$scores                             # scores
```

3.8.2 Cluster Analysis and Dendrogram

Cluster analysis will be discussed in Chapter 6 in detail. Here we introduce cluster analysis as an alternative nonlinear mapping technique for exploratory data analysis. The method allows gaining more insight into the relations between the objects if a linear method like PCA fails. This relation can be measured by the pairwise distances, and the distance measures presented in Section 2.4 are frequently used. Then a hierarchy of the objects is constructed according to their similarity. Here only an agglomerative, hierarchical clustering method is considered. At the beginning each object forms its own cluster. The process is started by combining the two most similar single-object clusters in one larger cluster. Using a similarity measure between clusters, the clusters with the smallest similarity can be determined and combined into a new larger cluster. This process is repeated until all objects end up in only one big cluster. This hierarchy of similarities can be displayed by a DENDROGRAM.

Figure 3.17 (left) shows a dendrogram of the glass vessels data containing samples of groups 2–4 with $n = 35$ and $m = 13$ (Janssen et al. 1998). The vertical axis (height) represents the similarity between the clusters, and the horizontal axis shows the objects in a special ordering to avoid line crossing in the dendrogram. Horizontal lines indicate when clusters are combined, and their vertical position shows the cluster similarity. The dendrogram shows that the objects within the three groups are combined in an early stage (at a small height), then the clusters consisting of objects from groups 3 and 4 are combined, and finally at a considerably larger height all objects are combined. This means that the objects form very clear groups, and that the groups correspond to the different types of glass vessels. Note that the information of group membership was not used during the cluster analysis. Therefore, this method is also called "unsupervised." The resulting dendrogram may heavily depend on the used distance measure and on the applied clustering algorithm. For comparison, the PCA score plot for the first two PCs is shown in Figure 3.17 (right); it represents 82.2% of the total variance. It also indicates a separation of the groups and a higher similarity between groups 3 and 4 compared to group 2. In this example PCA and dendrogram give similar results. In general PCA score plots and cluster analysis are complementing methods for exploratory data analysis.

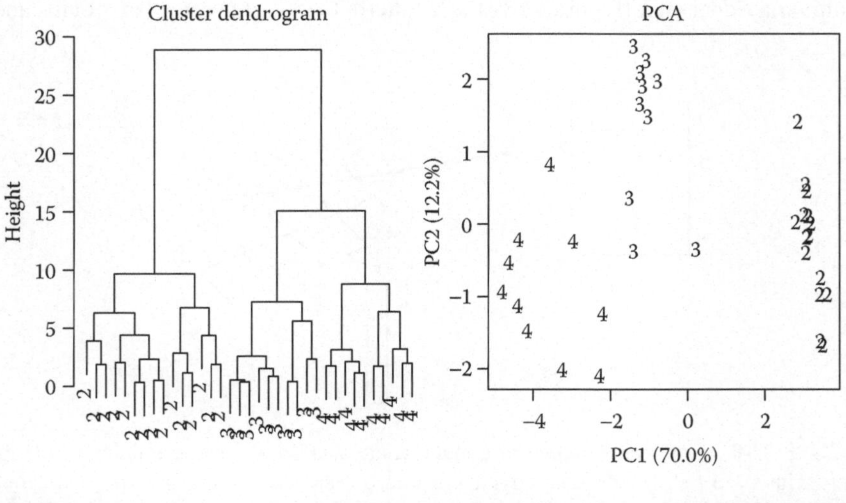

FIGURE 3.17 Glass vessels data containing groups 2–4: Dendrogram of hierarchical cluster analysis (left) and projection on the first two PCs (right) produce similar results. (Data from Janssen, K.H.A., De Raedt, I., Schalm, O., and Veeckman, J., *Microchim. Acta*, 15, 253, 1998.)

In ℝ, the dendrogram can be obtained by

```
ℝ:  X_dist <- dist(X)               # distance matrix (Euclidean)
    X_clust <- hclust(X_dist)       # hierarchical clustering
    plot(X_clust)                   # plots a dendrogram
```

3.8.3 Kohonen Mapping

Kohonen maps are named after the Finnish mathematician Teuvo Kohonen who invented this method and called it SELF-ORGANIZING MAPS (SOM) (Kohonen 1995, 2001). Kohonen mapping is a nonlinear method to represent high-dimensional data in a typically two-dimensional plot (map)—similar to, for instance, the score plot in PCA but not by a linear projection. The method is considered to belong to artificial neural networks, and has even been propagated as a model of the biological brain function; unfortunately it is often overloaded with biological nomenclature. Kohonen mapping and a number of applications in chemoinformatics have been described by Zupan and Gasteiger (1993, 1999). Main application of Kohonen mapping in chemometrics is exploratory data analysis (cluster analysis). Typically the objects—represented by multivariate data—are assigned to squares of a chessboard-like map. During an iterative process, areas in the map containing similar objects are automatically formed. The aim of Kohonen mapping is to assign similar objects to neighboring squares. An interesting version uses a toroid-like map (Zupan and Gasteiger 1999).

The basic theory of Kohonen maps—and only this will be treated here—is mathematically simple. A typical Kohonen map consists of a rectangular (often quadratic) array of fields (squares, cells, nodes, NEURONS) with a typical size of 5×5 (25 fields) to 100×100 (10,000 fields). Each field k is characterized by a vector $\boldsymbol{w}_k$, containing the weights $w_{k1}, w_{k2}, \ldots, w_{km}$, with m being the number of variables of a multivariate data set $\boldsymbol{X}$ (Figure 3.18); the lengths of the weight vectors are, for instance,

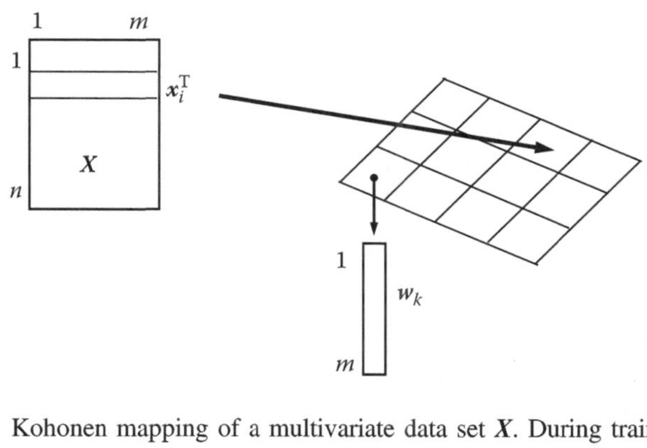

FIGURE 3.18 Kohonen mapping of a multivariate data set $\boldsymbol{X}$. During training of the map (here with 4×3 fields/nodes/neurons) the neuron is searched containing a weight vector $\boldsymbol{w}_k$ that is most similar to object vector $\boldsymbol{x}_i$. The winner neuron and its neighbors are adapted to become even more similar to $\boldsymbol{x}_i$. The object vectors are treated sequentially and repeatedly, resulting in a self-organizing of the map with similar objects located in neighboring neurons.

normalized to the value 1. The map is trained by an iterative procedure starting with random numbers for the components of the weight vectors (normalized to length 1). The object vectors are ordered randomly and are sequentially used by the training algorithm. For each object vector $\boldsymbol{x}_i$, distance measures (similarity measures, etc.), d_{ik}, to all weight vectors $\boldsymbol{w}_k$ are computed to find the most similar weight vector. For instance, the squared Euclidean distance is computed by

$$d_{ik}^2 = (\boldsymbol{x}_i - \boldsymbol{w}_k)^{\mathrm{T}} \cdot (\boldsymbol{x}_i - \boldsymbol{w}_k) \tag{3.31}$$

The winner weight vector, $\boldsymbol{w}_{\mathrm{c}}$ (c for central), is adjusted to make it even more similar—but not identical—to $\boldsymbol{x}_i$. The adjusted weight vector $\boldsymbol{w}_{c,\mathrm{NEW}}$ can be computed by

$$\boldsymbol{w}_{\mathrm{c,NEW}} = (1 - \tau)\boldsymbol{w}_{\mathrm{c}} + \tau(\boldsymbol{x}_i - \boldsymbol{w}_k) \tag{3.32}$$

with τ being the learning factor ($0 \leq \tau \leq 1$). Not only the winner weight vector is corrected, but also the weight vectors of neighboring neurons—usually the correction is smaller if the spatial distance from neuron c gets larger. The size of the considered neighborhood influences the number of empty fields (with no object assigned to) in the final map. The new weight vectors replace the original ones. One speaks about an epoch (a training cycle) if all objects have been treated once (in random sequence). The training cycles are repeated many times until convergence is reached (only minor changes of the weight vectors). It is essential to decrease the learning factor with the number of epochs (defined by a monotonic function). Also the size of the considered neighborhood of a central neuron can be diminished with the number of epochs.

Usually, the training process automatically assigns parts of the map to groups of similar objects (self-organization). After training, each object is assigned to one of the fields of the map (defined by one integer coordinate for the horizontal position, and one for the vertical position of the field). Different groups of objects usually build more or less separated areas in the map (with the weight vectors corresponding to prototype objects); some parts of the map may remain empty. The result of Kohonen mapping can be graphically represented with the number of objects per neuron indicated, for instance, by a color scale, a bar graph, a pie plot, or simply by numbers. New objects can be mapped by using the final weight vectors; the most similar weight vector determines the cluster of objects to which the new object is classified—or if falling into an empty region of the map, indicates a "really new" object.

Kohonen maps are advantageously used for exploratory data analysis if the linear method PCA fails (Melssen et al. 1993; Zupan and Gasteiger 1999). Disadvantages of this approach are (1) different but poorly documented algorithms are implemented in various software products; (2) the large number of adjustable parameters may be confusing; (3) results depend on initial values; and (4) for large data sets, extensive memory requirements and long computation times are restricting. An extension of the method—named counterpropagation network—has been developed for multivariate calibration or classification and the investigation of relationships between sets of variables (Zupan and Gasteiger 1999), for instance, between IR data and 3D chemical structures (Hemmer and Gasteiger 2000).

TABLE 3.2
Result of Kohonen Mapping for the Reduced Glass Vessels Data Containing Three Groups of Samples

Cell Number	x	y	Number of Objects Class 2	Class 3	Class 4	Sum
1	1	1	0	0	0	0
2	2	1	1	0	0	1
3	3	1	0	7	0	7
4	4	1	0	0	0	0
5	1	2	4	0	0	4
6	2	2	0	0	0	0
7	3	2	0	0	0	0
8	4	2	0	1	1	2
9	1	3	3	0	0	3
10	2	3	7	0	0	7
11	3	3	0	1	0	1
12	4	3	0	0	6	6
13	1	4	0	0	0	0
14	2	4	0	1	0	1
15	3	4	0	0	1	1
16	4	4	0	0	2	2

Source: Data from Janssen, K.H.A., De Raedt, I., Schalm, O., and Veeckman, J., *Microchim. Acta*, 15, 253, 1998.

Note: A map with 4×4 fields was taken resulting in 16 cells (rows of the table). x and y are the field positions in horizontal and vertical direction. The numbers of objects assigned to the cells are given for the object classes 2, 3, 4, and in total.

As an example we consider the groups 2–4 of the glass vessels data ($n = 35$, $m = 13$ (Janssen et al. 1998)). Table 3.2 lists the results of Kohonen mapping. A map with 4×4 fields was taken resulting in 16 cells corresponding to the rows in the table. The columns x and y refer to the horizontal and vertical position in the map, starting at the lower left corner. The numbers in Columns "Class 2," "Class 3," and "Class 4" are the frequencies how many objects from the corresponding groups are assigned to the cells. For example, to cell number 2 (position $x = 2$ and $y = 1$) 1 object of group 2 is assigned and none of groups 3 and 4. The last column in the table is the sum of assigned objects in each cell.

ℝ:
```
library(som)               # package for Self Organizing Maps
library(chemometrics)      # needed for plotting the results
Xs <- scale(X)             # autoscaling (mean 0, variance 1)
Xn <- Xs/sqrt(apply(Xs^2,1,sum))
                       # normalize length of row vectors to 1
X_SOM <- som(Xn,xdim=4,ydim=4)   # SOM for 4x4 fields
plotsom(X_SOM,grp,type="num")    # plot results with numbers
plotsom(X_SOM,grp,type="bar")    # plot results with bars
```

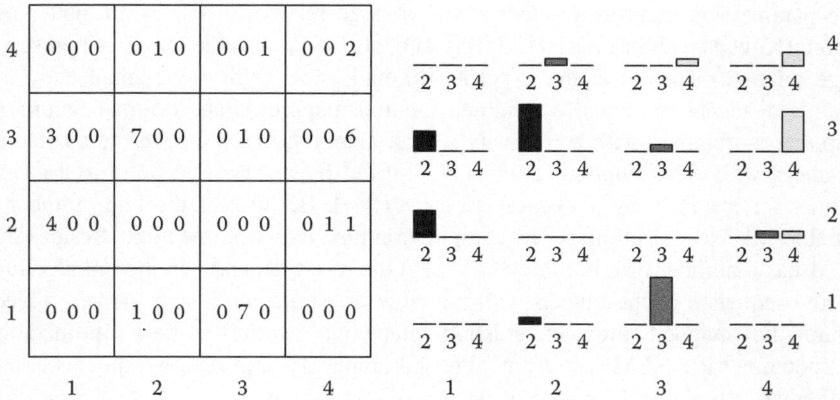

FIGURE 3.19 Result of Kohonen mapping for the reduced glass vessels data containing three groups of samples. A map with 4×4 fields was taken resulting in 16 cells. The numbers in the left plot are the number of objects in the three groups that have been assigned to the cells. In the right plot the numbers are visualized by bars. (Data from Janssen, K.H.A., De Raedt, I., Schalm, O., and Veeckman, J., *Microchim. Acta*, 15, 253, 1998.)

Figure 3.19 visualizes the results of Kohonen mapping presented in Table 3.2. The left plot shows the map with 4×4 fields. The numbers in each cell are the numbers of objects from each group that were assigned to the cell, see Table 3.2. The right plot in Figure 3.19 is a visualization of the numbers from the left plot in terms of barplots. The heights of the bars represent the numbers of assigned objects of each group. This plot gives a good visual impression about the ability of the Kohonen map to separate the groups. Overall, the separation is very successful, only the field (4, 2) results in an overlap.

3.8.4 Sammon's Nonlinear Mapping

Nonlinear mapping (NLM) as described by Sammon (1969) and others (Sharaf et al. 1986) has been popular in chemometrics. Aim of NLM is a two-(eventually a one- or three-) dimensional scatter plot with a point for each of the n objects preserving optimally the relative distances in the high-dimensional variable space. Starting point is a distance matrix for the m-dimensional space applying the Euclidean distance or any other monotonic distance measure; this matrix contains the distances of all pairs of objects, d_{ik}. A two-dimensional representation requires two map coordinates for each object; in total $2n$ numbers have to be determined. The starting map coordinates can be chosen randomly or can be, for instance, PCA scores. The distances in the map are denoted by d_{ik}^*. A mapping error ("stress," loss function) E_{NLM} can be defined as

$$E_{\text{NLM}} = \frac{1}{\sum\limits_{k<i} \left(d_{ik}^*\right)^p} \sum_{k<i} \frac{\left(d_{ik}^* - d_{ik}\right)^2}{\left(d_{ik}^*\right)^p} \quad \text{for } i = 1, \ldots, n \tag{3.33}$$

The parameter p controls whether small or large relative distances are better preserved (Kowalski and Bender 1973; Sharaf et al. 1986). A value of $p=-2$ preserves large relative distances at the expense of small ones. With $p=2$ small and large relative distances are equally weighted (neutral mapping). The original Sammon's mapping is obtained with $p=1$; it is a local mapping that preserves small relative distances well. The mapping error is minimized by an iterative optimization procedure, for instance, by a steepest ascent method. Because of the high number of variables ($2n$) to be optimized the computation time may become huge. Besides this, NLM has a number of other drawbacks: (1) the result depends on the initial values, (2) the sequence of the objects, (3) the value of parameter p, and (4) new objects cannot be mapped into an existing map but require a new optimization. In chemometrics NLM has been almost completely replaced by the faster and mathematically easier Kohonen mapping.

Sammon's NLM is one form of MULTIDIMENSIONAL SCALING (MDS). There exist a number of other MDS methods with the common aim of mapping the similarities or dissimilarities of the data. The different methods use different distance measures and loss functions (see Cox and Cox 2001).

For an illustration, the reduced glass vessels data set (only groups 2–4, $n=35$, $m=13$; Janssen et al. 1998) is used. NLM is applied on the Euclidean distance matrix. Figure 3.20 (left) shows the NLM result for $p=1$. This presentation allows a very good separation of the three groups. Figure 3.20 (right) presents the results of NLM for $p=-2$ (preserving large distances at the expense of small ones). Now, the main parts of the groups are even better separated but the groups 3 and 4 show an outlier.

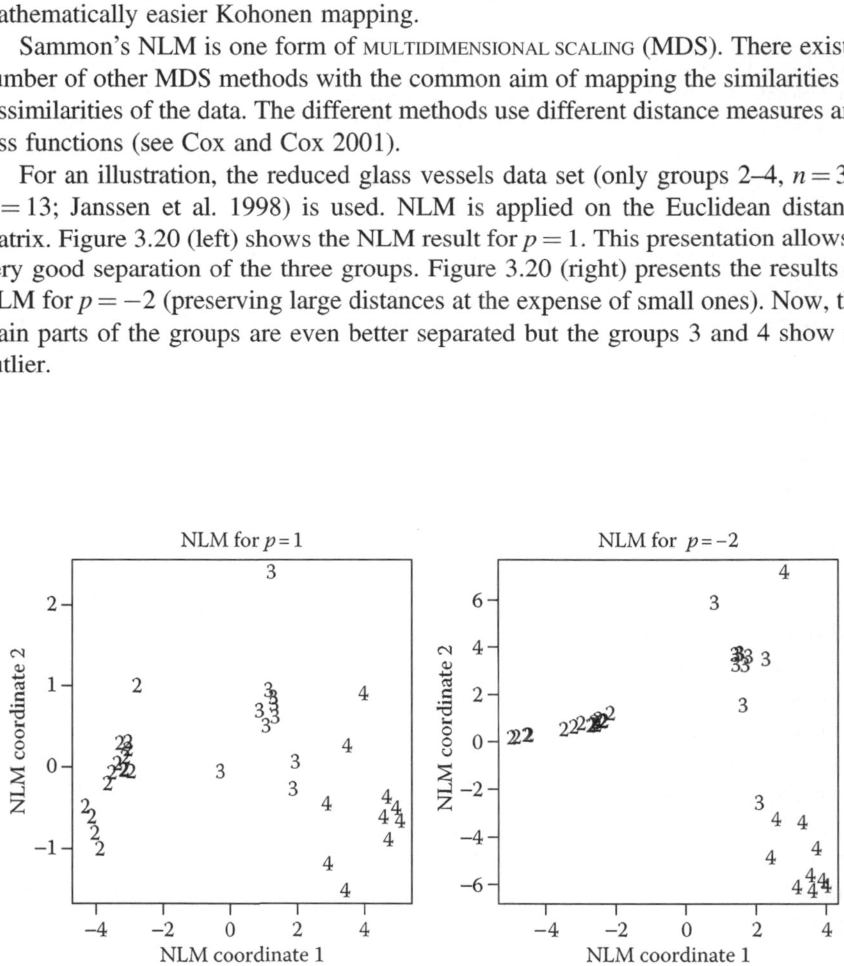

FIGURE 3.20 Sammon's NLM of groups 2–4 of the glass vessels data. Left: NLM for $p=1$; right: NLM for $p=-2$. (Data from Janssen, K.H.A., De Raedt, I., Schalm, O., and Veeckman, J., *Microchim. Acta*, 15, 253, 1998.)

```
ℝ: X_dist <- dist(scale(X))         # distance matrix of the
                                    # autoscaled data
   library(MASS)
   sam1 <- isoMDS(X_dist,p=1)       # NLM for p=1
   plot(sam1$points)
   sam2 <- isoMDS(X_dist,p=-2)      # NLM for p=-2
   plot(sam2$points)
```

3.8.5 Multiway PCA

A data matrix can be considered as a two-way array, with the objects and variables forming the two different "ways." In some applications it is necessary to extend this scheme to multiway arrays. For example, a three-way array (a rectangular block of data, Figure 3.21) occurs if the variables of objects are measured at various time points; applications in chemistry are data from hyphenated methods (chromatography combined with spectroscopy). There are several possibilities to analyze such data; only basic ideas are presented here. The interested reader is referred to Smilde et al. (2004).

The simplest versions are UNFOLDING METHODS where the multiway array is unfolded to a matrix. In above example the objects measured at different time points can be arranged as blocks of rows in a single matrix. Usual decomposition methods like PCA can then be applied. Of course, in this case the three-way nature of the data is ignored.

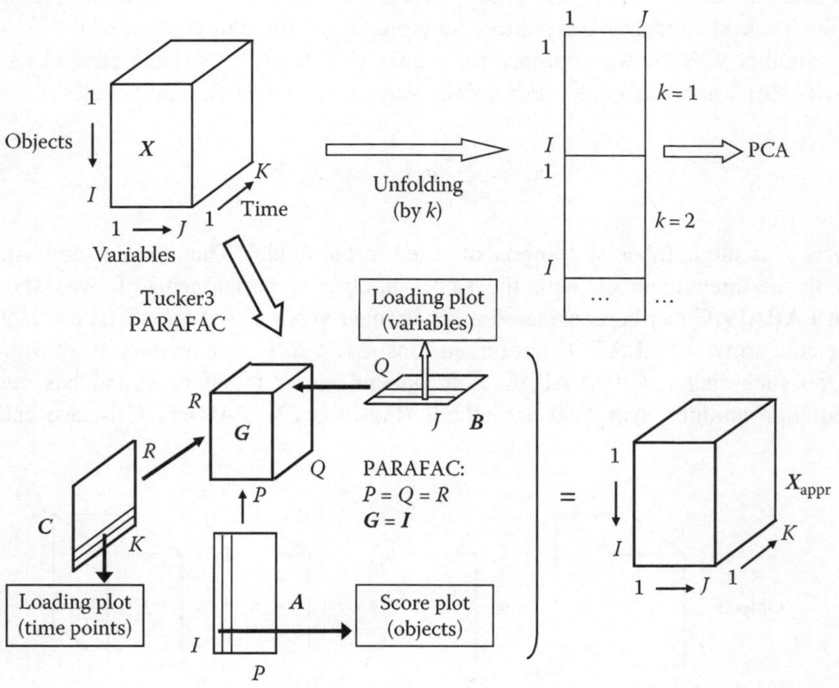

FIGURE 3.21 Three-way methods: Unfolding Tucker3 and PARAFAC. Resulting loading plots and score plot are shown for using the first two components.

Another possibility is the TUCKER3 model where a decomposition of the array into sets of scores and loadings is performed that should describe the data in a more condensed form than the original data array. For the sake of simplicity we will describe the model for a three-way array, but it is easy to extend the idea to multiway data. Let x_{ijk} denote an element of a three-way array $\boldsymbol{X}$ of dimension $I \times J \times K$. Basic assumption is that the data are influenced by a relatively small set of driving forces (factors). Then the Tucker3 model is defined as

$$x_{ijk} = \sum_{p=1}^{P} \sum_{q=1}^{Q} \sum_{r=1}^{R} a_{ip} b_{jq} c_{kr} g_{pqr} + e_{ijk} \tag{3.34}$$

where

e_{ijk} is an element of the error term (three-way array $\boldsymbol{E}$)
a_{ip}, b_{jq}, and c_{kr} are elements of loading matrices $\boldsymbol{A}, \boldsymbol{B}, \boldsymbol{C}$ with dimensions $I \times P$, $J \times Q$, $K \times R$, respectively (Figure 3.21)
g_{pqr} is an element of the core-array $\boldsymbol{G}$ with dimension $P \times Q \times R$

Thus for each mode a factorization (decomposition into scores and loadings) is done, expressed by three matrices and a three-way core-array $\boldsymbol{G}$. The matrices $\boldsymbol{A}$, $\boldsymbol{B}$, $\boldsymbol{C}$, and $\boldsymbol{G}$ are computed by minimizing the sum of squared errors; the optimum number of factors, P, Q, R, can be estimated by cross validation. In a similar manner the Tucker2 model can be defined which reduces only two of the three modes, as well as the Tucker1 model which reduces only one of the three modes.

Another well-known approach for multiway data analysis is the parallel factor (PARAFAC) analysis model. For a three-way array, the PARAFAC model is

$$x_{ijk} = \sum_{r=1}^{R} a_{ir} b_{jr} c_{kr} + e_{ijk} \tag{3.35}$$

where R is the number of components used in the model. Thus the decomposition results in three matrices with the same number of components. It was shown that PARAFAC can be considered a constrained version of Tucker3 (Kiers 1991). The core array in PARAFAC has dimensions $R \times R \times R$; it is an identity array with 1s in the superdiagonal. PARAFAC is in general easier to interpret and has better prediction abilities than Tucker methods (Figure 3.22). PARAFAC is also called

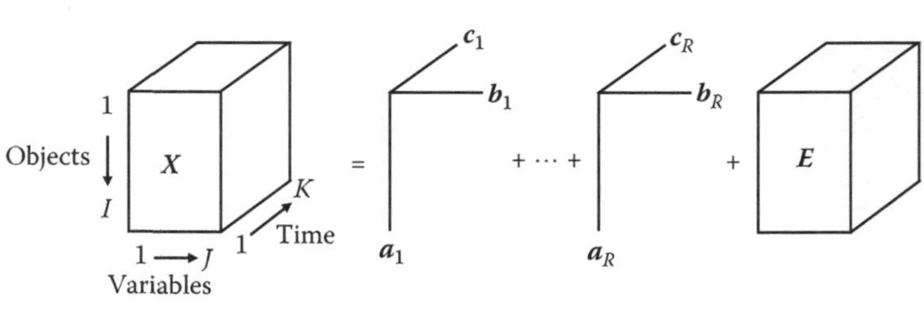

FIGURE 3.22 PARAFAC model for three-way data $\boldsymbol{X}$. R components are used to approximate $\boldsymbol{X}$ by a trilinear model as defined in Equation 3.35.

trilinear because if two modes are fixed (e.g., the a's and the b's) then the third mode is linear (the c's).

3.9 EXAMPLES

3.9.1 Tissue Samples from Human Mummies and Fatty Acid Concentrations

The concentration profiles of fatty acids are characteristic for many types of biological stuff, and such data have been frequently evaluated by multivariate data analysis. In a pioneering work by Forina and Armanino (1982), it was shown that Italian olive oil can be successfully classified according to the geographical origin using the concentrations of eight fatty acids. Recent works in food chemistry using chemometrics for fatty acid profiles deal, for instance, with apple juice (Blanco-Gomis et al. 2002), rainbow trouts (Barrado et al. 2003), sausages (Stahnke et al. 2002), and olive oils (Mannina et al. 2003). Medical applications deal with the classification of human tissues (Greco et al. 1995), insulin resistance (Manco et al. 2000), and bacteria (Xu et al. 2000).

The data used in this example are from anthropology. A set of $n = 34$ tissue samples from human mummies and references have been characterized by the concentrations of eight fatty acids ($m = 8$). These data have been selected from a larger data set worked out and evaluated in a project about the Tyrolean Iceman (Makristathis et al. 2002; Varmuza et al. 2005).

The used samples can be divided into six groups as follows (Table 3.3). Group 1 consists of two samples from the Tyrolean Iceman (Spindler et al. 1996), a well-preserved mummy from Neolithic time, found in 1991 on a glacier near the Austrian–Italian border in South Tyrol, Italy, with a burial time of 5200 years. Group 2 consists of nine samples from two corpses found on Austrian glaciers, with burial times of 29 and 57 years. Group 3 consists of three samples from a corpse that was 50 years in the cold water 50 m deep in the mountain lake Achensee, Austria. Group 4 consists of two samples from a freeze-dried Inca child mummy (Ice Maiden) found near the summit of Mount Ampato (6288 m), Peru, with a burial time of about 500 years. Group 5 includes only one sample from a mummy found in the extreme dry Ilo desert, Peru, with a burial time of 1000 years. Group 6 consists of 17 fresh tissue samples from three

TABLE 3.3
Tissue Samples from Human Mummies

Group Number	Group Name	n	Location	Burial Time (Years)
1	Iceman	2	Italy; glacier; 3200 m	5200
2	Glacier	9	Austria; glaciers; 2700–2800 m	29–57
3	Lake	3	Austria; Achensee; 50 m depth	50
4	Ampato	2	Peru; Mt. Ampato; 6200 m	500
5	Desert	1	Peru; Ilo desert	1000
6	Fresh	17	Austria	—

TABLE 3.4
Fatty Acids Used to Characterize Tissue Samples from Human Mummies Listed in Sequence of Eluation in GC Analysis

				Mean of Sample Groups (*n*)		
No.	Code	Name	Chemical Formula	Glacier (9)	Lake (3)	Fresh (17)
1	14:0	Myristic acid	$C_{14}H_{28}O_2$	5.3	18.0	3.7
2	15:0	Pentadecanoic acid	$C_{15}H_{30}O_2$	0.0	1.1	0.0
3	16:1	Palmitoleic acid	$C_{16}H_{30}O_2$	0.3	0.0	6.8
4	16:0	Palmitic acid	$C_{16}H_{32}O_2$	33.4	41.0	26.0
5	18:2	Linoleic acid	$C_{18}H_{32}O_2$	0.0	0.0	13.9
6	18:1	Oleic acid	$C_{18}H_{34}O_2$	4.2	3.4	36.5
7	16:0, 10 OH	10-Hydroxy palmitic acid	$C_{16}H_{32}O_3$	2.9	0.0	0.0
8	18:0, 10 OH	10-Hydroxy stearic acid	$C_{18}H_{36}O_3$	36.4	20.3	0.0

Note: The concentrations, in percent of total fatty acids, are used as variables.

recently deceased people. The samples are very diverse in origin, age, environmental conditions, and altitude of finding place; however, all mummy samples were naturally mummified.

The concentrations of eight selected fatty acids (Table 3.4) have been determined by gas chromatography as previously described (Makristathis et al. 2002); they are given in percent of all fatty acids captured by the used analytical technique. The variable matrix $\boldsymbol{X}$ contains 34 rows for the samples, and 8 columns with the fatty acid concentrations.

The primary aim of data evaluation is an exploratory investigation for obtaining an overview and insight into the data, and eventually to recognize relationships. For this purpose PCA is the method of choice. Furthermore, the capabilities of fatty acid profile data should be examined for information about the origin, age, and fate of mummies. Of special interest is the modification of fatty acids in corpses that have spent a long time in ice or cold water. Under these conditions a gray wax-like mixture of lipids is formed from body fat; it is called ADIPOCERE. Mummies from groups 1–3 show this typical conservation. Adipocere is built by conversion of unsaturated fatty acids—predominant in fresh tissue—into saturated fatty acids, hydroxy fatty acids, and oxo fatty acids. These reaction products have higher melting points than the unsaturated fatty acids and appear as a wax. A multivariate data evaluation (considering all eight fatty acid concentrations instead of single ones) should aim in a satisfying data interpretation.

Figure 3.23 shows scatter plots from PCA comprising the first three PCs. Usually, the score plot for PC1 and PC2 gains the first interest, and one should first look how much of the total variance is preserved by the projection axes. PC1 preserves 52.3% and PC2 24.7%; the sum of 77.0% is high enough for a good representation of the eight-dimensional variable space in the two-dimensional projection. Eventually, also PC3 with 9.6% of total variance can be considered by the score plot PC3 versus PC1 giving additional information about the distances of the samples.

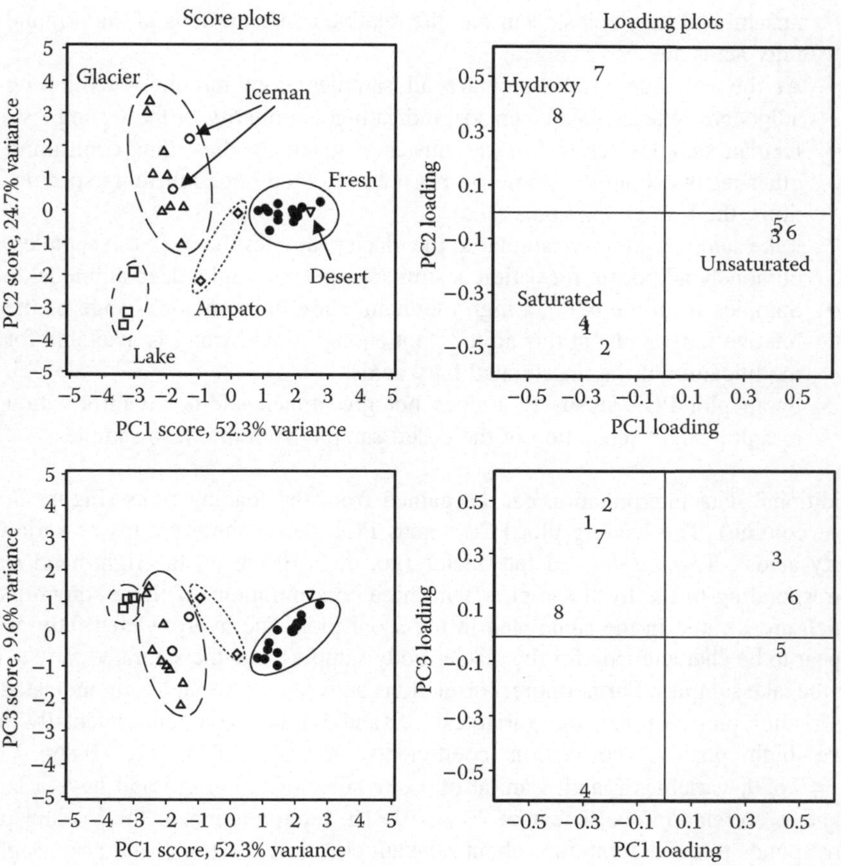

FIGURE 3.23 Scatter plots from PCA of fatty acid data characterizing tissue samples from human mummies and fresh reference samples.

The score plot PC2 versus PC1 shows a good clustering of the objects according to the groups defined in Table 3.3, although not all classes are completely separated. Note that no group information is used by PCA, and the sequence of the rows in the X-matrix is meaningless; the ellipses surrounding the object classes have been drawn manually. PCA is not a classification method (optimizing the class separation) but represents the distances between object points optimally. Any class separation obtained by PCA therefore indicates that the data contain class information—that may be enhanced by the application of real classification methods. We interpret the score plots as follows:

- Fresh reference samples form a compact cluster; the samples are all very similar (as characterized by their fatty acid concentrations).
- The 1000-year-old sample from the desert mummy is located within the cluster of the fresh samples. At a first glance this similarity is surprising, but a simple explanation is that without water no chemical reactions of the

unsaturated fatty acids occur and the relative concentrations of the original fatty acids are less changed.

- At the left side of the plot are all samples from mummies exhibiting adipocere (glacier, lake, iceman), indicating a similarity of these samples.
- Iceman samples are within the cluster of glacier bodies, thus confirming other results about the origin of this mummy (and not confirming speculations, the iceman may be a fake).
- Lake samples are more similar to the glacier samples than to fresh samples; obviously adipocere formation is similar in glaciers and a deep alpine lake.
- Samples from the 6200 m high mountain show only minor change of the relative fatty acids; at this altitude not enough liquid water is available for modification of the unsaturated fatty acids.
- Score plot PC3 versus PC1 does not give much additional information except a slight separation of the desert sample from the fresh samples.

Additional data interpretation can be gained from the loading plots (Figure 3.23, right column). The loading plot PC2 versus PC1 shows three groups of variables (fatty acids). The unsaturated fatty acids (no. 3, 5, 6) are on the right-hand side, corresponding to the fresh samples (with high concentrations of these compounds) which are located in the same area in the score plot. The hydroxylated fatty acids appear to be characteristic for the glacier body samples, and the saturated fatty acids for the lake samples. Furthermore, correlations between the variables are indicated in the loading plot. For instance, variables 3, 5, and 6 are close together; actually they have high positive correlation coefficients of $r_{35}=0.574$, $r_{36}=0.866$, and $r_{56}=0.829$; variables 6 and 8 are at opposite sides of the origin and have a large negative correlation coefficient of $r_{68}=-0.810$. Interpretation of the loading plot corresponds to the assumptions about relevant chemical reactions, and also roughly to the group means given in Table 3.4. Score plots and corresponding loading plots are a good visualization of relationships within the data and the sample classes.

The same data as used for PCA (concentrations of eight fatty acids in percent of all fatty acids) have been used for the construction of a dendrogram applying hierarchical clustering with complete linkages (Sections 3.8.2 and 6.4). The dendrogram confirms the clustering found in the PCA score plots and gives some additional insight. Fresh samples (6) and glacier samples (2) form distinct clusters; the iceman samples (1) and the lake samples (3) are in the glacier cluster with no significant separation (PCA shows more separation of the lake samples). The Ampato samples (4) and the desert sample (5) are within the fresh sample cluster but show some separation (Figure 3.24).

For comparison the same fatty acid data have been visualized by Sammon's NLM with parameter p set to 1 which makes a local mapping preserving well the small distances between objects (Figure 3.25, left). The NLM result shows a compact cluster of the fresh reference samples (class 6), and somewhat separated the desert sample (5). The horizontal map coordinate separates fresh samples from the high mountain samples (4) and the other samples (glacier samples, 2; lake samples, 3; and iceman samples, 1). The clustering has been indicated by manually drawn ellipses; it is very similar as obtained with the scores of the PC1 and PC2. Furthermore, Figure 3.25

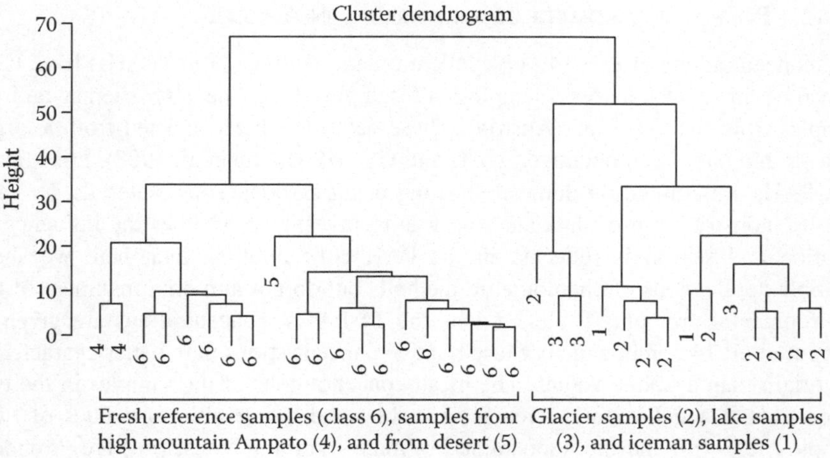

FIGURE 3.24 Dendrogram of fatty acid concentration data from mummies and reference samples. Hierarchical cluster analysis (complete linkage) with Euclidean distances has been applied.

(right) shows a Kohonen map of the same data; the main clustering is identical as obtained with PCA or NLM; however, fewer details are visible.

Application of multivariate statistics to fatty acid data from the Tyrolean Iceman and other mummies is a mosaic stone in the investigation of this mid-European ancestor, which is still a matter of research (Marota and Rollo 2002; Murphy et al. 2003; Nerlich et al. 2003). The iceman is on public display in the South Tyrol Museum of Archaeology in Bolzano, Italy, stored at −6 °C and 98% humidity, the conditions as they probably were during the last thousands of years.

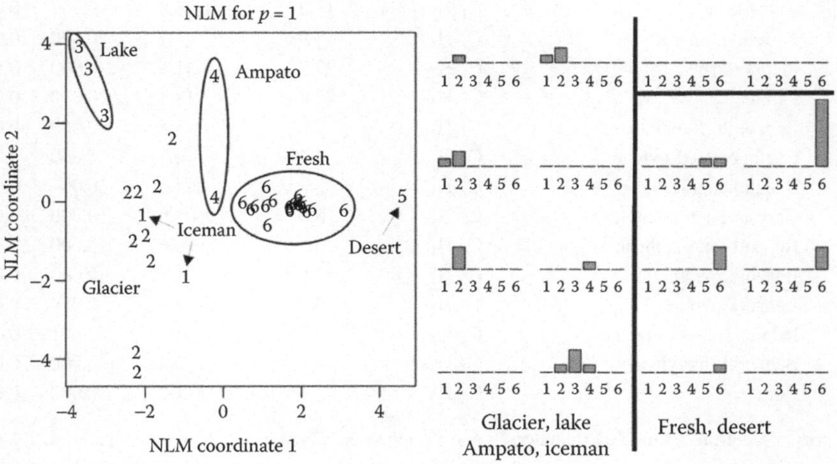

FIGURE 3.25 Sammon's NLM with $p = 1$ (left) and Kohonen map (right) of fatty acid concentration data from mummies and reference samples.

3.9.2 Polycyclic Aromatic Hydrocarbons in Aerosol

The concentrations of $m = 14$ polycyclic aromatic hydrocarbons (PAHs) have been measured in $n = 64$ aerosol samples, 47 samples from the city Vienna and 17 samples from the city Linz (Austria). These data have been selected from a larger data set and have been obtained by GC and GC/MS (Jaklin et al. 1988). It is known that PAHs are emitted by domestic heating using carbonaceous materials, by cars, and by industry. Aim of data analysis was to investigate whether the influence of location and season is reflected in the PAH concentration data; here we show possible applications of chemometric methods but do not aim on a treatment of the environmental problem. Table 3.5 lists the 14 PAHs; concentrations are given in percent of all 14 compounds because relative concentrations may better characterize the origin than absolute values. The mean concentrations of the samples in the two cities Vienna and Linz are given in the table, and the resulting p-values of two-sample t-tests (p_t) and the nonparametric Mann–Whitney U-test (p_U) are provided. These results already indicate that the groups are well separated; that means the origin of the PAH pollution is different.

An overview of the multivariate data set can first be obtained by PCA. Since the objects summed up to a constant value (compositional data, see Section 2.2.4), the data were first transformed with the isometric logratio (ILR) transformation.

TABLE 3.5
PAHs Measured in Aerosol Samples Collected in the Cities Vienna and Linz (Austria)

			Concentration (%)			
No.	Name	Chemical Formula	Mean Vienna	Mean Linz	p_t	p_U
1	Anthracene	$C_{14}H_{10}$	11.2	6.0	0.000	0.000
2	2-Methyl-anthracene	$C_{15}H_{12}$	4.9	1.1	0.000	0.000
3	Fluoranthene	$C_{16}H_{10}$	22.7	31.5	0.000	0.000
4	Pyrene	$C_{16}H_{10}$	23.4	18.4	0.000	0.000
5	Benzo[ghi]fluoranthene	$C_{18}H_{10}$	3.9	1.8	0.000	0.000
6	Cyclopenta[cd]pyrene	$C_{18}H_{10}$	3.4	0.9	0.000	0.000
7	Benz[a]anthracene	$C_{18}H_{12}$	3.3	4.3	0.007	0.003
8	Chrysene + triphenylene	$C_{18}H_{12}$	5.0	9.0	0.000	0.000
9	Benzo[b]fluoranthene + [j] + [k]	$C_{18}H_{10}$	6.1	10.6	0.000	0.000
10	Benzo[e]pyrene	$C_{20}H_{12}$	2.6	4.1	0.000	0.000
11	Benzo[a]pyrene	$C_{20}H_{12}$	3.0	3.6	0.078	0.086
12	Indeno[1,2,3-cd]pyrene	$C_{22}H_{12}$	2.7	3.3	0.021	0.021
13	Benzo[ghi]perylene	$C_{22}H_{12}$	4.4	3.5	0.000	0.000
14	Coronene	$C_{24}H_{12}$	3.4	1.8	0.000	0.000

Source: Data from Jaklin, J., Krenmayr, P., and Varmuza, K., *Fresenius Z. Anal. Chem.*, 331, 479, 1988.
Note: The concentrations, in percent of all 14 PAHs, are used as variables in exploratory data analysis. The mean concentrations for the samples in Vienna and Linz are given, as well as the p-values for the t-test and the U-test.

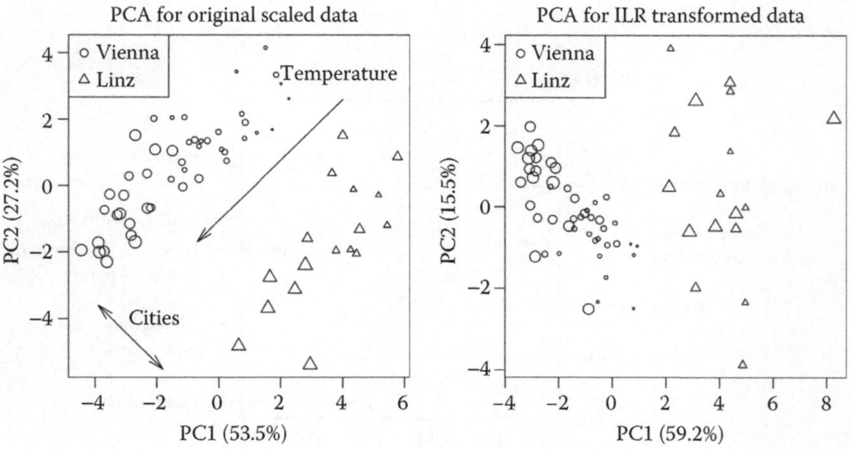

FIGURE 3.26 Plot of the first and second PCA scores for original scaled data (left) and the ILR transformed data (right). The different symbols correspond to the samples of Vienna and Linz, respectively, and the symbol size is proportional to the temperature. For both data sets PCA is able to separate the samples from the two cities. Also clusters of different temperatures are visible.

Figure 3.26 compares the PCA score plots obtained from the original data and the ILR transformed data; additionally the 24 h average temperature of the day on which the sample was collected is displayed by the size of the symbol. Both methods are able to separate the samples from the two cities. With ILR transformed data PC1 separates the two cities; the original scaled data require a rotation for a separation along a single component. Also clusters of different temperatures are visible; mean temperatures varied in Vienna between −5.1°C and 20.9°C, in Linz between 1.8°C and 18.1°C. The relationship between temperature and PAH concentration profile is more pronounced in Vienna than in Linz. Actually in the PCA plot of the original data two guiding factors can be seen, one factor (direction in the score plot) separates the cities, the other is responsible for temperature. A simple explanation for this data structure is the assumption of different main sources of pollution: in Vienna it is traffic, and in winter domestic heating, and in Linz it is the steel production industry.

The different origin of pollution can also be seen in the PCA loading plot, shown in Figure 3.27 for the original scaled data (corresponding to the score plot in Figure 3.26, left). In the PCA score plot, the samples from Linz are in the right-hand side; consequently the variables in approximately the same area in the loading plot are characteristic for this city. Actually, all seven PAHs in the right-hand side have been marked by the *t*-test and the *U*-test to have significantly higher concentrations in Linz than in Vienna. These compounds have their main origin in coke production accompanied to steel production. On the other hand, in Vienna the contribution of traffic is important shown by PAHs which are characteristic for exhaust fumes of cars, for instance, anthracene, cyclopenta[cd]pyrene, and coronene. A loading plot

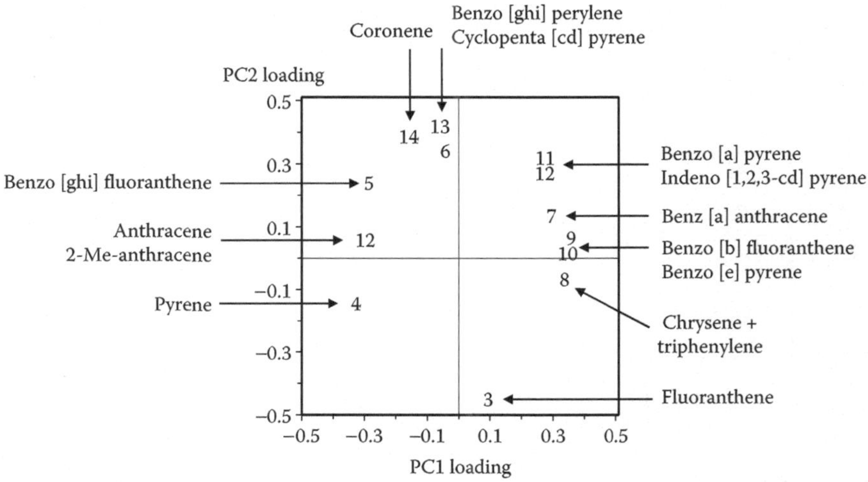

FIGURE 3.27 Loading plot of PC1 and PC2 computed from the original scaled data, corresponding to the score plot in Figure 3.26, left. The variable numbers 1–14 are as given in Table 3.5. PAHs in the left upper part are characteristic for Vienna, in the right part for Linz.

for the ILR transformed data would not be interpretable because the transformed variables have no direct relation to the original variables.

For comparison also a dendrogram (Figure 3.28) and a nonlinear mapping (NLM) (Figure 3.29) have been performed on the PAH data. Results from these methods show a clear separation of the samples from Linz and Vienna, but not much more details. The clusters in the NLM plots are very similar to the clusters in the PCA score plots. Thus, preserving the distances using two dimensions—the goal of

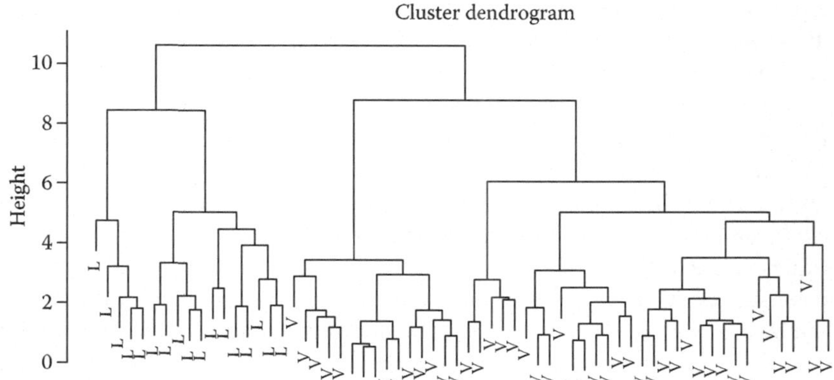

FIGURE 3.28 Dendrogram resulting from hierarchical cluster analysis of the original scaled data. The dendrogram of the ILR transformed data is very similar. Vienna (V) and Linz (L) form clearly separated clusters.

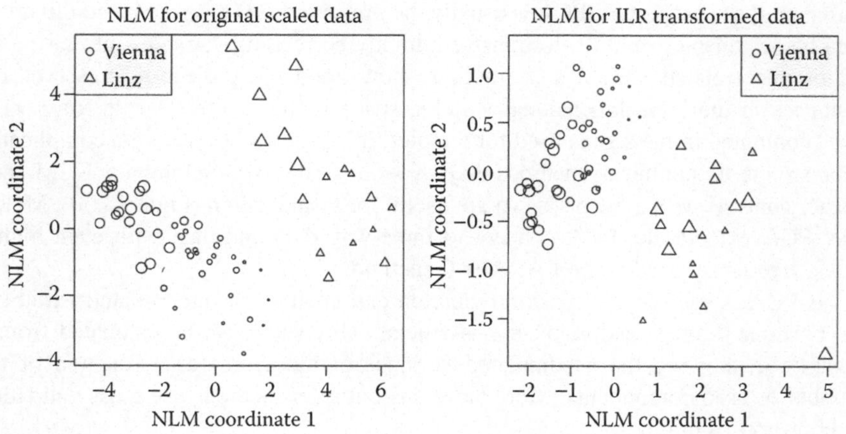

FIGURE 3.29 Sammon's NLM ($p = 1$) using the original scaled data (left) and the ILR transformed data (right).

PCA and NLM—can give very similar results for a linear method (like PCA) and in a nonlinear method (like NLM). Note that neither the dendrogram nor the NLM plots allow a direct interpretation of the PAHs responsible for the origin of pollution.

3.10 SUMMARY

PCA transforms a data matrix $\boldsymbol{X}(n \times m)$—containing data for n objects with m variables—into a matrix of lower dimension $\boldsymbol{T}(n \times a)$. In the matrix $\boldsymbol{T}$ each object is characterized by a relative small number, a, of PCA scores (PCs, latent variables). Score t_i of the ith object $\boldsymbol{x}_i$ is a linear combination of the vector components (variables) of vector $\boldsymbol{x}_i$ and the vector components (loadings) of a PCA LOADING VECTOR $\boldsymbol{p}$; in other formulation the score is the result of a scalar product $\boldsymbol{x}_i^{\mathrm{T}} \cdot \boldsymbol{p}$. The score vector $\boldsymbol{t}_k$ of PCA component k contains the scores for all n objects; $\boldsymbol{T}$ is the SCORE MATRIX for n objects and a components; $\boldsymbol{P}$ is the corresponding LOADING MATRIX (see Figure 3.2).

PCA scores and loadings have UNIQUE PROPERTIES as follows:

- PCA score 1 (PC1, first principal component) is the linear latent variable with the maximum possible variance. The direction of PC2 is orthogonal to the direction of PC1 and again has maximum possible variance of the scores. Subsequent PCs follow this rule.
- All PCA loading vectors are orthogonal to each other; PCA is a rotation of the original orthogonal coordinate system resulting in a smaller number of axes.
- PCA scores are uncorrelated latent variables.
- A scatter plot using the scores of the first two PCs preserves best the distances (similarities) of the objects using two dimensions and a linear method.

Based on these properties, PCA is usually the first choice (1) to visualize multivariate data by scatter plots and (2) to transform highly correlating variables into a smaller set of uncorrelated variables. A measure how good the projection preserves the distances in the high-dimensional variable space is the PERCENT OF THE TOTAL VARIANCE contained in the scores used for the plot. If PCA fails because of a complicated data structure, nonlinear methods like KOHONEN MAPPING, Sammon's NLM, and cluster analysis with a DENDROGRAM are useful for visualization (Figure 3.30). MULTIWAY PCA extends the PCA concept to three-way data and data with even higher ways; frequently used is the PARAFAC method.

If PCA is used for dimension reduction and creation of uncorrelated variables, the OPTIMUM NUMBER OF COMPONENTS is crucial. This value can be estimated from a SCREE PLOT showing, the accumulated variance of the scores as a function of the number of used components. More laborious but safer methods use cross validation or bootstrap techniques.

Outliers may heavily influence the result of PCA. DIAGNOSTIC PLOTS help to find OUTLIERS (leverage points and orthogonal outliers) falling outside the hyper-ellipsoid which defines the PCA model. Essential is the use of robust methods that are tolerant against deviations from multivariate normal distributions.

Methods of ROBUST PCA are less sensitive to outliers and visualize the main data structure; one approach for robust PCA uses a robust estimation of the covariance matrix, another approach searches for a direction which has the maximum of a robust variance measure (PROJECTION PURSUIT).

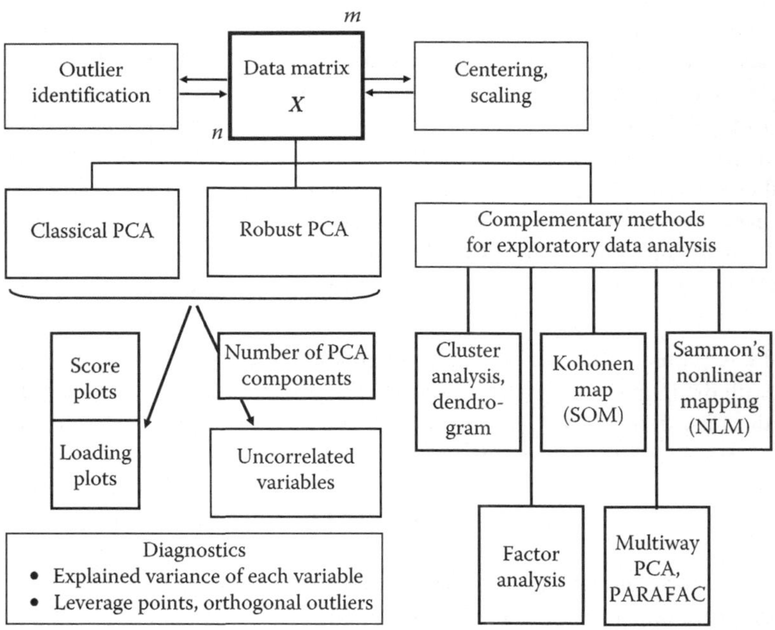

FIGURE 3.30 PCA and related methods.

All algorithms used for PCA are iterative; the most used is SVD. In chemometrics the NIPALS algorithm is popular, and a classic standard method is JACOBI ROTATION. Application of PCA does not require special mathematical knowledge because usually score and loading plots are evaluated visually.

REFERENCES

Barrado, E., Jimenez, F., Prieto, F., Nuevo, C.: *Food Chem.* 81, 2003, 13–20. The use of fatty-acid profiles of the lipids of the rainbow trout (*Oncorhynchus mykiss*) to differentiate tissue and dietary feed.

Basilevsky, A.: *Statistical Factor Analysis and Related Methods. Theory and Applications*. Wiley, New York, 1994.

Blanco-Gomis, D., Mangas, A. J. J., Margolles, C. I., Arias, A. P.: *J. Agric. Food. Chem.* 50, 2002, 1097–1100. Characterization of cider apples on the basis of their fatty acid profiles.

Cox, M. F., Cox, M. A. A.: *Multidimensional Scaling*. Chapman & Hall, London, United Kingdom, 2001.

Croux, C., Filzmoser, P., Oliveira, M. R.: *Chemom. Intell. Lab. Syst.* 87, 2007, 218–225. Algorithms for projection-pursuit robust principal component analysis.

Forina, M., Armanino, C.: *Ann. Chim.* 72, 1982, 127–141. Eigenvector projection and simplified non-linear mapping of fatty acid content of Italian olive oils.

Greco, A. V., Mingrone, G., Gasbarrini, G.: *Clin. Chim. Acta* 239, 1995, 13–22. Free fatty acid analysis in ascitic fluid improves diagnosis in malignant abdominal tumors.

Harman, H. H.: *Modern Factor Analysis*. University of Chicago Press, Chicago, IL, 1976.

Hemmer, M. C., Gasteiger, J.: *Anal. Chim. Acta* 420, 2000, 145–154. Prediction of three-dimensional molecular structures using information from infrared spectra.

Hubert, M., Rousseeuw, P. J., Vanden Branden, K.: *Technometrics* 47, 2005, 64–79. ROBPCA: A new approach to robust principal components.

Jaklin, J., Krenmayr, P., Varmuza, K.: *Fresenius Z. Anal. Chem.* 331, 1988, 479–485. Polycyclic aromatic compounds in the atmosphere of Linz (Austria) (in German).

Janssen, K. H. A., De Raedt, I., Schalm, O., Veeckman, J.: *Microchim. Acta* 15(suppl.), 1998, 253–267. Compositions of 15th–17th century archaeological glass vessels excavated in Antwerp.

Johnson, R. A., Wichern, D. W.: *Applied Multivariate Statistical Analysis*. Prentice Hall, Upper Saddle River, NJ, 2002.

Kiers, H. A. L.: *Psychometrika* 56, 1991, 449–454. Hierarchical relations among three-way methods.

Kohonen, T.: *Self-Organizing Maps*. Springer, Berlin, Germany, 1995.

Kohonen, T.: *Self-Organizing Maps*. Springer, Berlin, Germany, 2001.

Kowalski, B. R., Bender, C. F.: *J. Am. Chem. Soc.* 95, 1973, 686–693. Pattern recognition. II. Linear and nonlinear methods for displaying chemical data.

Makristathis, A., Schwarzmeier, J., Mader, R. M., Varmuza, K., Simonitsch, I., Chavez, J. C., Platzer, W., Unterdorfer, H., Scheithauer, R., Derevianko, A., Seidler, H.: *J. Lipid Res.* 43, 2002, 2056–2061. Fatty acid composition and preservation of the Tyrolean iceman and other mummies.

Malinowski, E. R.: *Factor Analysis in Chemistry*. Wiley, New York, 2002.

Manco, M., Mingrone, G., Greco, A. V., Capristo, E., Gniuli, D., De Gaetano, A., Gasbarrini, G.: *Metabolism* 49, 2000, 220–224. Insulin resistance directly correlates with increased saturated fatty acids in skeletal muscle triglycerides.

Mannina, L., Dugo, G., Salvo, F., Cicero, L., Ansanelli, G., Calcagni, C., Segre, A.: *J. Agric. Food Chem.* 51, 2003, 120–127. Study of the cultivar-composition relationship in Sicilian olive oils by GC, NMR, and statistical methods.

Maronna, R., Martin, D., Yohai, V.: *Robust Statistics: Theory and Methods*. Wiley, Toronto, Onatario, Canada, 2006.

Marota, I., Rollo, F.: *Cell. Mol. Life Sci.* 59, 2002, 97–111. Molecular paleontology.

Melssen, W. J., Smits, J. R. M., Rolf, G. H., Kateman, G.: *Chemom. Intell. Lab. Syst.* 18, 1993, 195–204. Two-dimensional mapping of IR spectra using a parallel implemented self-organizing feature map.

Miyashita, Y., Itozawa, T., Katsumi, H., Sasaki, S. I.: *J. Chemom.* 4, 1990, 97–100. Comments on the NIPALS algorithm.

Murphy, W. A., zur Nedden, D., Gostner, P., Knapp, R., W., R., Seidler, H.: *Radiology* 226, 2003, 614–629. The Iceman: Discovery and imaging.

Nerlich, A. G., Bachmeier, B., Zink, A., Thalhammer, S., Egarter-Vigl, E.: *Lancet* 362, 2003, 334. Oetzi had a wound on his right hand.

Sammon, J. W.: *IEEE Trans. Comput.* C-18 1969, 401–409. A nonlinear mapping for data structure analysis.

Seasholtz, M. B., Pell, R. J., Gates, K.: *J. Chemom.* 4, 1990, 331–334. Comments on the power method.

Sharaf, M. A., Illman, D. L., Kowalski, B. R.: *Chemometrics*. Wiley, New York, 1986.

Smilde, A., Bro, R., Geladi, P.: *Multi-Way Analysis with Applications in the Chemical Sciences*. Wiley, Chichester, United Kingdom, 2004.

Spindler, K., Wilfing, H., Rastbichler-Zissernig, E., zur Nedden, D., Nothdurfter, H.: *Human Mummies—the Man in the Ice*. Springer, Vienna, Austria, 1996.

Stahnke, L. H., Holck, A., Jensen, A., Nilsen, A., Zanardi, E.: *J. Food Sci.* 67, 2002, 1914–1921. Maturity acceleration of Italian dried sausage by *Staphylococcus carnosus*—relationship between maturity and flavor compounds.

Vandeginste, B. G. M., Massart, D. L., Buydens, L. C. M., De Jong, S., Smeyers-Verbeke, J.: *Handbook of Chemometrics and Qualimetrics: Part B*. Elsevier, Amsterdam, The Netherlands, 1998.

Varmuza, K., Makristathis, A., Schwarzmeier, J., Seidler, H., Mader, R. M.: *Mass Spectrom. Rev.* 24, 2005, 427–452. Exploration of anthropological specimens by GC–MS and chemometrics.

Xu, M., Basile, F., Voorhees, K. J.: *Anal. Chim. Acta* 418, 2000, 119–128. Differentiation and classification of user specified bacterial groups by in situ thermal hydrolysis and methylation of whole bacterial cells with tert-butyl bromide chemical ionization ion trap mass spectrometry.

Zupan, J., Gasteiger, J.: *Neural Networks for Chemists*. VCH, Weinheim, Germany, 1993.

Zupan, J., Gasteiger, J.: *Neural Networks in Chemistry and Drug Design*. Wiley-VCH, Weinheim, Germany, 1999.

4 Calibration

4.1 CONCEPTS

A fundamental task in science and technology is modeling a PROPERTY y by one or several VARIABLES x. The property is considered as the interesting fact of a system, but often cannot be determined directly or only with high cost; in contrary the x-data are often easily available but not the primary aim of an investigation.

Depending on how well known and how strictly defined the relationship between x and y is, we can distinguish different levels of creating and applying models that predict y from x.

- The relationship is described by a fundamental scientific law (a first principle), formulated as a relative simple mathematical equation with all parameters known. An example is for instance the time, y, a falling stone needs for a given height, x; the gravity constant, g, is known and y can be easily calculated by $(2x/g)^{0.5}$—if special effects like air friction are ignored.
- The relationship is well described by a relatively simple mathematical equation—usually based on physical/chemical knowledge—but the parameters are not known. An example is the often cited Beer's law (better Bouguer–Lambert–Beer's law), a mathematical model for photometry based on fundamental, idealized physical principles. The concentration of a light-absorbing substance, c, is given by $A/(a \cdot l)$ with l being the path length, a the absorption coefficient, and A the absorbance defined by $\log(I_0/I)$ with I_0 the incident light intensity, and I the light intensity after passing the sample. Light intensities and path length can be measured easily but the absorption coefficient is in general not known; it has to be determined from a set of standard solutions with known concentrations and by application of a regression method—a so-called calibration procedure. This method becomes very powerful if many wavelengths in the infrared (IR) or near infrared (NIR) range are used, and it is one of the main applications of chemometrics (multivariate calibration).
- In many cases of practical interest, no theoretically based mathematical equations exist for the relationships between x and y; we sometimes know but often only assume that relationships exist. Examples are for instance modeling of the boiling point or the toxicity of chemical compounds by variables derived from the chemical structure (molecular descriptors). Investigation of quantitative structure–property or structure–activity relationships (QSPR/QSAR) by this approach requires multivariate calibration methods. For such purely empirical models—often with many variables—the

prediction performance has to be estimated very carefully. On the other hand, the obtained model parameters can eventually enhance the understanding of the relationship between x and y.

Models of the form $y = f(x)$ or $y = f(x_1, x_2, \ldots, x_m)$ can be linear or nonlinear; they can be formulated as a relatively simple equation or can be implemented as a less evident algorithmic structure, for instance in artificial neural networks (ANN), tree-based methods (CART), local estimations of y by radial basis functions (RBF), k-NN like methods, or splines. This book focuses on linear models of the form

$$y = b_0 + b_1 x_1 + b_2 x_2 + \cdots + b_j x_j + \cdots + b_m x_m + e \tag{4.1}$$

where

b_0 is called the INTERCEPT
b_1 to b_m the REGRESSION COEFFICIENTS
m the number of variables
e the RESIDUAL (error term)

Often mean-centered data are used and then b_0 becomes zero. Note that the model corresponds to a linear latent variable—as described in Section 2.6—which best predicts y (or in other words has maximum correlation with y). Nonlinearities can be introduced in linear models by nonlinear transformations of the variables (for instance squares or logarithms), or by adding new variables that are nonlinear functions of the original variables (for instance cross products that consider interactions of the variables). Prominent aim of this chapter is the description of methods for estimating appropriate values of b_0 and b_1 to b_m, and of methods enabling a realistic estimation of the prediction errors. The parameters of a model are estimated from a CALIBRATION SET (TRAINING SET) containing the values of the x-variables and y for n samples (objects). The resulting model is evaluated with a TEST SET (with known values for x and y). Because modeling and prediction of y-data is a defined aim of data analysis, this type of data treatment is called SUPERVISED LEARNING.

All regression methods aim at the MINIMIZATION OF RESIDUALS, for instance minimization of the sum of the squared residuals. It is essential to focus on minimal prediction errors for new cases—the test set—but not (only) for the calibration set from which the model has been created. It is relatively easy to create a model—especially with many variables and eventually nonlinear features—that very well fits the calibration data; however, it may be useless for new cases. This effect of OVERFITTING is a crucial topic in model creation. Definition of appropriate criteria for the PERFORMANCE OF REGRESSION MODELS is not trivial. About a dozen different criteria—sometimes under different names—are used in chemometrics, and some others are waiting in the statistical literature for being detected by chemometricians; a basic treatment of the criteria and the methods how to estimate them is given in Section 4.2.

Regression can be performed directly with the values of the variables (ordinary least-squares regression, OLS) but in the most powerful methods, such as principal component regression (PCR) and partial least-squares regression (PLS), it is done via a small set of INTERMEDIATE LINEAR LATENT VARIABLES (the COMPONENTS). This approach has important advantages:

TABLE 4.1
Type of *x*- and *y*-Data for Regression Models and Appropriate Methods

Number of *x*-Variables	Number of *y*-Variables	Name	Methods
1	1	Simple	Simple OLS, robust regression
Many	1	Multiple	PLS, PCR, multiple OLS, robust regression, Ridge regression, Lasso regression
Many	Many	Multivariate	PLS2, CCA

- Data with highly correlating x-variables can be used (correlating x-variables can even be considered as useful "duplicate" measurements).
- Data sets with more variables than samples can be used.
- The complexity of the model can be controlled by the number of components and thus overfitting can be avoided and maximum prediction performance for test set data can be approached.

Depending on the type of data, different methods are available (Table 4.1).

- For only one x-variable and one y-variable—simple x/y-regression—the basic equations are summarized in Section 4.3.1. Ordinary least-squares (OLS) regression is the classical method, but also a number of robust methods exist for this purpose.
- Data sets with many x-variables and one y-variable are most common in chemometrics. The classical method multiple OLS is rarely applicable in chemometrics because of highly correlating variables and the large number of variables (Section 4.3.2). Work horses are PLS regression (Section 4.7) and PCR (Section 4.6).
- If more than one y-variable has to be modeled, a separate model can be developed for each y-variable or methods can be applied that work with an $\boldsymbol{X}$- and a $\boldsymbol{Y}$-matrix, such as PLS2 (Section 4.7.1), or CANONICAL CORRELATION ANALYSIS (CCA) (Section 4.8.1).

Closely related to the creation of regression models by OLS is the problem of VARIABLE SELECTION (FEATURE SELECTION). This topic is therefore presented in Section 4.5, although variable selection is also highly relevant for other regression methods and for classification.

An EXAMPLE WITH ARTIFICIAL DATA demonstrates the capabilities of multiple regression (more than one x-variable) in comparison to simple regression using only a single x-variable. The data set (Table 4.2) contains 25 samples; for each sample the variables x_1, x_2, and x_3 are given, as well as a property y. Figure 4.1 shows that none of the variables is capable to model y; x_1 and x_2 show at least a weak correlation with y; x_3 obviously is only noise. Next, we apply multiple regression using all three variables—and for simplicity a calibration set with all samples. The used method is OLS regression and the result is a linear latent variable (a direction in

TABLE 4.2
Artificial Data Set with Three *x*-Variables and One *y*-Variable, Showing the Advantage of Multiple Regression in Comparison to Univariate Regression

	x_1	x_2	x_3	y
	1.00	1.00	0.41	11.24
	1.00	2.00	−0.11	12.18
	1.00	3.00	−3.87	15.77
	1.00	4.00	0.33	16.20
	1.00	5.00	1.56	30.33
	2.00	1.00	−6.60	15.90
	2.00	2.00	1.57	20.48
	2.00	3.00	5.57	25.92
	2.00	4.00	2.87	26.76
	2.00	5.00	−1.90	27.05
	3.00	1.00	−0.30	13.47
	3.00	2.00	6.09	25.22
	3.00	3.00	6.95	26.72
	3.00	4.00	−7.20	33.60
	3.00	5.00	−1.71	41.30
	4.00	1.00	−4.64	24.12
	4.00	2.00	−0.43	32.44
	4.00	3.00	3.30	28.24
	4.00	4.00	2.48	34.11
	4.00	5.00	−5.75	37.64
	5.00	1.00	−3.08	27.11
	5.00	2.00	2.07	32.54
	5.00	3.00	7.23	32.05
	5.00	4.00	0.83	40.19
	5.00	5.00	−12.19	44.43
s	1.44	1.44	4.71	9.27
$\bar{x}$	3.00	3.00	−0.26	27.00

Notes: s, standard deviation; $\bar{x}$, mean.

the three-dimensional variable space) for which the scores have maximum correlation coefficient with y. We obtain a model for $\hat{y}$, the predicted y

$$\hat{y} = 1.418 + 4.423x_1 + 4.101x_2 - 0.0357x_3 \tag{4.2}$$

This linear combination of the variables correlates much better than any of the single variables (Figure 4.2, left). The STANDARDIZED REGRESSION COEFFICIENTS

$$_0b_j = b_j \times s_j \tag{4.3}$$

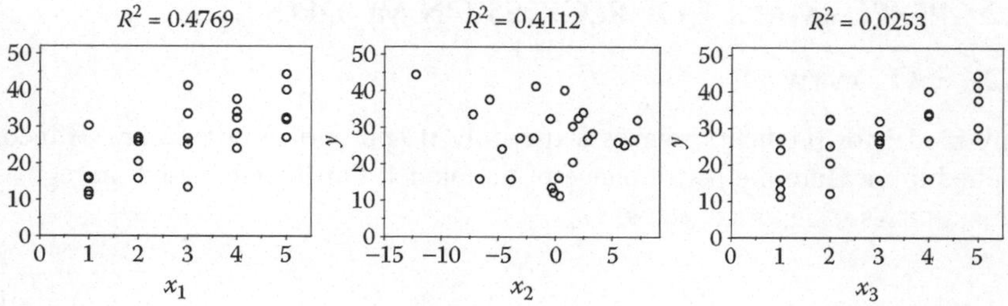

FIGURE 4.1 Univariate regression is not successful; none of the single variables x_1, x_2, x_3 (data set in Table 4.2) is useful to predict the property y. R^2 is the squared Pearson correlation coefficient.

with s_j for the standard deviation of variable j are measures for the importance of the variables in the regression model. The standardized regression coefficients of the variables x_1, x_2, and x_3 in Equation 4.2 are 6.37, 5.91, and 0.33, respectively, indicating that x_1 and x_2 have the main influence on y; x_3 has almost no influence. The OLS-model using only x_1 and x_2 is

$$\hat{y} = 1.353 + 4.433x_1 + 4.116x_2 \tag{4.4}$$

and it has similar regression coefficients and similar performance as the model including the noise variable (Figure 4.2, right).

The data of this examples have been simulated as follows: x_1 and x_2 have been systematically varied as shown in Table 4.2; x_3 contains random numbers from a normal distribution $N(0, 5)$. Property y is calculated as a theoretical value $5x_1 + 4x_2$ with noise added from a normal distribution $N(0, 3)$.

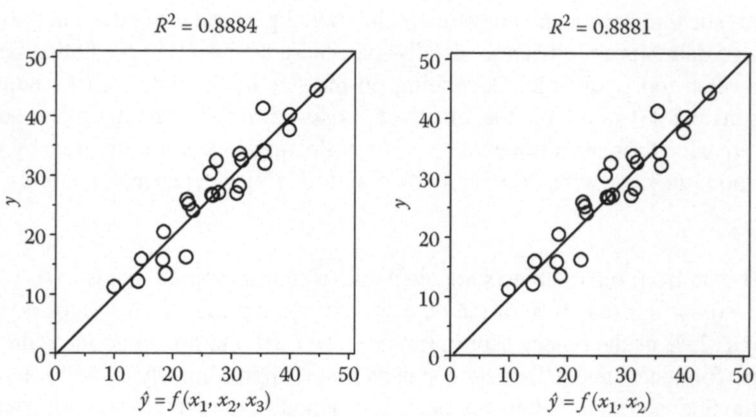

FIGURE 4.2 Linear combinations of the x-variables (data set in Table 4.2) are useful for the prediction of property y. For the left plot, x_1, x_2, and x_3 have been used to create an OLS-model for y, Equation 4.2; for the right plot x_1 and x_2 have been used for the model, Equation 4.4. R^2 is the squared Pearson correlation coefficient. Both models are very similar; the noise variable x_3 does not deteriorate the model.

4.2 PERFORMANCE OF REGRESSION MODELS

4.2.1 OVERVIEW

Any model for prediction makes sense only if appropriate criteria are defined and applied to measure the performance of the model. For models based on regression, the RESIDUALS (PREDICTION ERRORS) e_i

$$e_i = y_i - \hat{y}_i \tag{4.5}$$

are the basis for performance measures, with y_i for the given (experimental, "true") value and $\hat{y}_i$ the predicted (modeled) value of an object i. The different approaches to estimate model performance (the GENERALIZATION ERROR) use different strategies for the selection of objects used for model generation and for model test, and define different mathematical measures derived from the prediction errors. One aim of performance measures is MODEL SELECTION; that means to find out the approximate best model. The other aim is MODEL ASSESSMENT in which for a final model expected prediction errors for new cases are estimated. For model selection sometimes simple, fast to compute, criteria are used or have to be used; for instance for variable selection by a genetic algorithm (GA) typically several 100,000 models have to be compared. For model assessment, computing time should not play an important role, because a good estimation of the performance for new cases justifies a high effort.

An often-used performance measure estimates the STANDARD DEVIATION OF THE PREDICTION ERRORS (STANDARD ERROR OF PREDICTION, SEP, Section 4.2.3). In general, a performance measure that only characterizes how well the model fits given data is not acceptable; consequently a realistic estimate of the performance for new cases is essential. Using the same objects for calibration and test should be strictly avoided; such an approach can estimate the model performance adequately only for very large data sets and "friendly data"—otherwise the resulting prediction performance is often too optimistic. Depending on the size of the data set (the number of objects available)—and on the effort of work—different strategies are possible; however, not all are recommendable. The following levels are ordered by typical applications to data with decreasing size and also by decreasing reliability of the results.

1. If data from many objects are available, a split into three sets is best: into a TRAINING SET (ca. 50% of the objects) for creating models, a VALIDATION SET (ca. 25% of the objects) for optimizing the model to obtain good prediction performance, and a TEST SET (PREDICTION SET, approximately 25%) for testing the final model to obtain a realistic estimation of the prediction performance for new cases. The three sets are treated separately. Applications in chemistry rarely allow this strategy because of a too small number of objects available.
2. The data set is split into a CALIBRATION SET used for model creation and optimization and a TEST SET (PREDICTION SET) to obtain a realistic estimation

of the prediction performance for new cases. The calibration set is divided into a training set and a validation set by cross validation (CV) (Section 4.2.5) or bootstrap (Section 4.2.6); first the optimum complexity (for instance optimum number of PLS components) of the model is estimated (Section 4.2.2) and then a model is built from the whole calibration set applying the found optimum complexity; this model is applied to the test set.

3. CV or bootstrap is used to split the data set into different calibration sets and test sets. A calibration set is used as described above to create an optimized model and this is applied to the corresponding test set. All objects are principally used in training set, validation set, and test set; however, an object is never simultaneously used for model creation and for test. This strategy (double CV or double bootstrap or a combination of CV and bootstrap) is applicable to a relatively small number of objects; furthermore, the process can be repeated many times with different random splits resulting in a high number of test-set-predicted values (Section 4.2.5).
4. All data are used as calibration set. CV or bootstrap is applied to determine the optimum model complexity. Because no test set data are used, the obtained prediction performance (obtained by CV from calibration set) is usually too optimistic. This approach cannot be recommended.
5. From all data a model is created that best fits the data, and no optimization of the model complexity for best prediction is performed. This model is applied to the same data; the resulting prediction errors are usually far too optimistic, and consequently this approach should be avoided.

Mostly the split of the objects into training, validation, and test sets is performed by simple random sampling. However, more sophisticated procedures—related to experimental design and theory of sampling—are available. In chemometrics the Kennard–Stone algorithm is sometimes used. It claims to set up a calibration set that is representative for the population, to cover the x-space as uniformly as possible, and to give more weight to objects outside the center. This aim is reached by selecting objects with maximum distances (for instance Euclidean distances) in the x-space (Kennard and Stone 1969, Snee 1977, Vandeginste et al. 1998).

The methods CV (Section 4.2.5) and bootstrap (Section 4.2.6), which are necessary for small data sets, are called RESAMPLING strategies. They are applied to obtain a reasonable high number of predictions, even with small data sets. It is evident that the larger the data set and the more reliable (more friendly in terms of modeling) the data are, the better the prediction performance can be estimated. Thus a qualitative UNCERTAINTY RELATION of the general form is presumed:

$$\text{(size of data)} * \text{(friendliness of data)} * \text{(uncertainty of performance measure)} = \text{constant}$$

An informative description of the prediction errors is a visual representation, for instance by a histogram or a probability density curve; these plots, however, require a reasonable large number of predictions. For practical reasons, the error distribution can be characterized by a single number, for instance the standard deviation.

A number of performance criteria are not primarily dedicated to the users of a model but are applied in model generation and optimization. For instance, the MEAN SQUARED ERROR (MSE) or similar measures are considered for optimization of the number of components in PLS or PCA. For variable selection, the models to be compared have different numbers of variables; in this case—and especially if a fit criterion is used—the performance measure must consider the number of variables; appropriate measures are the ADJUSTED SQUARED CORRELATION COEFFICIENT, $_{\mathrm{ADJ}}R^2$, or the AKAIKE'S INFORMATION CRITERION (AIC); see Section 4.2.3.

Unfortunately, definitions, nomenclature, and abbreviations used for performance criteria are sometimes confusing (Frank and Todeschini 1994; Kramer 1998). For instance in the abbreviations MSEC, PRESS, RMSEP, SEC, SEE, SEP, E means error or estimate, R means residual or root, and S means squared or standard or sum. To make it not too complicated, at least in these examples, C is always calibration, M is mean, and P is prediction.

4.2.2 OVERFITTING AND UNDERFITTING

The more complex (the "larger") a model is, the better is it capable to fit given data (the calibration set). The prediction error for the calibration set in general decreases with increasing complexity of the model (Figure 4.3). Thus an appropriate highly complicated model can fit almost any data with almost zero deviations (residuals) between experimental (true) y and modeled (predicted) y. It is evident that such models are not necessarily useful for new cases, because they are probably OVERFITTED;

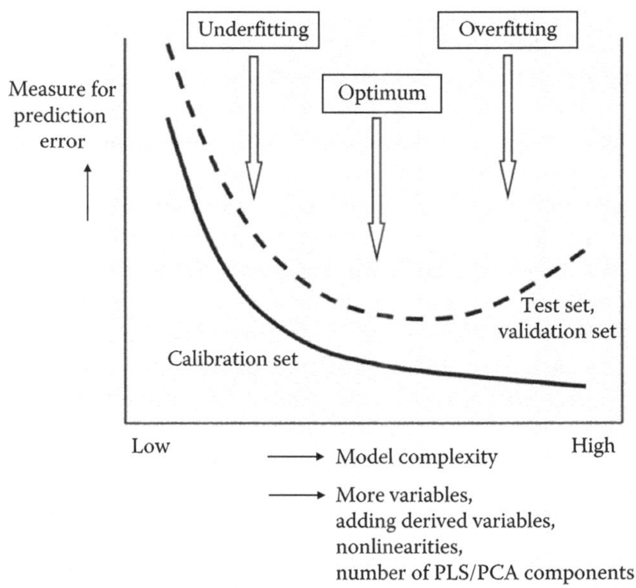

FIGURE 4.3 Model complexity versus prediction error for calibration set and for test or validation set (schematically).

that means they are very well adapted to the calibration data but do not possess sufficient GENERALIZATION. In general, the prediction errors for new cases (test set, prediction set, objects not used in model generation) show a minimum at a medium model complexity. Prediction errors for new cases are high for "small" models (UNDERFITTING, low complexity, too simple models) but also for overfitted models.

For regression models, the complexity becomes higher by

- Increasing the number of variables (OLS, Section 4.3)
- Adding derived variables (nonlinear functions of the original variables)
- Implementing nonlinearities into the model
- Increasing the number of "components" in PLS and PCR (Sections 4.6 and 4.7)

Determination of the optimum complexity of a model is an important but not always an easy task, because the minimum of measures for the prediction error for test sets is often not well marked. In chemometrics, the complexity is typically controlled by the number of PLS or PCA components, and the optimum complexity is estimated by CV (Section 4.2.5). Several strategies are applied to determine a reasonable optimum complexity from the prediction errors which may have been obtained by CV (Figure 4.4). CV or bootstrap allows an estimation of the prediction error for each object of the calibration set at each considered model complexity.

- The GLOBAL MINIMUM would be the complexity which has the minimum prediction error for a validation set (within the investigated range of complexities). This approach often results in overfitting.
- A LOCAL MINIMUM at a low complexity avoids overfitting and is most often used. Several heuristic algorithms are implemented in commercial software.

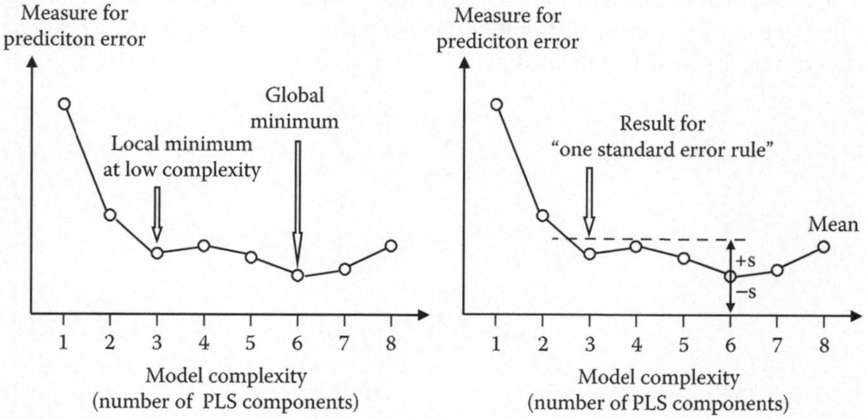

FIGURE 4.4 Determination of optimum complexity of regression models (schematically). Measure for prediction errors for instance $RMSE_{CV}$ in arbitrary linear units. Left, global and local minimum of a measure for prediction performance. Right, one standard error rule.

The common idea is to choose the model complexity where the curve of the measure for the prediction errors "flattens out."

- A ONE STANDARD ERROR RULE is described in Hastie et al. (Hastie et al. 2001). It is assumed that several values for the measure of the prediction error at each considered model complexity are available (this can be achieved, e.g., by CV or by bootstrap, Sections 4.2.5 and 4.2.6). Mean and standard error (standard deviation of the means, s) for each model complexity are computed, and the most parsimonious model whose mean prediction error is no more than one standard error above the minimum mean prediction error is chosen. Figure 4.4 (right) illustrates this procedure. The points are the mean prediction errors and the arrows indicate mean plus/minus one standard error. A horizontal line is drawn at the global minimum of the mean prediction errors plus one standard error. We choose that model that is as small as possible, but where the mean prediction error is still below this line.

4.2.3 Performance Criteria

The basis of all performance criteria are prediction errors (residuals), $y_i - \hat{y}_i$, obtained from an independent test set, or by CV or bootstrap, or sometimes by less reliable methods. It is crucial to document from which data set and by which strategy the prediction errors have been obtained; furthermore, a large number of prediction errors is desirable. Various measures can be derived from the residuals to characterize the prediction performance of a single model or a model type. If enough values are available, visualization of the error distribution gives a comprehensive picture. In many cases, the distribution is similar to a normal distribution and has a mean of approximately zero. Such distribution can well be described by a single parameter that measures the spread. Other distributions of the errors, for instance a bimodal distribution or a skewed distribution, may occur and can for instance be characterized by a tolerance interval.

The classical standard deviation of prediction errors is widely used as a measure of the spread of the error distribution, and is called in this application STANDARD ERROR OF PREDICTION (SEP) defined by

$$\mathrm{SEP} = \sqrt{\frac{1}{z-1}\sum_{i=1}^{z}(y_i - \hat{y}_i - \mathrm{bias})^2} \tag{4.6}$$

with

$$\mathrm{bias} = \frac{1}{z}\sum_{i=1}^{z}(y_i - \hat{y}_i) \tag{4.7}$$

where

y_i are the given (experimental, "true") values
$\hat{y}_i$ are the predicted (modeled) values
z is the number of predictions

Note that z can be larger than the number of objects, n, if for instance repeated CV or bootstrap has been applied. The BIAS is the arithmetic mean of the prediction errors and should be near zero; however, a systematic error (a nonzero bias) may appear if, for instance, a calibration model is applied to data that have been produced by another instrument. In the case of a normal distribution, about 95% of the prediction errors are within the tolerance interval $\pm 2 \cdot \text{SEP}$. The measure SEP and the tolerance interval are given in the units of y, and are therefore most useful for model applications.

SEP without any index is mostly used for prediction errors obtained from a test set (for clarity better named SEP_{TEST}). SEP_{CV} is calculated from prediction errors obtained in CV, for instance during optimization of the number of components in PLS or PCR. SEP_{CV} is usually smaller (more optimistic) than SEP_{TEST}. If Equation 4.6 is applied to predictions of the calibration set, the result is called STANDARD ERROR OF CALIBRATION (SEC); SEC is usually a too optimistic estimation of the prediction errors for new cases.

The MEAN SQUARED ERROR (MSE) is the arithmetic mean of the squared errors,

$$\text{MSE} = \frac{1}{z}\sum_{i=1}^{z}(y_i - \hat{y}_i)^2 \tag{4.8}$$

MSEC (or MSE_{CAL}) refers to results from a calibration set, MSECV (or MSE_{CV}) to results obtained in CV, and MSEP (or MSE_{TEST}) to results from a prediction/test set. MSE minus the squared bias gives the squared SEP,

$$\text{SEP}^2 = \text{MSE} - \text{bias}^2 \tag{4.9}$$

The ROOT MEAN SQUARED ERROR (RMSE) is the square root of MSE, and can again be given for calibration (RMSEC or RMSE_{CAL}), CV (RMSECV or RMSE_{CV}) or for prediction/test (RMSEP, or $\text{RMSE}_{\text{TEST}}$). In the case of a negligible bias, RMSEP and SEP are almost identical, as well as MSE_{TEST} and SEP^2.

$$\text{RMSE} = \sqrt{\frac{1}{z}\sum_{i=1}^{z}(y_i - \hat{y}_i)^2} \tag{4.10}$$

MSE is preferably used during the development and optimization of models but is less useful for practical applications because it has not the units of the predicted property. A similar widely used measure is PREDICTED RESIDUAL ERROR SUM OF SQUARES (PRESS), the sum of the squared errors; it is often applied in CV.

$$\text{PRESS} = \sum_{i=1}^{z}(y_i - \hat{y}_i)^2 = z \cdot \text{MSE} \tag{4.11}$$

Instead of the classical estimation of the standard deviation, SEP, robust methods can be applied as described in Section 1.6.4, for instance the spread measures s_{IQR}

or s_{MAD}. The TOLERANCE INTERVAL is independent from the shape of the distribution and can be defined for instance by the 2.5% and 97.5% percentiles of the empirical error distribution; 95% of the prediction errors can be expected within these limits. Correlation measures between experimental y and predicted y are frequently used to characterize the model performance. Mostly used is the squared PEARSON CORRELATION COEFFICIENT; more robust measures (Section 2.3.2) may be considered but are rarely used.

4.2.4 CRITERIA FOR MODELS WITH DIFFERENT NUMBERS OF VARIABLES

In variable selection models with a different number of variables have to be compared, and the applied performance criteria must consider the number of variables in the compared models. The model should not contain a too small number of variables because this leads to poor prediction performance. On the other hand, it should also not contain a too large number of variables because this results in overfitting and thus again poor prediction performance (see Section 4.2.2). In the following, m denotes the number of regressor variables (including the intercept if used) that are selected for a model.

The following criteria are usually directly applied to the calibration set to enable a fast comparison of many models as it is necessary in variable selection. The criteria characterize the fit and therefore the (usually only few) resulting models have to be tested carefully for their prediction performance for new cases. The measures are reliable only if the model assumptions are fulfilled (independent normally distributed errors). They can be used to select an appropriate model by comparing the measures for models with various values of m.

The ADJUSTED R-SQUARE, ${}_{\mathrm{ADJ}}R^2$, is defined by

$$ {}_{\mathrm{ADJ}}R^2 = 1 - \frac{n-1}{n-m-1}(1-R^2) \tag{4.12} $$

where

n is the number of objects

R^2 is called COEFFICIENT OF DETERMINATION, expressing the proportion of variance that is explained by the model

As the model complexity increases, R^2 becomes larger. In linear regression, R^2 is the squared correlation coefficient between y and $\hat{y}$, and ${}_{\mathrm{ADJ}}R^2$ is called the ADJUSTED SQUARED CORRELATION COEFFICIENT. Thus ${}_{\mathrm{ADJ}}R^2$ is a modification of the R^2 that penalizes larger models. A model with a large value of ${}_{\mathrm{ADJ}}R^2$ is preferable. Another, equivalent representation for ${}_{\mathrm{ADJ}}R^2$ is

$$ {}_{\mathrm{ADJ}}R^2 = 1 - \frac{\mathrm{RSS}/(n-m-1)}{\mathrm{TSS}/(n-1)} \tag{4.13} $$

with residual sum of squares (RSS) for the sum of the squared residuals

$$\mathrm{RSS} = \sum_{i=1}^{n} (y_i - \hat{y}_i)^2 \tag{4.14}$$

and total sum of squares (TSS) for the sum of the squared differences to the mean $\bar{y}$ of y,

$$\mathrm{TSS} = \sum_{i=1}^{n} (y_i - \bar{y})^2 \tag{4.15}$$

The following three performance measures are commonly used for variable selection by stepwise regression or by best-subset regression. An example in Section 4.5.8 describes use and comparison of these measures.

AKAIKE'S INFORMATION CRITERION (AIC) is given by (log denotes the logarithm with base e)

$$\mathrm{AIC} = n \log(\mathrm{RSS}/n) + 2m \tag{4.16}$$

For an increasing number of regressor variables, RSS becomes smaller. The AIC penalizes large models by the additive term $2m$. The model with a small value of AIC is preferable. Note that the AIC value itself is meaningless; it can only be used for comparing different models.

The BAYES INFORMATION CRITERION (BIC) is very similar to AIC and given by

$$\mathrm{BIC} = n \log(\mathrm{RSS}/n) + m \log n \tag{4.17}$$

If the number of objects $n > 7$, $\log n$ is larger than 2, and BIC gives more penalty to larger models than AIC. In other words, BIC usually selects smaller models than AIC. The model with a small value of BIC is preferable. As for the AIC, the BIC value itself is meaningless.

MALLOW'S CP (Cp) is defined by (Frank and Todeschini 1994; Mallows 1973; Massart et al. 1997)

$$Cp = \mathrm{RSS}/s^2 - n + 2m \tag{4.18}$$

and is mostly used as a stopping rule for stepwise regression or best-subset regression. In this definition, s^2 is the estimated error variance which can be obtained from regression with the full model (using all variables). If a full model cannot be computed (too many variables, collinearity, etc.), a regression can be performed on the relevant principal components (PC), and the error variance is estimated from the resulting residuals. For the "true" model, Cp is approximately m and otherwise greater than m. Thus a model where Cp is approximately m should be selected, and where the model with smallest m should be preferred.

4.2.5 CROSS VALIDATION

The most used resampling strategy in chemometrics to obtain a reasonable large number of predictions is cross validation (CV). CV is also often applied to optimize

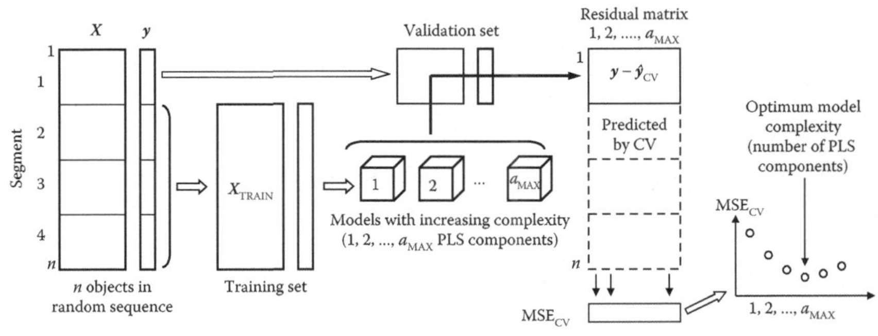

FIGURE 4.5 CV with four segments (leave-a-quarter-out) applied to estimation of the optimum complexity of the model.

the complexity of a model, for instance to estimate the optimum number of PLS or PCA components. However, the prediction errors obtained during model optimization are in general not appropriate to estimate the prediction performance for new cases. CV can also be applied to split the data into calibration sets and test sets.

The procedure of CV applied to model optimization can be described as follows. The available set with n objects is randomly split into s SEGMENTS (parts) of approximately equal size. The number of segments can be 2 to n, often values between 4 and 10 are used; Figure 4.5 demonstrates CV with four segments. One segment is left out as a validation set. The other $s-1$ sets are used as a training set to create models which have increasing complexity (for instance, models with 1, 2, 3, ..., a_{MAX} PLS components). The models are separately applied to the objects of the validation set resulting in predicted values connected to different model complexities. This procedure is repeated so that each segment is a validation set once. The result is a matrix with n rows and a_{MAX} columns containing predicted values $\hat{y}_{CV}$ (predicted by CV) for all objects and all considered model complexities. From this matrix and the known y-values, a residual matrix (matrix with prediction errors) is computed. An error measure (for instance, MSE_{CV}) is calculated from the residuals column-wise. According to Figures 4.3 and 4.4, the lowest MSE_{CV} or a similar criterion indicates the optimum model complexity.

A single CV as described gives n predictions. For many data sets in chemistry n is too small for a visualization of the error distribution. Furthermore, the obtained performance measure may heavily depend on the split of the objects into segments. It is therefore recommended to repeat the CV with different random splits into segments (REPEATED CV), and to summarize the results. Knowing the variability of MSE_{CV} at different levels of model complexities also allows a better estimation of the optimum model complexity, see "one standard error rule" in Section 4.2.2 (Hastie et al. 2001).

Different methods for CV are applied:

- CV with four segments is also called LEAVE-A-QUARTER-OUT CV.
- A very time-consuming and therefore rarely used strategy is LEAVE-OUT-ALL-POSSIBLE-SUBSETS of size v (LEAVE-v-OUT), especially if v is varied.

- If the number of segments is equal to the number of objects (each segment contains only one object), the method is called LEAVE-ONE-OUT CV or FULL CV. This method has advantages and disadvantages: (1) Randomization of the sequence of objects is senseless, therefore only one CV run is necessary resulting in n predictions. (2) The number of created models is n which may be time-consuming for large data sets. (3) Depending on the data, full CV may give too optimistic results, especially if pairwise similar objects are in the data set, for instance from duplicate measurements. (4) Full CV is easier to apply than repeated CV or bootstrap, and it is implemented in today's commercial software. (5) In many cases, full CV gives a reasonable first estimate of the model performance.

Different methods can be applied for the split into segments. Mode "111222333" denotes that the first n/s (rounded to integer) objects are put into segment 1, the next n/s objects into segment 2, and so on. Mode "123123123" puts object 1 into segment 1, object 2 into segment 2, and so on. Mode "random" makes a random split but usually without any user control. We recommend to sort the objects by a user-created random permutation of the numbers 1 to n, and then to apply mode "111222333"—and to repeat this several times.

In DOUBLE CV, the CV strategy is applied in an outer loop (OUTER CV) to split all data into test sets and calibration sets, and in an inner loop (INNER CV) to split the calibration set into training sets and validation sets (Figure 4.6). The inner loop is used to optimize the complexity of the model (for instance, the optimum number of PLS components as shown in Figure 4.5); the outer loop gives predicted values $\hat{y}_{\mathrm{TEST}}$ for all n objects, and from these data a reasonable estimation of the prediction performance for new cases can be derived (for instance, the $\mathrm{SEP}_{\mathrm{TEST}}$). It is important

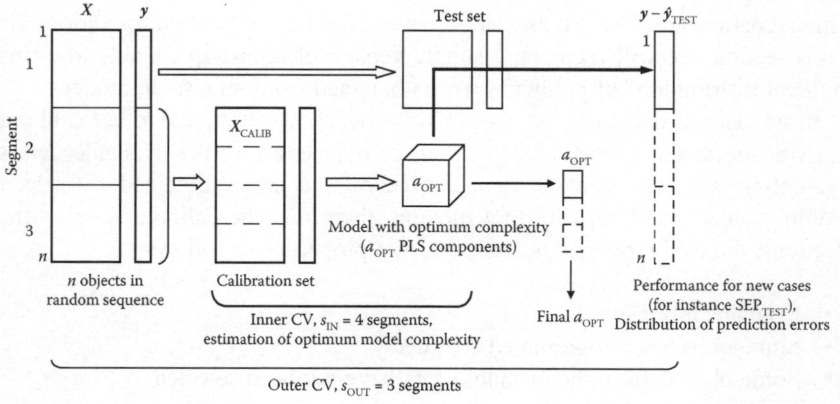

FIGURE 4.6 Double CV with three segments (s_{OUT}) in the outer CV and four segments (s_{IN}) in the inner CV. In the outer CV, test sets are defined and the prediction performance for new cases is estimated (for instance the standard error of prediction, $\mathrm{SEP}_{\mathrm{TEST}}$). In the inner CV, the optimum complexity of the model is estimated from a calibration set as shown in Figure 4.5.

to strictly avoid any optimization using the results of the test sets, because this would give a too optimistic prediction performance.

The number of segments in the outer and inner loop (s_{OUT} and s_{IN}, respectively) may be different. Each loop of the outer CV results in an optimum complexity (for instance, optimum number of PLS components, a_{OPT}). In general, these s_{OUT} values are different; for a final model the median of these values or the most frequent value can be chosen (a smaller complexity would avoid overfitting, a larger complexity would result in a more detailed model but with the risk of overfitting). A final model can be created from all n objects applying the final optimum complexity; the prediction performance of this model has been estimated already by double CV. This strategy is especially useful for PLS and PCR.

Also for double CV, the obtained performance measure depends on the split of the objects into segments, and therefore it is advisable to repeat the process with different random splits into segments (REPEATED DOUBLE CV), and to summarize the results.

4.2.6 Bootstrap

The word "bootstrap" has different—but related—meanings: A bootstrap is for instance a small strap or loop at the back of a boot that enables to pull the boot on. "Pulling yourself up by your own bootstraps" means to lever off to great success from a small beginning. In software technology, the bootstrap is a small initial program that allows loading a larger program—usually an operating system (booting a system). Within multivariate data analysis, bootstrap is a resampling method that can be used as an alternative to CV, for instance, to estimate the prediction performance of a model or to estimate the optimum complexity (Efron 1983; Efron and Tibshirani 1993, 1997). The bootstrap is a simple but rather time-consuming procedure that is applied to reach a high goal as, for instance, a good estimation of prediction performance. In general, bootstrap can be used to estimate the distribution of model parameters; many bootstrap schemes have been described for various aims. In this section, we will focus on a simple version of bootstrap suitable to estimate empirical distributions of prediction errors obtained from regression models.

Basic ideas of bootstrap are RESAMPLING WITH REPLACEMENT, and to use calibration sets with the same number of objects, n, as objects are in the available data set. Resampling with replacement means, a calibration set is obtained by selecting randomly objects and copying (not moving) them into the calibration set. After n selections, the calibration set is ready and has properties as follows:

- Contains n objects
- Some objects are represented repeatedly
- Some objects from the available data have not been selected

Using the calibration set, a model is created; in the case of PLS or PCR, additionally an optimization of the number of components has to be done—for instance by CV, or by an inner bootstrap within the calibration set. The resulting optimized model is then applied to the objects not contained in the calibration set giving predicted values and the corresponding prediction errors. The whole procedure is repeated z_{BOOT} times, usually 100 to some 1000 times.

The probability for an object to be in a calibration set is about 0.63 as can be shown easily: The probability a particular object is selected is $1/n$, and the probability for not being selected is $1 - 1/n$. The probability an object is not selected in n drawings is $(1 - 1/n)^n$ and the probability it is selected is $1 - (1 - 1/n)^n$. The latter expression is for large n approximately 0.63. The number of obtained predictions in z_{BOOT} repetitions is therefore approximately $0.37 \cdot z_{BOOT} \cdot n$. Note that the number of available predictions per object varies (theoretically between 0 and z_{BOOT}). From the corresponding residuals, performance measures can be derived as defined in Section 4.2.3.

Advantages of the bootstrap are a simple strategy, and having always n objects in the calibration set. Disadvantages are time-consuming computations, not all objects may have been considered equally, and sometimes too optimistic results. Therefore improved versions of bootstrap algorithms have been suggested that give a more realistic estimation of the prediction errors (Hastie et al. 2001). The bootstrap is increasingly used in chemometrics and is better than a single CV, however, gives similar results and requires about the same order of computational effort as for instance repeated double CV.

4.3 ORDINARY LEAST-SQUARES REGRESSION

4.3.1 SIMPLE OLS

Simple linear regression relates a property (response, dependent variable) y and a single independent variable (feature, measurement) x by a linear model. This approach has still many applications in analytical chemistry. In the frame of multivariate data analysis, we use this simple regression for instance to describe the relationship between experimental and predicted y (predicted by a multivariate model). Of course, regression with one x-variable is a special case of regression with many x-variables (multiple regression, Section 4.3.2); nevertheless, we treat simple regression separately in this section, summarize the basic principles and equations, and give relevant $\mathbb{R}$-code.

Let $\boldsymbol{x}$ be a vector containing the n values (objects) $x_1, x_2, \ldots, x_n$ of an independent variable x. The vector $\boldsymbol{y}$ contains the corresponding n values (responses) $y_1, y_2, \ldots, y_n$ of a dependent variable y. Thus, the values of the y-variable depend on the values of the x-variable, or, in statistical terms, y is a random variable and x consists of fixed values. A linear model that relates x and y can be defined as

$$\boldsymbol{y} = b_0 + b\boldsymbol{x} + \boldsymbol{e} \tag{4.19}$$

with b and b_0 being the regression parameters (REGRESSION COEFFICIENTS), b_0 is the INTERCEPT and b is the SLOPE (Figure 4.7). Since the data will in general not follow a perfect linear relation, the vector $\boldsymbol{e}$ contains the RESIDUALS (errors) $e_1, e_2, \ldots, e_n$. The task is now to find estimates of the regression coefficients in order to find a reliable linear relation between the x and y values, which practically means to minimize a function of the errors. Assume we have found such estimates for b_0 and b.

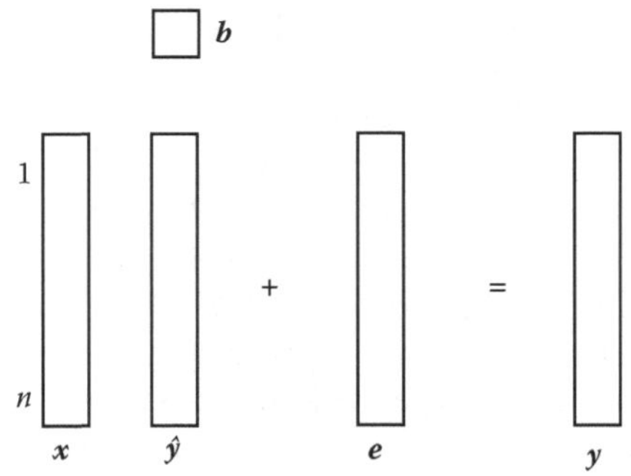

FIGURE 4.7 Simple OLS regression for mean-centered data: $\hat{\boldsymbol{y}} = b\boldsymbol{x}$.

(For simplicity, we denote the estimates by the same letters.) Then the predicted (modeled) property $\hat{y}_i$ for sample i and the prediction error e_i are calculated by

$$\hat{y}_i = b_0 + bx_i \tag{4.20}$$

$$e_i = y_i - \hat{y}_i \tag{4.21}$$

where $i = 1, \ldots, n$. There are many different possibilities how to optimally estimate the regression parameters. One very popular way is to take the ORDINARY LEAST-SQUARES (OLS) APPROACH, which minimizes the sum of the squared residuals $\sum e_i^2$ to estimate the model parameters b and b_0. This method has many advantages, one of them being explicit equations for the estimates of b and b_0:

$$b = \frac{\sum_{i=1}^{n} (x_i - \bar{x})(y_i - \bar{y})}{\sum_{i=1}^{n} (x_i - \bar{x})^2} \tag{4.22}$$

$$b_0 = \bar{y} - b \cdot \bar{x} \tag{4.23}$$

with $\bar{x}$ for the arithmetic mean of all x_i, and $\bar{y}$ for the arithmetic mean of all y_i. Note that the sum in the numerator in Equation 4.22 is proportional to the covariance between x and y, and the sum in the denominator is proportional to the variance of x. For mean-centered data ($\bar{x} = 0$, $\bar{y} = 0$), the intercept term is $b_0 = 0$ and

$$b = \frac{\sum_{i=1}^{n} x_i y_i}{\sum_{i=1}^{n} x_i^2} = \frac{\boldsymbol{x}^{\mathrm{T}}\boldsymbol{y}}{\boldsymbol{x}^{\mathrm{T}}\boldsymbol{x}} \tag{4.24}$$

For new x values, y is predicted by

$$\hat{y} = b_0 + bx \tag{4.25}$$

Note that the described model best fits the given (calibration) data, but is not necessarily optimal for predictions (see Section 4.2).

The least-squares approach can become very unreliable if outliers are present in the data (see Section 4.4). In this case, it is more advisable to minimize another function of the errors which results in more robust regression estimates. Although with the OLS approach the Equations 4.22 and 4.23 can always be applied, it is advisable to use the following assumptions for obtaining reliable estimates:

- Errors are only in y but not in x
- Residuals are uncorrelated and normally distributed with mean 0 and constant variance σ^2

Both assumptions are mainly needed for constructing confidence intervals and tests for the regression parameters, as well as for prediction intervals for new observations in x. The assumption of normal distribution additionally helps avoid skewness and outliers, mean 0 guarantees a linear relationship. The constant variance, also called HOMOSCEDASTICITY, is also needed for inference (confidence intervals and tests). This assumption would be violated if the variance of y (which is equal to the residual variance σ^2, see below) is dependent on the value of x, a situation called HETEROSCEDASTICITY, see Figure 4.8.

Besides estimating the regression coefficients, it is also of interest to estimate the variation of the measurements around the fitted regression line. This means that the RESIDUAL VARIANCE σ^2 has to be estimated which can be done by the classical estimator

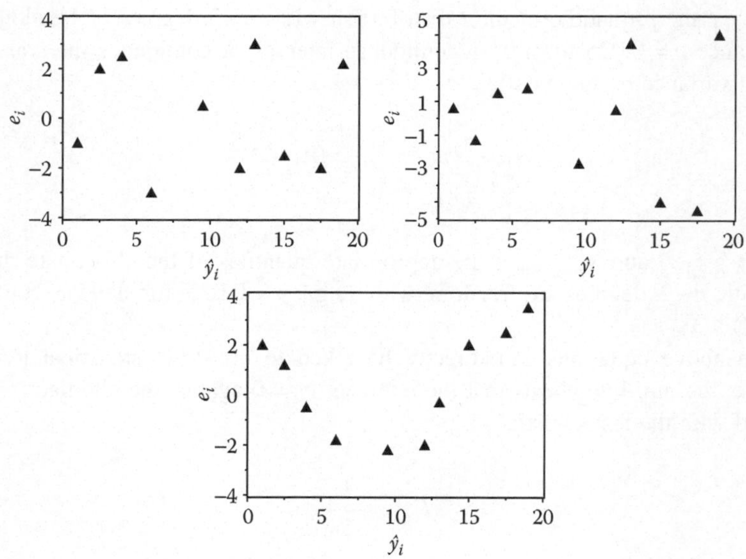

FIGURE 4.8 Examples of residual plots from linear regression. In the upper left plot, the residuals are randomly scattered around 0 (eventually normally distributed) and fulfill a requirement of OLS. The upper right plot shows heteroscedasticity because the residuals increase with y (and thus they also depend on x). The lower plot indicates a nonlinear relationship between x and y.

$$s_e^2 = \frac{1}{n-2} \sum_{i=1}^{n} (y_i - \hat{y}_i)^2 = \frac{1}{n-2} \sum_{i=1}^{n} e_i^2 \tag{4.26}$$

The denominator $n-2$ is used here because two parameters are necessary for a fitted straight line, and this makes s_e^2 an unbiased estimator for σ^2. The estimated residual variance is necessary for constructing confidence intervals and tests. Here the above model assumptions are required, and confidence intervals for intercept, b_0, and slope, b, can be derived as follows:

$$b_0 \pm t_{n-2;p} s(b_0) \tag{4.27}$$

$$b \pm t_{n-2;p} s(b) \tag{4.28}$$

with the standard deviations of b_0 and b

$$s(b_0) = s_e \sqrt{\frac{\sum_{i=1}^{n} x_i^2}{n \sum_{i=1}^{n} (x_i - \bar{x})^2}} \tag{4.29}$$

$$s(b) = \frac{s_e}{\sqrt{\sum_{i=1}^{n} (x_i - \bar{x})^2}} \tag{4.30}$$

and $t_{n-2;\,p}$ the p-quantile of the t-distribution with $n-2$ degrees of freedom, with for instance $p=0.025$ for a 95% confidence interval. A confidence interval for the residual variance σ^2 is given by

$$\frac{(n-2)s_e^2}{\chi^2_{n-2;1-p}} < \sigma^2 < \frac{(n-2)s_e^2}{\chi^2_{n-2;p}} \tag{4.31}$$

where $\chi^2_{n-2;1-p}$ and $\chi^2_{n-2;p}$ are the appropriate quantiles of the chi-square distribution with $n-2$ degrees of freedom with (e.g., $p=0.025$ for a 95% confidence interval).

The above equations can directly be taken to construct statistical tests. For example, the null hypothesis that the intercept $b_0=0$ against the alternative $b_0 \neq 0$ is tested with the test statistic

$$T_{b_0} = \frac{b_0}{s(b_0)} \tag{4.32}$$

where the estimated value for b_0 has to be used in the denominator. The null hypothesis is rejected at the significance level of α if $|T_{b_0}| > t_{n-2;1-\alpha/2}$. In this case, an intercept term is needed in the regression model. The test for $b=0$ is equivalent.

$$T_b = \frac{b}{s(b)} \tag{4.33}$$

Often it is of interest to obtain a confidence interval for the prediction at a new x value. Even more general, a confidence band for predicted values $\hat{y}$ as a function of x is given by Massart et al. (1997, p. 195)

$$\hat{y} \pm s_e \sqrt{2F_{2,n-2;p}} \sqrt{\frac{1}{n} + \frac{(x-\bar{x})^2}{\sum_{i=1}^{n}(x_i-\bar{x})^2}} \tag{4.34}$$

with $F_{2,n-2;p}$ the p-quantile of the F-distribution with 2 and $n-2$ degrees of freedom. Best predictions are possible in the mid part of the range of x where most information is available (Figure 4.9).

An example for OLS regression is shown in Figure 4.9; data for x and y are the same as in the next $\mathbb{R}$ example. The solid line is the OLS line given by the OLS estimates with intercept b_0 and slope b. The points seem to be scattered randomly around the regression line. Additionally, the dashed hyperbolic lines show the 95% confidence band. The true regression line (for the true parameters b_0 and b) will fall into this band with probability 95%.

```
R: x=c(1.5,2,2.5,2.9,3.4,3.7,4,4.2,4.6,5,5.5,5.7,6.6)
   y=c(3.5,6.1,5.6,7.1,6.2,7.2,8.9,9.1,8.5,9.4,9.5,11.3,11.1)
   res <- lm(y ~ x)  # linear model for y on x. The symbol ~
     # allows to construct a formula for the relation
   summary(res)        # gives the output of Fig. 4.10
```

Figure 4.10 shows the $\mathbb{R}$ output for OLS regression for the data shown in Figure 4.9. Under "Residuals," a summary for the residuals is given (minimum, first quartile, median, third quartile, maximum). Under "Coefficients," we see the

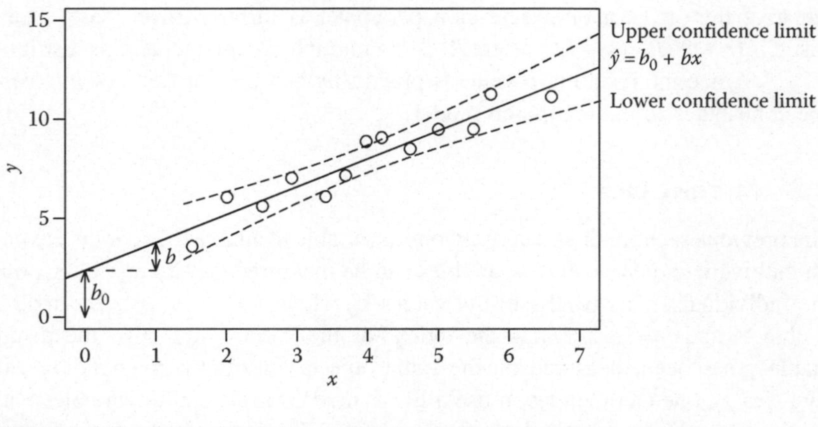

FIGURE 4.9 Confidence band for predicted values in linear regression.

```
Call:
lm(formula = y ~ x)
Residuals:
    Min       1Q  Median       3Q      Max
-0.9724 -0.5789 -0.2855  0.8124  0.9211
Coefficients:
              Estimate    Std.Error    t value    Pr(>|t|)
(Intercept)   2.3529      0.6186       3.804      0.00293 **
x             1.4130      0.1463       9.655      1.05e-06 ***
---
Signif. codes:  0 '***' 0.001 '**' 0.01 '*' 0.05 '.' 0.1 ' ' 1
Residual standard error: 0.7667 on 11 degrees of freedom
Multiple R-Squared: 0.8945,     Adjusted R-squared: 0.8849
F-statistic: 93.22 on 1 and 11 DF,  p-value: 1.049e-06
```

FIGURE 4.10 Output of the ℝ function "summary" for linear models. A linear regression model was fit using OLS to the data shown in Figure 4.9.

regression parameters and corresponding test results (STATISTICAL INFERENCE). The first line refers to the intercept term b_0 and the second to the slope b (for variable x). The column "Estimate" shows the estimated parameters, i.e., $b_0 = 2.3539$ and $b = 1.4130$. The column "Std. Error" provides the standard errors according to Equations 4.29 and 4.30, and the final columns include information for tests on the parameters (see Equations 4.32 and 4.33). Since the p-values, shown in column "Pr(>|t|)," are much smaller than a reasonable significance level α, e.g., $\alpha = 0.05$, both intercept and slope are important in our regression model. Below in Figure 4.10, there is information on the residual standard error s_e. The "Multiple R-Squared" in the case of univariate x and y is the same as the squared Pearson correlation coefficient between x and y, and is a measure of model fit (see Section 4.2.3). The "Adjusted R-squared" is similar to the multiple R-squared measure but penalizes the use of a higher number of parameters in the model (see Section 4.2.4). Finally, the F-statistic on the last line tests whether all parameters are zero, against the alternative that at least one regression parameter is different from zero. The test statistic is $(n-2)\,R^2/(1-R^2)$ where R^2 is the multiple R-squared, and its distribution is $F_{1,n-2}$. Since the resulting p-value is practically zero, at least one of intercept or slope contributes to the regression model.

4.3.2 Multiple OLS

In the previous section, there was only one x-variable available to model y. However, with multivariate data several x-variables can be measured, say $x_1, x_2, \ldots, x_p$, on the same individuals. This results in the values x_{ij} where $i = 1, \ldots, n$ is the index for the objects and $j = 1, \ldots, m$ is the index for the variables. If also the response variable y has been measured on the same objects with the corresponding values $y_1, y_2, \ldots, y_n$, one can then again use a linear model to relate all x-variables with y. This results in the set of equations

$$\begin{aligned} y_1 &= b_0 + b_1 x_{11} + b_2 x_{12} + \cdots + b_m x_{1m} + e_1 \\ y_2 &= b_0 + b_1 x_{21} + b_2 x_{22} + \cdots + b_m x_{2m} + e_2 \\ &\vdots \\ y_n &= b_0 + b_1 x_{n1} + b_2 x_{n2} + \cdots + b_m x_{nm} + e_n \end{aligned} \tag{4.35}$$

Thus, y is related to a linear combination of the x-variables, plus an additive error term. The difference to simple regression is that for each additional x-variable a new regression coefficient is needed, resulting in the unknown coefficients $b_0, b_1, \ldots, b_m$ for the m regressor variables. It is more convenient to formulate Equation 4.35 in matrix notation. Therefore, we use the vectors $\boldsymbol{y}$ and $\boldsymbol{e}$ like in Equation 4.19, but define a matrix $\boldsymbol{X}$ of size $n \times (m+1)$ which includes in its first column n values of 1, and in the remaining columns the values $x_{ij} (i=1, \ldots, n; j=1, \ldots, m)$. Thus, the first column of $\boldsymbol{X}$ will take care of the intercept term, and Equation 4.35 can be written as

$$\boldsymbol{y} = \boldsymbol{X}\boldsymbol{b} + \boldsymbol{e} \tag{4.36}$$

with the regression coefficients $\boldsymbol{b} = (b_0, b_1, \ldots, b_m)^{\mathrm{T}}$. This model is called the MULTIPLE LINEAR REGRESSION MODEL (Figure 4.11). For mean-centered x data, the column with ones can be omitted.

Similar to the univariate case, also for the multiple regression case, the residuals are calculated by

$$\boldsymbol{e} = \boldsymbol{y} - \hat{\boldsymbol{y}} \tag{4.37}$$

with $\hat{\boldsymbol{y}}$ being the predicted y values. There are several ways to estimate the regression coefficients, one of them being OLS estimation minimizing the sum of squared residuals $\boldsymbol{e}^{\mathrm{T}}\boldsymbol{e}$. This results in an explicit solution for the regression coefficients,

$$\boldsymbol{b} = (\boldsymbol{X}^{\mathrm{T}}\boldsymbol{X})^{-1}\boldsymbol{X}^{\mathrm{T}}\boldsymbol{y} \tag{4.38}$$

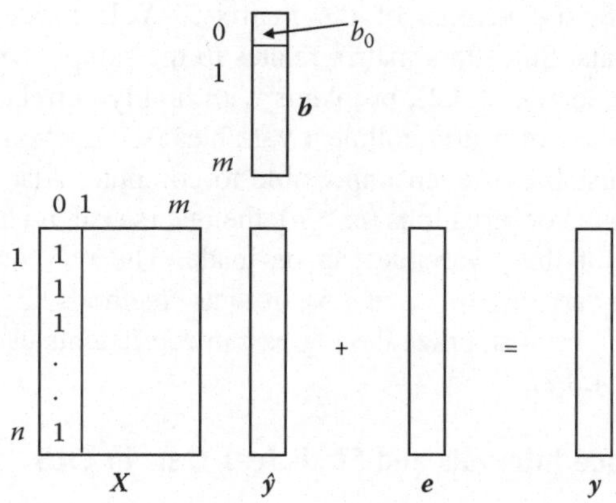

FIGURE 4.11 Multiple OLS regression: $\hat{\boldsymbol{y}} = \boldsymbol{X}\boldsymbol{b}$.

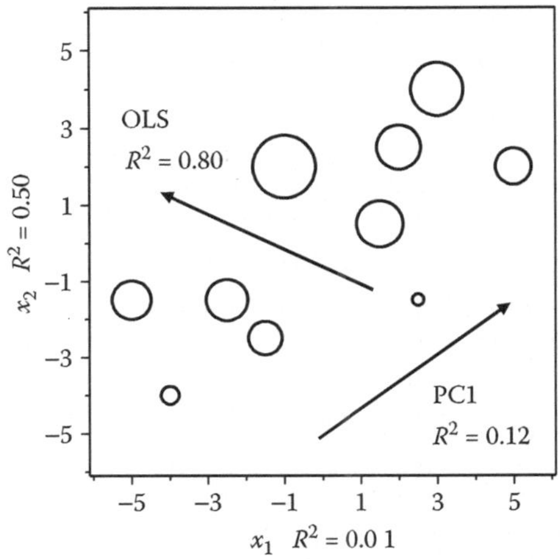

FIGURE 4.12 Linear latent variable with maximum variance of scores (PCA) and maximum correlation coefficient between y and scores (OLS). Scatter plot of a demo data set with 10 objects and two variables (x_1, x_2, mean-centered); the diameter of the symbols is proportional to a property y; R^2 denotes the squared correlation coefficients between y and x_1, y and x_2, y and PC1 scores, y and $\hat{y}$ from OLS.

and thus the fitted y values are

$$\hat{\boldsymbol{y}} = \boldsymbol{X}\boldsymbol{b} \tag{4.39}$$

The regression coefficients (including the intercept) form a vector $\boldsymbol{b}$ which can be considered as a loading vector for a latent variable in the x-space; actually it is the latent variable which gives the MAXIMUM PEARSON CORRELATION COEFFICIENT between y and $\hat{y}$ (Figure 4.12). A loading vector in PCA is usually normalized to unit length; however, the vector with regression coefficients is scaled for prediction of y.

There is an important difference to the univariate case. As can be seen from Equation 4.38, the inverse of the matrix $\boldsymbol{X}^{\mathrm{T}}\boldsymbol{X}$ is needed to compute the regression coefficients. Since this matrix relates to the sample covariance matrix of the x-variables (see Section 2.3.2), problems with highly correlated x-variables can be expected. In the case of nearly collinear variables (see Section 2.3.1), the inverse will become very instable or even impossible to compute. Also, if there are more regressor variables than observations ($m > n$), the inverse can no longer be computed and no predictions of the y-variable can be made. The way out is to reduce the number of regressor variables (Section 4.5), or to use methods like PCR (Section 4.6) or PLS (Section 4.7), or to stabilize the regression coefficients using Ridge or Lasso regression (Section 4.8.2).

4.3.2.1 Confidence Intervals and Statistical Tests in OLS

Also in multiple regression, confidence intervals for the parameters can be derived. From a practical point of view, it is, however, more important to test if single regression coefficients are zero, i.e., if single x-variables contribute to the explanation of the y-variable. Such a test can be derived if the following assumptions are fulfilled:

- Errors $\boldsymbol{e}$ are independent n-dimensional normally distributed
- With mean vector $\mathbf{0}$
- And covariance matrix $\sigma^2\boldsymbol{I}_n$

Similar to Equation 4.26, an unbiased estimator for the residual variance σ^2 is

$$s_e^2 = \frac{1}{n-m-1}\sum_{i=1}^{n}(y_i - \hat{y}_i)^2 = \frac{1}{n-m-1}(\boldsymbol{y} - \boldsymbol{X}\boldsymbol{b})^{\mathrm{T}}(\boldsymbol{y} - \boldsymbol{X}\boldsymbol{b}) \tag{4.40}$$

where $\boldsymbol{b}$ contains the OLS estimated regression coefficients. The null hypothesis $b_j = 0$ against the alternative $b_j \neq 0$ can be tested with the test statistic

$$z_j = \frac{b_j}{s_e\sqrt{d_j}} \tag{4.41}$$

where d_j is the jth diagonal element of $(\boldsymbol{X}^{\mathrm{T}}\boldsymbol{X})^{-1}$. The distribution of z_j is t_{n-m-1}, and thus a large absolute value of z_j will lead to a rejection of the null hypothesis.

Also an F-test can be constructed to test the null hypothesis $b_0 = b_1 = \cdots = b_m = 0$ against the alternative $b_j \neq 0$ for any $j = 0, 1, \ldots, m$. The test statistic can be derived by a variance decomposition of y (ANOVA): The idea is to decompose the deviations of y_i from their arithmetic mean $\bar{y}$ into a part $(\hat{y}_i - \bar{y})$ that is explained by the regression and into an unexplained part $(y_i - \hat{y}_i)$. Using the abbreviations

- Total sum of squares $\mathrm{TSS} = \sum (y_i - \bar{y})^2$
- Residual sum of squares $\mathrm{RSS} = \sum (y_i - \hat{y}_i)^2$
- Regression sum of squares $\mathrm{RegSS} = \sum (\hat{y}_i - \bar{y})^2$

it is easy to verify that $\mathrm{TSS} = \mathrm{RegSS} + \mathrm{RSS}$. These different sums can also be used to express the proportion of total variability of y that is explained by the regression:

$$R^2 = 1 - \mathrm{RSS}/\mathrm{TSS} = \mathrm{RegSS}/\mathrm{TSS} \tag{4.42}$$

This measure is the R-squared measure that was already mentioned in the univariate case above. Coming back to the test for all coefficients being zero, the test statistic is

$$F = \frac{\mathrm{RegSS}/m}{\mathrm{RSS}/(n-m-1)} = \frac{n-m-1}{m} \cdot \frac{R^2}{1-R^2} \tag{4.43}$$

and its distribution is $F_{m,n-m-1}$.

This test has in a very similar form another important application. Suppose we have given two nested regression models, M_0 and M_1, of different size,

$$M_0: \quad y = b_0 + b_1x_1 + \cdots + b_{m_0}x_{m_0} + e$$

$$M_1: \quad y = b_0 + b_1x_1 + \cdots\cdots + b_{m_1}x_{m_1} + e$$

where $m_0 < m_1$ and M_1 contains additional variables. The question is whether model M_1 is able to explain the y-variable better than model M_0. With the above idea of variance decomposition, the test statistic is

$$F = \frac{(\text{RSS}_0 - \text{RSS}_1)/(m_1 - m_0)}{\text{RSS}_1/(n - m_1 - 1)} \tag{4.44}$$

with distribution $F_{m_1-m_0,n-m_1-1}$. Here, RSS_0 and RSS_1 are the residual sum of squares for models M_0 and M_1, respectively. Thus, if $\text{RSS}_0 - \text{RSS}_1$ is large, the larger model M_1 can explain the y-variable in a better way; that means if $F > F_{m_1-m_0,n-m_1-1;1-\alpha}$ then H_0 (the smaller model M_0 is "true") is rejected at the significance level α. This test can be applied to any nested models as long as the assumptions for OLS are fulfilled. Comparisons of nested models containing different numbers of variables are performed by the criteria AIC, BIC or ${}_{\text{ADJ}}R^2$ as described in Section 4.2.4.

As an example for MLR, we consider the data from Table 4.2 (Section 4.1) where only variables x_1 and x_2 were in relation to the y-variable but not x_3. Nevertheless, a regression model using all x-variables is fitted and the result is presented in Figure 4.13. The statistical tests for the single regression coefficients clearly show that variable x_3 can be omitted from the model.

```
ℝ:  res <- lm(y ~ x1+x2+x3)  # linear model for y on x1,x2,x3 for
                              # the data of Table 4.2.
    summary(res)             # gives the output of Figure 4.13
```

```
Call:
lm(formula = y ~ x1+x2+x3)
Residuals:
    Min       1Q   Median      3Q     Max
-6.0340 -2.1819  0.1936  2.0693  6.0457
Coefficients:
             Estimate    Std. Error   t value    Pr(>|t|)
(Intercept)  1.41833     2.10969      0.672      0.509
x1           4.42317     0.46960      9.419      5.47e-09 ***
x2           4.10108     0.47209      8.687      2.14e-08 ***
x3          -0.03574     0.14519     -0.246      0.808
---
Signif. codes:  0 '***' 0.001 '**' 0.01 '*' 0.05 '.' 0.1 ' ' 1
Residual standard error: 3.309 on 21 degrees of freedom
Multiple R-Squared: 0.8884,     Adjusted R-squared: 0.8725
F-statistic: 55.72 on 3 and 21 DF,  p-value: 3.593e-10
```

FIGURE 4.13 Output of the ℝ function "summary" for linear models. A linear regression model was fit using OLS to the data presented in Table 4.2.

4.3.2.2 Hat Matrix and Full Cross Validation in OLS

In the context of OLS, the so-called HAT-MATRIX $\boldsymbol{H}$ plays an important role. $\boldsymbol{H}$ combines the observed and the predicted y-values by the equation

$$\hat{\boldsymbol{y}} = \boldsymbol{H}\boldsymbol{y} \tag{4.45}$$

and so it "puts the hat" on y. From Equation 4.38, it can be seen immediately that the hat-matrix is defined as

$$\boldsymbol{H} = \boldsymbol{X}(\boldsymbol{X}^{\mathrm{T}}\boldsymbol{X})^{-1}\boldsymbol{X}^{\mathrm{T}} \tag{4.46}$$

and that it only depends on the x-values. Of particular interest are the diagonal elements h_{ii} of the $n \times n$ matrix $\boldsymbol{H}$ because they reflect the influence of each value y_i on its own prediction $\hat{y}_i$. Large values of the diagonal elements would indicate a large influence, and therefore the values h_{ii} are often used for regression diagnostics (see Section 4.4.2).

There is a further application of the diagonal elements of the hat-matrix in full CV. By using the values h_{ii} from the OLS model estimated with all n objects, one can avoid to estimate OLS models for all subsets of size $n-1$, because there exists the relation

$$\mathrm{MSE}_{\text{FULL-CV}} = \frac{1}{n}\sum_{i=1}^{n}\left(\frac{y_i - \hat{y}_i}{1 - h_{ii}}\right)^2 \tag{4.47}$$

Here, $\hat{y}_i$ are the estimated y values using OLS for all n observations. Therefore, for estimating the prediction error with full CV, it suffices to perform an OLS regression with all n observations. This yields the same estimation of the prediction error as leave-one-out CV, but saves a lot of computing time.

There exists even a further simplification which makes it possible to directly use the sum of squared residuals RSS. In this so-called GENERALIZED CV (GCV), the values h_{ii} in Equation 4.47 are replaced by trace $(\boldsymbol{H})/n = \Sigma h_{ii}/n$, leading to a good approximation of the MSE:

$$\begin{aligned}\mathrm{MSE}_{\text{FULL-CV}} &\approx \frac{1}{(1-\mathrm{trace}(\boldsymbol{H})/n)^2}\cdot\frac{1}{n}\sum_{i=1}^{n}(y_i - \hat{y}_i)^2 \\ &= \frac{1}{(1-\mathrm{trace}(\boldsymbol{H})/n)^2}\cdot\frac{\mathrm{RSS}}{n}\end{aligned} \tag{4.48}$$

This approximation can not only be used in the context of multiple OLS, but also for methods where the estimated values are obtained via a relation $\hat{\boldsymbol{y}} = \boldsymbol{H}^* \boldsymbol{y}$ with $\boldsymbol{H}^*$ depending only on the x-variables (e.g., in Ridge regression—Section 4.8.2).

4.3.3 MULTIVARIATE OLS

While multiple linear regression aims at relating a single y-variable with several x-variables, MULTIVARIATE LINEAR REGRESSION relates several y-variables with several x-variables. Having available n observations for a number q of y-variables

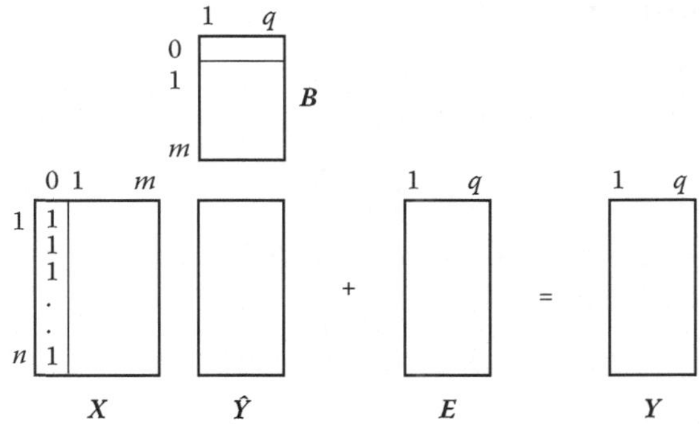

FIGURE 4.14 Multivariate OLS regression: $\hat{Y} = XB$.

and a number m of x-variables results in the data matrices Y of size $n \times q$ and X of size $n \times (m+1)$ (including the intercept), respectively. The regression model can be written as

$$Y = XB + E \tag{4.49}$$

with the $(m+1) \times q$ matrix B of regression coefficients and the $n \times q$ matrix of errors E (Figure 4.14). This model, however, can also be written in terms of the single y-variables. If y_j, b_j, and e_j denote the jth columns of Y, B and E, respectively, the regression model for the jth y-variable is

$$y_j = Xb_j + e_j \tag{4.50}$$

According to Equation 4.38, the resulting OLS estimator for the regression coefficients b_j is

$$b_j = (X^T X)^{-1} X^T y_j \tag{4.51}$$

for $j = 1, \ldots, q$. It is easy to see that the estimated coefficients can be combined to a matrix B of estimated regression coefficients, resulting in

$$B = (X^T X)^{-1} X^T Y \tag{4.52}$$

and in the fitted y values

$$\hat{Y} = XB \tag{4.53}$$

Matrix B consists of q loading vectors (of appropriate lengths), each defining a direction in the x-space for a linear latent variable which has maximum Pearson's correlation coefficient between y_j and $\hat{y}_j$ for $j = 1, \ldots, q$. Note that the regression coefficients for all y-variables can be computed at once by Equation 4.52, however, only for noncollinear x-variables and if $m < n$. Alternative methods to relate $\boldsymbol{X}$ and $\boldsymbol{Y}$ are PLS2 (Section 4.7) and CCA, see Section 4.8.1. These methods use linear latent variables for the x-space and the y-space; furthermore PLS2 also works with many and highly correlating variables.

4.4 ROBUST REGRESSION

4.4.1 OVERVIEW

If the model assumptions (normal distribution, homoscedasticity, etc., see Section 4.3) are fulfilled, the OLS regression estimator is also known as the BEST LINEAR UNBIASED ESTIMATOR (BLUE). So, among all unbiased estimators (on average the "true" regression parameters are estimated), the OLS estimator is the most precise one. However, if the assumptions do not hold, OLS estimation can lead to very poor prediction. Therefore it is recommended to check the model assumptions, e.g., by inspecting QQ-plots of the residuals (plotting quantiles of the standard normal distribution against the quantiles of the residuals, see Section 4.4.2), or by plotting the observed y-variable against the residuals. Diagnostics based on OLS estimation can, however, in certain situations be completely misleading. Figure 4.15 shows an example where 10 points follow a linear trend, but one object severely deviates from this trend. In the left plot, this deviating object is in the usual range of the x-variable but its value on the y-variable is outlying. In the right plot, the object is an outlier with respect to the x-variable, but in the usual range of the y-variable. OLS estimation results in an acceptable fit in the first case, but in a completely erroneous fit in the second case. However, the dashed line resulting from a robust regression method gives a good fit for the 10 "normal" points in both cases. This regression line would practically be identical with an OLS regression line only for the 10 "normal" points. Thus, only a single object may have a strong effect on OLS estimation, and the effect is more severe in the right plot with the x outlier. Such outliers are called

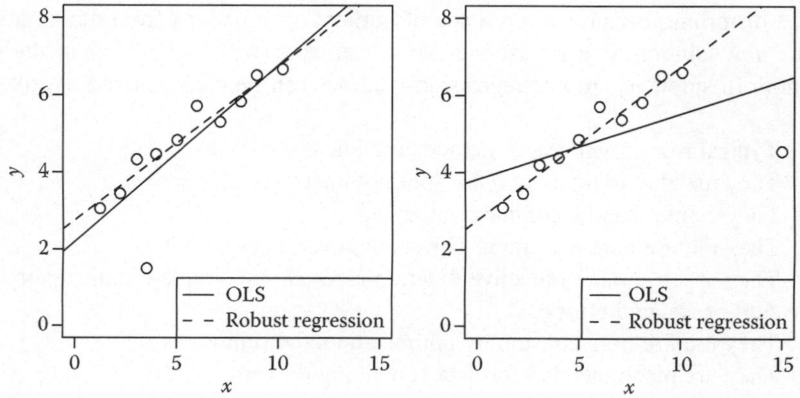

FIGURE 4.15 Comparison of OLS and robust regression on data with an outlier in the y-variable (left) and in the x-variable (right).

LEVERAGE POINTS because they can lever the OLS regression line. It is clear that statistical tests based on the OLS results are not useful. Even diagnostic plots using OLS residuals would be useless because the residuals can be large for "normal" points but small for outliers. One could argue that for the data used in Figure 4.15 it would be easy to visually identify the outlier and remove this object from the data. However, in the multiple or multivariate regression setup, this is no longer possible.

For robust regression, the OBJECTIVE FUNCTION is changed. While for OLS regression, the sum of all squared residuals is minimized; in robust regression, another function of the residuals is minimized. Three methods for robust regression are mentioned here:

- REGRESSION M-ESTIMATES minimize $\Sigma\rho(e_i/\sigma)$ to obtain the estimated regression coefficients, where the choice of the function ρ determines the robustness of the estimation, and σ is a robust estimation of the standard deviation of the residuals (Maronna et al. 2006). Note that both the residuals e_i and the residual scale σ depend on the regression coefficients, and several procedures have been proposed to estimate both quantities.
- REGRESSION MM-ESTIMATES use a special way of estimating these quantities, and the resulting regression coefficients have the advantage of attaining both high robustness and high precision (Maronna et al. 2006).
- LEAST TRIMMED SUM OF SQUARES (LTS) REGRESSION (Rousseeuw 1984) minimizes the sum of the smallest h squared residuals. Depending on the desired robustness, h can be varied between half and all the observations.

In general, robust regression requires an iterative optimization algorithm for obtaining the regression coefficients. Often it is only possible to find an approximate solution and not the global minimum of the objective function. Moreover, a new computation of the regression coefficients could result in a slightly different solution because random sampling is used within the optimization. This should not be disturbing because in presence of outliers or violations from model assumptions *any* solution of a robust regression estimator will be better than the OLS solution. In summary, robust regression methods can be characterized as follows:

- Typical robust regression methods are linear methods.
- They use the original variables (but not latent variable scores).
- They cannot handle collinear variables.
- They require data with more objects than variables ($n > 2m$).
- They apply another objective function as OLS (not simple minimization of sum of squared errors).
- They require time-consuming optimization algorithms.
- They are recommended for data containing outliers.

As an example for robust regression, we consider data from incineration of biomass. The problem is to model the softening temperature (SOT) of ash by the elemental composition of the ash. Data from 99 ash samples (Reisinger et al. 1996)—originating from different biomass—comprise the experimental SOT (630–1410 °C, used as the dependent y-variable) and the experimentally determined eight mass concentrations of the elements P, Si, Fe, Al, Ca, Mg, Na, and K (used as independent x-variables, with the sum of the oxides normalized to 100). Since the distribution of most variables is skewed, we additionally include the logarithmically transformed data in the regression model, resulting in 16 regressor variables. The ℝ-code for robust MM-regression is as follows.

```
ℝ:  library(robustbase)  # base library for robust statistics
    data(ash,package="chemometrics")           # load ash-data
    set.seed(4)       # set this seed to obtain this result
    reslmrob <- lmrob(SOT ~.,data=ash,compute.rd=TRUE)
             # robust linear regression
             # compute also robust Mahalanobis distances
    summary(reslmrob)       # gives the output of Figure 4.16
    plot(reslmrob)          # diagnostic plots of Figure 4.17
```

The summary statistics of Figure 4.16 for robust regression looks very similar to the usual summary statistics from OLS regression (see Figure 4.13 and explanations given there). We can see which variables significantly contribute to the robust regression model; marked by one or two asterisks are log(Fe_2O_3), log(MgO), log(Na_2O), and log(K_2O). Moreover, there is information of the weights used for downweighting outlying observations. The observations 5, 12, 57, and 96 received a weight close to zero; hence they are strong outliers.

4.4.2 Regression Diagnostics

In Section 4.3.2, the model assumptions for multiple regression were stated, which are important for confidence intervals and statistical tests. For robust regression, these assumptions can be violated, but they must be fulfilled at least for the majority of the objects. It is thus recommended to check the model assumptions with diagnostic plots that offer information about outliers and model fit. Figure 4.17 shows diagnostic plots for the example used in the previous section. The plot in the upper left part of Figure 4.17 shows the robust standardized residuals (robust residuals divided by their robustly estimated standard deviation) versus the robust Mahalanobis distances. The latter is computed only for the x-variables, and is thus equivalent to outlier detection for the x-variables (see Section 2.5) using robust Mahalanobis distances. The idea is to identify leverage points, as shown in Figure 4.15 (right). A vertical dotted line indicates the critical chi-square quantile, which is taken as cutoff value for separating regular points from the outliers (see Section 2.5). The two horizontal dotted lines at ± 2.5 separate regular observations from outliers in the y-variable. Outliers in y are objects with absolute larger standardized residuals than 2.5. For this data set, we have many objects which are

```
Call:
lmrob(formula = SOT ~ ., data = d3, compute.rd = TRUE)
Weighted Residuals:
     Min       1Q    Median        3Q       Max
-376.358  -37.044     8.401    59.296   625.884
Coefficients:
               Estimate    Std. Error    t value     Pr(>|t|)
(Intercept)   542120.93    1046942.79      0.518     0.60598
P2O5           -5401.47      10466.41     -0.516     0.60719
SiO2           -5412.53      10470.38     -0.517     0.60659
Fe2O3          -5422.14      10470.20     -0.518     0.60595
Al2O3          -5409.15      10469.99     -0.517     0.60680
CaO            -5408.32      10469.01     -0.517     0.60682
MgO            -5395.54      10471.00     -0.515     0.60774
Na2O           -5398.30      10470.31     -0.516     0.60753
K2O            -5419.97      10469.26     -0.518     0.60606
log(P2O5)        -35.95         31.22     -1.151     0.25298
log(SiO2)         52.99         50.82      1.043     0.30020
log(Fe2O3)        72.05         24.53      2.937     0.00430 **
log(Al2O3)        14.15         13.61      1.040     0.30149
log(CaO)         -28.93         47.76     -0.606     0.54633
log(MgO)        -129.33         56.15     -2.303     0.02379 *
log(Na2O)        -29.13         13.73     -2.121     0.03694 *
log(K2O)          72.99         22.30      3.273     0.00156 **
---
Signif. codes:  0 '***' 0.001 '**' 0.01 '*' 0.05 '.' 0.1 ' ' 1
```

```
Robust residual standard error: 77.32
Convergence in 45 IRWLS iterations
Robustness weights:
 4 observations c(5,12,57,96) are outliers with |weight| < 1.01e-06;
 4 weights are ~= 1; the remaining 91 ones are summarized as
   Min. 1st Qu.  Median    Mean 3rd Qu.    Max.
 0.0193  0.9010  0.9702  0.8864  0.9921  0.9999
```

FIGURE 4.16 Summary statistics of the result of robust regression for the ash data.

nonoutlying for the *y*-variable but outlying in the *x*-space. Although these objects are leverage points, they are even stabilizing the regression hyperplane because they are along the direction of the linear trend of the data. Thus, only the objects marked with "+" are potentially influencing OLS regression, as was demonstrated in Figure 4.15.

The remaining diagnostic plots shown in Figure 4.17 are the QQ-plot for checking the assumption of normal distribution of the residuals (upper right), the values of the *y*-variable (response) versus the fitted *y* values (lower left), and the residuals versus the fitted *y* values (lower right). The symbols "+" for outliers were used for the same objects as in the upper left plot. Thus it can be seen that the

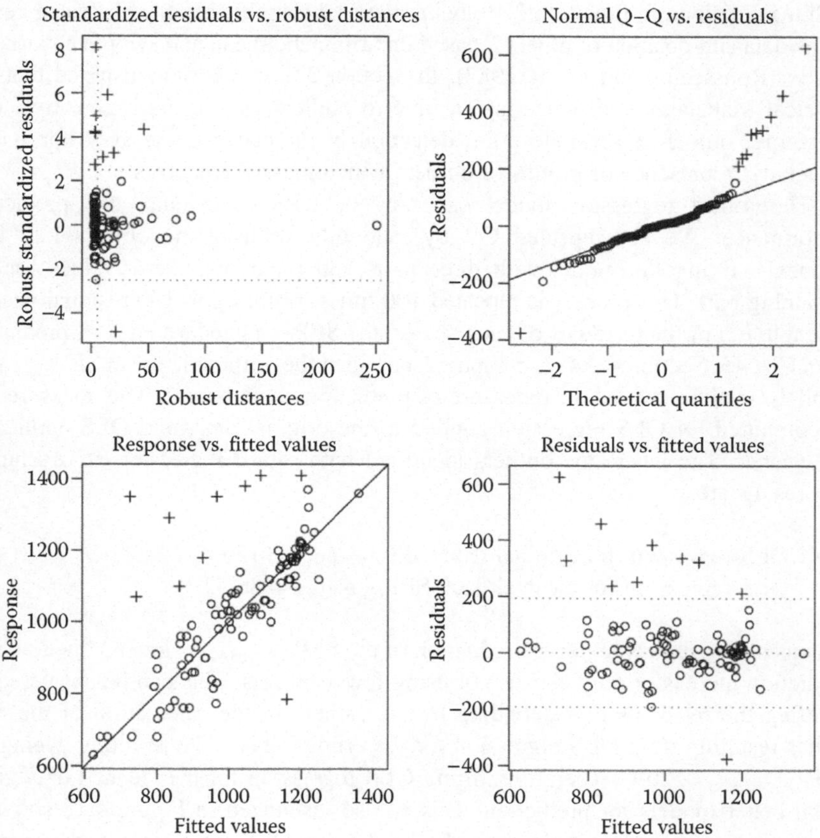

FIGURE 4.17 Diagnostic plots from robust regression on the ash data.

assumption of normal distribution for the residuals holds for the regular (nonoutlying) observations, that the linear model is useful for these objects, and that the residuals show about the same scattering along the whole range of the fitted values (homoscedasticity).

Note that also for OLS regression diagnostic plots are used. Besides QQ-plots and plots of residuals or response versus fitted values, a plot for the identification of leverage points is made. For the latter plot, the diagonal elements h_{ii} of the hat-matrix $\boldsymbol{H}$ are plotted (see Section 4.3.2.2). Since the values h_{ii} reflect the influence of each value y_i on its own prediction $\hat{y}_i$, large values refer to potential leverage points having large influence on the OLS regression estimates. It can, however, be shown that the diagonal elements of $\boldsymbol{H}$ are related to the squared Mahalanobis distance through

$$h_{ii} = \frac{d^2(\boldsymbol{x}_i)}{n-1} + \frac{1}{n} \tag{4.54}$$

with $d(\boldsymbol{x}_i)$ defined in Equation 2.19, being the Mahalanobis distance between center of the data and position of object i, based on arithmetic mean and sample covariance matrix (Rousseeuw and Leroy 1987). In Section 2.5, it was demonstrated that the classical Mahalanobis distance is sensitive to outliers, and so h_{ii} is sensitive, too. Therefore, outlier or leverage point detection by h_{ii} can become very unreliable, especially in presence of groups of outliers which mask themselves.

The robust regression model can now be used to evaluate the prediction performance. We use repeated CV by randomly splitting the objects into four segments, fitting the model with three parts and predicting the y values for the remaining part. This process is repeated 100 times, resulting in 100 residual vectors of length n. For the residuals of each vector, the SEP_{CV} (standard error of prediction from CV, see Section 4.2.4) is computed, and then the arithmetic mean $SEP_{CV\text{-}MEAN}$ of all 100 values is used as a measure of prediction performance. This measure can be computed for OLS regression applied to the original data, and OLS applied to the cleaned data where the outliers identified with robust regression are excluded. The results are

- OLS regression for original data: $SEP_{CV\text{-}MEAN} = 156\,°C$
- OLS regression for cleaned data: $SEP_{CV\text{-}MEAN} = 88\,°C$

The question arises whether the reduction of the $SEP_{CV\text{-}MEAN}$ is indeed due to better prediction models or only because of using fewer objects. This can be answered by omitting the residuals corresponding to the outliers in the calculation of the SEP values resulting from OLS regression for the original data. The resulting average is $SEP_{CV\text{-}MEAN} = 120°C$. Therefore, fitting OLS regression for the cleaned data gives much better models for prediction. This is also visualized in Figure 4.18 showing the response versus the fitted values. In the left plot OLS regression on the original

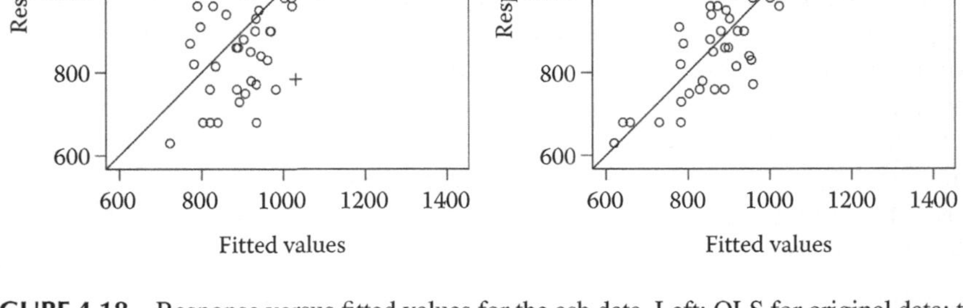

FIGURE 4.18 Response versus fitted values for the ash data. Left: OLS for original data; the symbol "+" indicated outliers identified with robust regression. Right: OLS regression for cleaned data where outliers are excluded.

data is performed, and the plot symbols are due to the outlier identification of robust regression (Figure 4.17). The right plot shows the results from OLS regression on the cleaned data. Even if the outliers marked with "+" would be excluded in the left plot, the fit is much poorer than in the right plot.

4.4.3 Practical Hints

When should we apply OLS and when robust regression? A recommendation could be to start with robust regression (e.g., MM-regression) and to inspect the regression diagnostics plots. If no outliers are visible in the plot of the robust distances versus the standardized residuals (upper left plot in Figure 4.17), switch to OLS and evaluate the regression model, e.g., with the SEP_{CV}. If outliers are visible, either remove the outliers and proceed with OLS and the evaluation (SEP_{CV}), or do the evaluation with robust regression on the uncleaned data. The latter approach can be computationally expensive because for each CV or bootstrap sample robust regression has to be applied. Moreover, a robust performance measure has to be used, like a trimmed SEP_{CV} where a fixed percentage (e.g., 20%) of the largest residuals is trimmed off. The advantage is that regression procedures like MM-regression do not use 0/1 weights for outlier rejection, but they smoothly downweight outliers and thus include more information to achieve a higher precision.

In situations where robust regression cannot be applied (collinearity, $n < 2m$) the x-variables could be summarized by robust principal components (Section 3.5). Then the procedure as mentioned above can be applied (see Section 4.6).

4.5 VARIABLE SELECTION

4.5.1 Overview

In the previous Sections 4.3.2 and 4.4, for multiple regression all available variables $x_1, x_2, \ldots, x_m$ were used to build a linear model for the prediction of the y-variable. This approach is useful as long as the number m of regressor variables is small, say not more than 10. However, in many situations in chemometrics, one has to deal with several hundreds of regressor variables. This may result in problems, because OLS regression (or robust regression) is no longer computable if the regressor variables are highly correlated, or if the number of objects is lower than the number of variables. Although the widely used regression methods PCR (Section 4.6) and PLS (Section 4.7) can handle such data without problems, there are arguments against using all available regressor variables:

- Use of all variables will produce a better fit of the model for the training data because the residuals become smaller and thus the R^2 measure increases (see Section 4.2). However, we are usually not interested in maximizing the fit for the training data but in maximizing the prediction performance for the test data. Thus a reduction of the regressor variables can avoid the effects of overfitting and lead to an improved prediction performance.

- A regression model with a high (e.g., hundreds) number of variables is practically impossible to interpret. The interpretation is feasible if no more than about a dozen variables are used in the model.
- Using a smaller number of regressor, variables can also reduce the computation time considerably. Although this argument might not seem to be relevant nowadays because OLS estimation is very fast to compute, it might be more relevant if careful but laborious evaluation schemes are used (e.g., repeated double CV or bootstrap techniques), or if more complex algorithms (e.g., for robust regression) are applied.

In chemistry, we often have a large number of variables obtained by automatic instruments (such as spectrometers or chromatographs) but a rather small number of samples. It is evident that for instance the absorbances at neighboring wavelengths are often highly correlated. Also mixture components (e.g., fatty acids) may have highly correlating concentrations. The basic philosophy in chemometrics is to consider such correlating variables as "parallel" measurements; parallel measurements have advantages in reducing noise and therefore the primary goal is not elimination of correlating variables. Thus, variable selection (often called FEATURE SELECTION) is intensively and controversy discussed in chemometrics (Anderssen et al. 2006; Baumann 2003; Forina et al. 2004; Frank and Friedman 1993; Nadler and Coifman 2005). Automatically omitting variables that contain essentially the same information as others could reduce the stability of a model. On the other hand, variables that contain essentially noise but have no relation to the *y*-variable will very likely reduce the prediction performance. However, it is not that simple to identify these variables, and sometimes variables with a very small correlation with the *y*-variable can be still useful in a multivariate model.

The most reliable approach would be an EXHAUSTIVE SEARCH among all possible variable subsets. Since each variable could enter the model or be omitted, this would be $2^m - 1$ possible models for a total number of m available regressor variables. For 10 variables, there are about 1000 possible models, for 20 about one million, and for 30 variables one ends up with more than one billion possibilities—and we are still not in the range for m that is standard in chemometrics. Since the goal is best possible prediction performance, one would also have to evaluate each model in an appropriate way (see Section 4.2). This makes clear that an expensive evaluation scheme like repeated double CV is not feasible within variable selection, and thus mostly only fit-criteria (AIC, BIC, adjusted R^2, etc.) or fast evaluation schemes (leave-one-out CV) are used for this purpose. It is essential to use performance criteria that consider the number of used variables; for instance simply R^2 is not appropriate because this measure usually increases with increasing number of variables.

Since an exhaustive search—eventually combined with exhaustive evaluation—is practically impossible, any variable selection procedure will mostly yield suboptimal variable subsets, with the hope that they approximate the global optimum in the best possible way. A strategy could be to apply different algorithms for variable selection and save the best candidate solutions (typically 5–20 variable subsets). With this low number of potentially interesting models, it is possible to perform a detailed evaluation (like repeated double CV) in order to find one or several variables subsets best suitable for prediction. One could combine the different models, or accept several solutions that perform equally well but have a different interpretation.

An important point is the evaluation of the models. While most methods select the best model at the basis of a criterion like adjusted R^2, AIC, BIC, or Mallow's *Cp* (see Section 4.2.4), the resulting optimal model must not necessarily be optimal for prediction. These criteria take into consideration the residual sum of squared errors (RSS), and they penalize for a larger number of variables in the model. However, selection of the final best model has to be based on an appropriate evaluation scheme and on an appropriate performance measure for the prediction of new cases. A final model selection based on fit-criteria (as mostly used in variable selection) is not acceptable.

The performance of different variable selection methods usually depends on the data set and thus on the inherent data structure. Unfortunately, there is no general rule or guideline for the choice of the method that is ideally suited for the data at hand. There are just limitations of certain methods, for example a limitation of BEST-SUBSET REGRESSION for the number of variables in the data (see Section 4.5.4). In any case, if the data show no structure but are just randomly distributed, any variable selection method will deliver no better results than a simple MONTE CARLO selection (random selection of variable subsets and testing the models), which is usually faster. A stochastic search for a good variable subset—together with some strategy—can be performed by genetic algorithms (see Section 4.5.6).

4.5.2 UNIVARIATE AND BIVARIATE SELECTION METHODS

There exist various strategies for variable selection that are not using the multivariate data information but are only based on simple coefficients computed for pairs of variables or even just for single variables. The advantage of such methods is the low computational cost, a great disadvantage is that the resulting variable subsets are often far from being optimal. Therefore, these methods are used as preselection tools, either for eliminating variables with very poor quality, or for selecting a set of potentially useful variables.

Criteria for the ELIMINATION of regressor variables are for example:

- A considerable percentage of the variable values is missing or below a small threshold.
- All or nearly all variable values are equal. This usually implies that the variance of the variable is low, and therefore the information content is poor. This is especially important for categorical or binary variables.
- The variable includes many and severe outliers. Outliers in data can distort least-square-based regression methods, and usually it is not possible to replace outliers by meaningful data values.
- Compute the correlation between pairs of regressor variables. If the correlation is high (positive or negative), exclude that variable having the larger sum of (absolute) correlation coefficients to all remaining regressor variables. The variable with the smaller sum covers more of additional or new information that is not already included in the other regressor variables.

Criteria for the IDENTIFICATION of potentially useful regressor variables are for example:

- High variance of the variable; this, however, depends on the data scale, and is thus only appropriate if the variables are measured at a similar scale.
- High (absolute) correlation coefficient with the y-variable; however, note that also variables with a weak correlation can become important if they are able to explain variability of y that is not captured by other regressor variables. A better strategy could thus be based on an iterative procedure, where only in the first step the variable with highest correlation with y is selected. Then a model is computed using the selected variable and the residuals are computed. In the next step the variable with the (absolute) highest correlation coefficient with the residuals is selected, and so on. However, this approach already considers the multivariate data information.

In practice, it often turns out that a significant number of variables can be deleted due to the above criteria, or that some variables are considered as a "must" for inclusion in a variable subset. This can speed up other algorithms for variable selection considerably.

4.5.3 STEPWISE SELECTION METHODS

A stepwise variable selection method adds or drops one variable at a time. Basically, there are three possible procedures (Miller 2002):

- FORWARD SELECTION: Start with the empty model (or with preselected variables) and add that variable to the model that optimizes a criterion; continue to add variables until a stopping rule becomes active.
- BACKWARD ELIMINATION: Start with the full model (or with preselected variables) and drop that variable from the model that optimizes a criterion; continue to drop variables until a stopping rule becomes active.
- BOTH DIRECTIONS: Combine the above strategies; the decision on whether a variable should be added or dropped (and which variable) is based on a criterion.

Criteria for the different strategies were mentioned in Section 4.2.4. For example, if the AIC measure is used for stepwise model selection, one would add or drop that variable which allows the biggest reduction of the AIC. The process is stopped if the AIC cannot be further reduced. This strategy has been applied in the example shown in Section 4.9.1.6.

An often-used version of stepwise variable selection (stepwise regression) works as follows: Select the variable with highest absolute correlation coefficient with the y-variable; the number of selected variables is $m_0 = 1$. Add each of the remaining x-variables separately to the selected variable; the number of variables in each subset is $m_1 = 2$. Calculate F as given in Equation 4.44,

$$F = \frac{(\mathrm{RSS}_0 - \mathrm{RSS}_1)/(m_1 - m_0)}{\mathrm{RSS}_1/(n - m_1 - 1)} \tag{4.55}$$

with RSS being the sum of the squared residuals $\sum_{i=1}^{n}(y_i - \hat{y}_i)^2$ (Equation 4.14). Consider the added variable which gives the highest F, and if the decrease of RSS is significant, take this variable as the second selected one. Significance is obtained if, e.g., $F > F_{m_1-m_0,n-m_1-1;0.95}$. Forward selection of variables would continue in the same way until no significant change occurs; a disadvantage of this strategy is that a selected variable cannot be removed later on. Usually the better strategy is to continue with a backward step. All variables in the current subset (at the moment only two) are tested whether elimination causes a significant increase of RSS or not (again by using the F-test as described above). The result of this step is the elimination of one variable or not. Now, another forward step is done, followed by another backward step, and so on, until no significant change of RSS occurs or a defined maximum number of variables is reached.

Note that starting from the full model is not possible in case of highly correlated variables, or for data sets with more variables than objects. Moreover, the different strategies will usually result in different variable subsets. However, applying one strategy will always give a unique solution because there is no randomization included in the algorithm. On the other hand, since only one variable can be included or excluded in each step, the optimal solution (that would be found by an exhaustive search) will in general be unattainable, and the approximate solution can be poor.

```
R:  lmfull <- lm(y~.,data=dat)  # linear model with all
                                # variables
    lmstep <- step(lmfull,direction="both")
                  # drop/add variables in each step
    summary(lmstep)  # summary output for the final model
```

4.5.4 Best-Subset Regression

In Section 4.5.1, we already mentioned that even for a moderate number of regressor variables an exhaustive search for the best variable subset is infeasible for interesting data sets. There is, however, a strategy that allows excluding complete branches in the tree of all possible subsets, and thus finding the best subset for data sets with up to about 30–40 variables.

The basic idea of the algorithm can be explained using a simple, hypothetical example. Suppose we have three regressor variables x_1, x_2, and x_3. Then the tree of all possible subsets is shown in Figure 4.19. We assume that the model selection is for instance based on the AIC measure. Starting from the full model, we may compute an AIC of 10 for the model including x_1 and x_2, and an AIC of 20 for a model with x_2 and x_3. One can show that reducing the latter model by one variable would result in an AIC of at least 18. This comes from the definition of the AIC in Equation 4.16, $\mathrm{AIC} = n\log(\mathrm{RSS}/n) + 2m$. Reducing m by one decreases the measure by 2, and RSS will be increased in general. In the other branch, we see that reducing

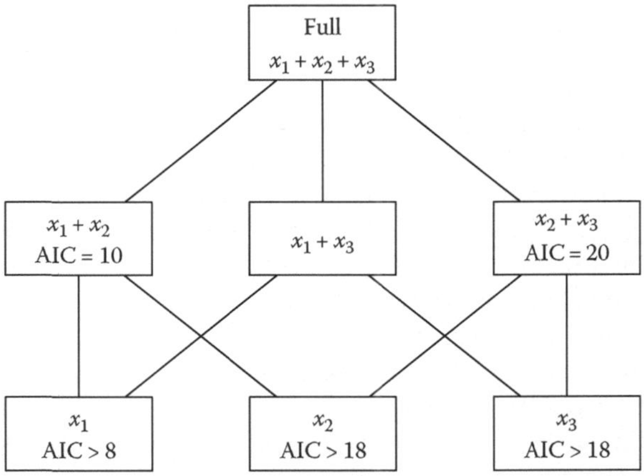

FIGURE 4.19 Tree of all possible models for three regressor variables with AIC values for a hypothetical example. If the AIC is used for model selection the right branch (models with $x_2 + x_3$ or x_2 or x_3) of the tree can be excluded.

the model with x_1 and x_2 by one variable results in a much smaller AIC of at least 8. Since we want to select that model which gives the smallest value of the AIC, the complete branch with $x_2 + x_3$ can be ignored, because any submodel in this branch is worse (AIC > 18) than the model $x_1 + x_2$ with AIC $= 10$. In this example, only the models x_1 and $x_1 + x_3$ could lead to an improved AIC value, and thus they need to be investigated.

Best-subset regression is also called ALL-SUBSETS REGRESSION, although not all subsets are tested. There are various different strategies and algorithms that allow excluding branches in the tree, like the LEAPS AND BOUNDS algorithm ("rapidly going/jumping up") (Furnival and Wilson 1974), or regression-tree methods (Hofmann et al. 2007). The criteria used for the identification of the best subsets are all based on the RSS, like the adjusted R^2, AIC, BIC, or Mallow's *Cp* (see Section 4.2.4). As discussed earlier (Section 4.5.1), all these criteria are fit-criteria, and thus they do not directly evaluate the performance of the regression models for new cases. Evaluating the prediction performance of each submodel would, however, be infeasible, and thus one hopes that these criteria lead to a good (not necessarily the best) model. Once this final model (or a couple of potentially good final models) has (have) been found, a careful evaluation of the prediction performance is required (Section 4.5.1). The algorithms for best-subset regression are usually limited to a maximal size of about 40 regressor variables.

```
R:  library(leaps)
    bestsub <- regsubsets(X,nvmax=15)
                  # models with at most 15 variables are
                  # considered
    summary(bestsub)   # summary output of the results
    plot(bestsub)      # plot subsets versus the criterion
```

4.5.5 Variable Selection Based on PCA or PLS Models

A simple strategy for variable selection is based on the information of other multivariate methods like PCA (Chapter 3) or PLS regression (Section 4.7). These methods form new latent variables by using linear combinations of the regressor variables, $b_1x_1 + b_2x_2 + \cdots + b_mx_m$ (see Section 2.6). The coefficients (loadings, regression coefficients) $b_1, b_2, \ldots, b_m$ reflect the importance of an x-variable for the new latent variable. Of course, these coefficients are only comparable if the x-variables are scaled to the same variance or they are multiplied with the standard deviation of the variable (see Section 4.1). Coefficients that are close to zero point at variables of less importance. Thus, the (absolute) size of the coefficients could be used as a criterion for variable selection. Note that the PCA loadings are only based on the information of the x-variables, while PLS loadings also account for the relation to the y-variable. Also note that the importance of a variable depends on the variable set in which the variable is included.

In Section 4.8.2, we will describe a method called Lasso regression. Depending on a tuning parameter, this regression technique forces some of the regression coefficients to be exactly zero. Thus the method can be viewed as a variable selection method where all variables with coefficients different from zero are in the final regression model.

4.5.6 Genetic Algorithms

Variable selection is an optimization problem. An optimization method that combines randomness with a strategy that is borrowed from biology is a technique using genetic algorithms—a so-called NATURAL COMPUTATION METHOD (Massart et al. 1997). Actually, the basic structure of GAs is ideal for the purpose of selection (Davis 1991; Hibbert 1993; Leardi 2003), and various applications of GAs for variable selection in chemometrics have been reported (Broadhurst et al. 1997; Jouan-Rimbaud et al. 1995; Leardi 1994, 2001, 2007). Only a brief introduction to GAs is given here, and only from the point of view of variable selection.

A particular selection of variables can be denoted by a vector consisting of binary components; a "1" indicates the variable is selected, a "0" it is not selected. Such a vector of length m (the total number of variables) defines one of the possible variable subsets and is simply a bit string. In the biologically inspired notation of GAs, such a vector is called a CHROMOSOME containing m GENES. A set of different chromosomes (each representing one of the possible variable subsets) is called a POPULATION (Figure 4.20). Usually one starts with randomly defined chromosomes, mostly restricted by a maximum number of "1" (selected variables) in a chromosome. The number of chromosomes—the size of the population—is typically 30–200.

For each chromosome (variable subset), a so-called FITNESS (RESPONSE, OBJECTIVE FUNCTION) has to be determined, which in the case of variable selection is a performance measure of the model created from this variable subset. In most GA applications, only fit-criteria that consider the number of variables are used (AIC, BIC, adjusted R^2, etc.) together with fast OLS regression and fast leave-one-out CV (see Section 4.3.2). Rarely, more powerful evaluation schemes are applied (Leardi 1994).

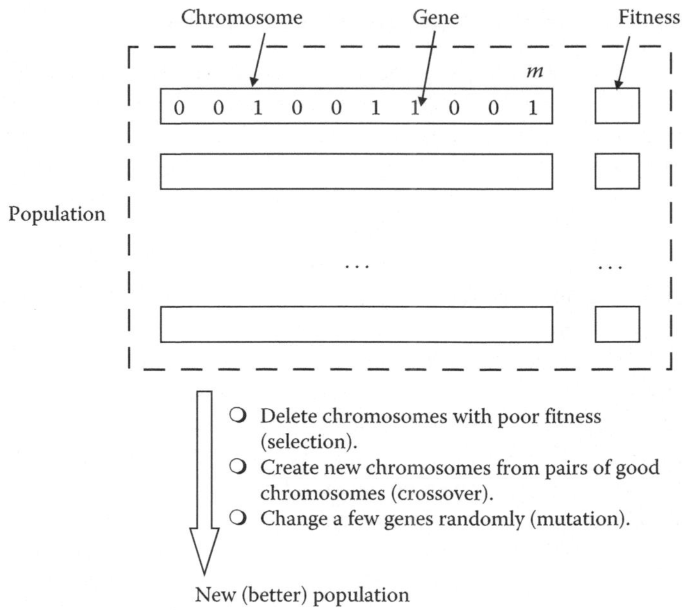

FIGURE 4.20 Scheme of a GA applied to variable selection. The first chromosome defines a variable subset with four variables selected from $m = 10$ variables. Fitness is a measure for the performance of a model built from the corresponding variable subset. The population of chromosomes is modified by genetically inspired actions with the aim to increase the fitness values.

In general, the population will contain chromosomes with different fitness, and the GA strategy is to produce better populations (EVOLVING OF THE POPULATION). A next population is obtained by biologically inspired actions as follows:

- Some of the worst chromosomes are deleted and replaced by new chromosomes (COMPETITION).
- New chromosomes are derived from pairs of good chromosomes, mostly by a so-called CROSSOVER (Figure 4.21). The idea is that a combination of two good chromosomes may produce an even better one.
- A small percentage of the genes are randomly changed by MUTATION. That means a few "0" are changed into "1" and vice versa. This random action should avoid to be trapped in local optima. The mutation rate may decrease during the training to achieve a better convergence.

Determination of the fitness for the new chromosomes completes one GENERATION of a GA training. The procedure is repeated until some termination criterion is reached (e.g., no increase of the fitness of the best chromosomes or the defined maximum number of generations reached). Typical GA trainings require some 100,000

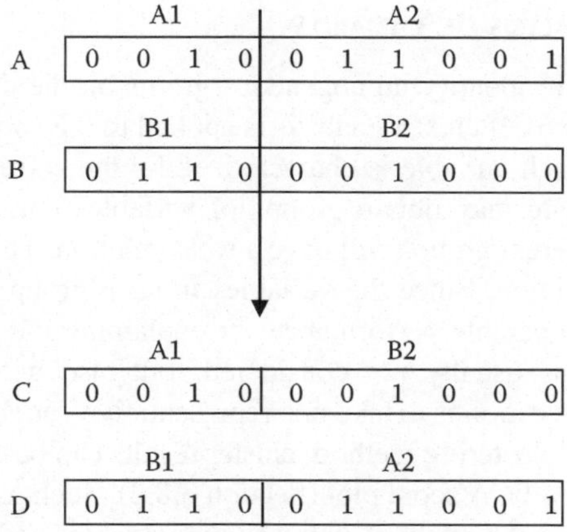

FIGURE 4.21 Crossover as used in GAs. Two chromosomes A and B are cut at a random position and the parts are connected in a crossover scheme resulting in two new chromosomes C and D. Various other rules for crossover have been suggested.

iterations (tested models), some 1000 generations, and computation times between minutes and hours. The final solution of variable selection is given by the chromosome with highest fitness—or often better by the first best chromosomes.

Several parameters have to be selected for a GA training, such as population size, initial population, selection mode, crossover parameters, mutation rate, and termination rule—all are influencing the final result. Repeating the training with different initializations and varied parameters is advisable despite the high computational effort necessary. The GA approach is especially useful for data sets between about 30 and 200 variables, and for selecting about 5–20 variables. Up to about 30 original variables, the method best-subset regression (Section 4.5.4) is capable to consider all possible variable subsets, and randomness is not necessary. For a large number of variables, the chance of finding a good solution in the huge search domain becomes small and computation time until convergence becomes very large. A preselection of the variables by fast methods is then necessary. If the philosophy is not to mix methods, a reasonable strategy is to use all (say 500) variables in several rather short GA runs, and to consider the most often selected 150–200 variables for extensive GA trainings (Yoshida et al. 2001). An implementation of GAs in ℝ is as following:

```
ℝ: library(subselect)
   Hmat <- lmHmat(X,y)  # for regression models
   gen <- genetic(Hmat$mat, kmin=5,kmax=15, H=Hmat$H,r=1,
                  crit="CCR12")
                  # models with 5-15 variables are considered
   gen$bestvalues  # best values of the criterion
   gen$bestsets    # best subsets
```

4.5.7 Cluster Analysis of Variables

Cluster analysis tries to identify homogeneous groups in the data, either objects or variables (see Chapter 6). If cluster analysis is applied to the correlation matrix of the regressor variables (each variable is characterized by the correlation coefficients to all other variables), one may obtain groups of variables that are strongly related, while variables in different groups will have a weak relation. This information can be used for variable selection. Since the variables in each group are similar, they are supposed to have comparable performance for explaining the y-variable; however, information about y is usually not considered (but also not necessary). Thus a strategy for variable selection is to take one representative variable from each cluster.

Depending on the clustering method, cluster results can be displayed in the form of a dendrogram or in a PCA score plot (Section 3.8.2). Such graphic allows a better visual impression about the relations between the variables and about the clustering structure. It is thus a suitable tool for selecting the variables.

4.5.8 Example

The concept described above will be demonstrated at an example from biomass technology. The HEATING VALUE OF BIOMASS is an important parameter for the design and the control of power plants. The so-called higher heating value (HHV) is the enthalpy of complete combustion of a fuel including the condensation enthalpy of formed water. Numerous empirical equations have been published to relate the heating value to the elemental composition of fuels other than biomass. In this example, data from $n = 122$ plant material samples (for instance wood, grass, rye, rape, reed) are used to develop regression models for a prediction of HHV (the y-variable) from the elemental composition (Friedl et al. 2005). The HHV data (kJ/kg) have been determined by bomb calorimetry. The used $m = 22$ x-variables consist of six basic variables, the mass % of carbon (C), hydrogen (H), nitrogen (N), sulfur (S), chlorine (Cl), and ash (A); furthermore of 16 derived variables, comprising the six squared terms C^2, H^2, N^2, S^2, Cl^2, A^2, the cross term C*H, the ratio C/H, and the eight logarithmic terms ln(C), ln(H), ln(N), ln(S), ln(Cl), ln(A), ln(C*H), ln(C/H). The nonlinear transformations of the basic variables have been used to model nonlinear relationships between HHV and the elemental composition.

The following methods for variable selection are used:

- *Correlation*: Select those 3, 5, 10, and 15 x-variables with the (absolute) highest Pearson correlation coefficient with the y-variable.
- *Variance*: Select those 3, 5, 10, and 15 x-variables with the highest variance.
- *Stepwise*: Perform a stepwise variable selection in both directions; start once from the empty model, and once from the full model; the AIC is used for measuring the performance.
- *Best-subset*: Best-subset regression; select the solution with the smallest BIC value.
- *GA* (genetic algorithm): Use the best three solutions.

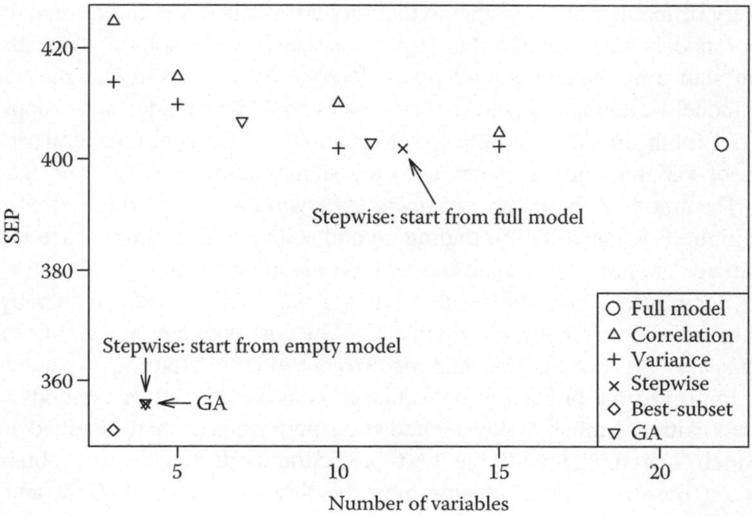

FIGURE 4.22 Performance of different heuristics for variable selection for the biomass data. The performance measure is the average standard error of prediction, SEP_{TEST}, resulting from repeated double CV, and it is compared with the number of variables in the subset. Best-subset regression leads to the best model, followed by solutions from stepwise regression and a GA. The simple heuristics using x-variables with the highest variance or with the highest Pearson correlation to y show about the same performance as the full model.

The variable subsets obtained from the different strategies are then tested with repeated double CV (see Section 4.2.5), where four segments are used for the outer loop (split all data into test sets and calibration sets), and seven segments are taken for the inner loop (determination of the optimum number of PLS components). Each run of the double CV gives a performance measure SEP^{i}_{TEST} (Section 4.2.3), and in total $i = 1, \dots, 100$ replications are carried out. The averages SEP_{TEST} of the resulting 100 values are shown in Figure 4.22. We also computed the standard errors around these averages, defined as the standard deviation of the SEP^{i}_{TEST} values, divided by $\sqrt{100}$, and they provide a measure of precision of SEP_{TEST}. They are in the range from 0.5 to 2, being smaller for smaller values of SEP_{TEST}, and are thus very reliable.

Although the results from this example cannot be generalized for any problem or data set, they reflect in some sense the typical behavior of the different variable selection methods. Since the number of variables was rather small ($m = 22$), *best-subset regression* can test all relevant subsets and should thus yield a very good solution. Figure 4.22 indeed shows that best-subset regression gave the best solution with the smallest value of SEP_{TEST}, closely followed by a solution from the GA and from stepwise regression. Note that best-subset is not necessarily the overall best solution, because a fit-measure is optimized (see discussion above). From the GA, the three solutions with the best fitness measure were used. Although all three solutions had almost the same value of the fitness function (fit-criterion), they show very different values for the prediction performance, leading to one good and two poor models for prediction. For *stepwise regression*, the solution depends on the start: the start from the empty model was clearly more successful than the start from the full model—although in both cases variables could be included and dropped from the model (both directions). Note that for data sets typical in chemometrics (high number of variables, highly correlated) the start is only possible from the empty model. The strategies based on *univariate or bivariate criteria* (highest correlation with y, highest variance) fail in finding a good solution. This shows once more that multivariate data have to be analyzed with multivariate concepts.

For comparison, also the result from the *full model* is shown, hereby using the same evaluation scheme by double CV. In this example, a variable selection by *best-subset regression*, GA, and *stepwise regression* (starting from the empty model) improved the prediction performance considerably; other methods resulted in models with a similar or lower prediction performance than obtained with the full model. The model with the best prediction performance was obtained by *best-subset regression* and contains only the three variables H, C*H and ln(N), with a $\mathrm{SEP}_{\mathrm{TEST}}$ of 351 kJ/kg corresponding to about 2% of the averaged HHV values used. The next best solutions (stepwise regression and GA) with a $\mathrm{SEP}_{\mathrm{TEST}}$ of 356 kJ/kg include in addition to the three variables H, C*H and ln(N) the variable Cl, which is surprising because the values of Cl are low and have a relatively high analytical error.

4.6 PRINCIPAL COMPONENT REGRESSION

4.6.1 OVERVIEW

Variable selection as introduced in Section 4.5 is one possibility of reducing the number of regressor variables and removing multicollinearity. This can lead to a regression model with a good interpretability, but the price to pay is a high computational effort, especially for a large number of x-variables. PCR also solves the problem of data collinearity and reduces the number of regressor variables, but the regressor variables are no longer the original measured x-variables but linear combinations thereof. The linear combinations that are taken for PCR are the principal component scores of the x-variables, so PCR is a combination of PCA (Chapter 3) and multiple linear regression (usually OLS, Section 4.3.2). Do not confuse "statistical PCR" with "biochemical PCR" (polymerase chain reaction); for the latter the Nobel Prize was given 1993 to one of the inventors. The scheme of PCR is visualized in Figure 4.23.

PCA decomposes a (centered) data matrix $\boldsymbol{X}$ into scores $\boldsymbol{T}$ and loadings $\boldsymbol{P}$, see Chapter 3. For a certain number a of PCs which is usually less than the rank of the data matrix, this decomposition is

$$\boldsymbol{X} = \boldsymbol{T}\boldsymbol{P}^{\mathrm{T}} + \boldsymbol{E} \tag{4.56}$$

with an error matrix $\boldsymbol{E}$ (see Equation 3.7). The score matrix $\boldsymbol{T}$ contains the maximum amount of information of $\boldsymbol{X}$ among all matrices that are orthogonal projections on

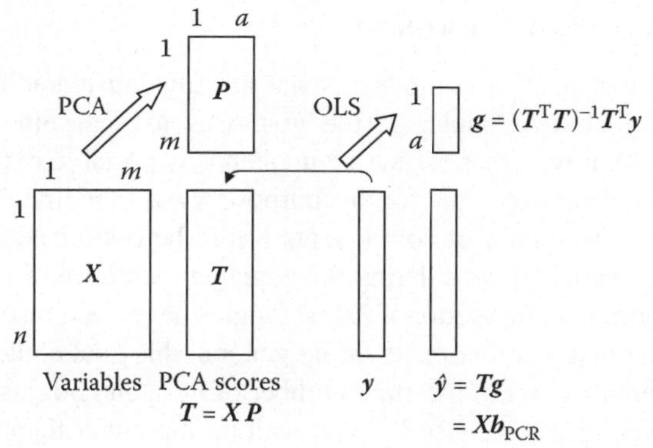

FIGURE 4.23 Scheme of PCR.

a linear combinations of the x-data. Note that $\boldsymbol{y}$ is not considered in this step. In a multiple linear regression model

$$\boldsymbol{y} = \boldsymbol{X}\boldsymbol{b} + \boldsymbol{e} \tag{4.57}$$

we replace the matrix $\boldsymbol{X}$ by the score matrix $\boldsymbol{T}$ and thus include major information of the x-data for regression on $\boldsymbol{y}$ (see Section 4.3.2). The resulting regression model is thus

$$\boldsymbol{y} = \boldsymbol{X}\boldsymbol{b} + \boldsymbol{e} = (\boldsymbol{T}\boldsymbol{P}^{\mathrm{T}})\boldsymbol{b} + \boldsymbol{e}_{\mathrm{T}} = \boldsymbol{T}\boldsymbol{g} + \boldsymbol{e}_{\mathrm{T}} \tag{4.58}$$

with the new regression coefficients $\boldsymbol{g} = \boldsymbol{P}^{\mathrm{T}}\boldsymbol{b}$ and the error term $\boldsymbol{e}_{\mathrm{T}}$. This indeed solves problems with data collinearity because the information of the highly correlated x-variables is compressed in few score vectors that are uncorrelated. Furthermore, the complexity of the regression model can be optimized by the number of used PCs (Section 4.6.2). OLS regression can now be used to estimate the regression coefficients, resulting in

$$\boldsymbol{g} = (\boldsymbol{T}^{\mathrm{T}}\boldsymbol{T})^{-1}\boldsymbol{T}^{\mathrm{T}}\boldsymbol{y} \tag{4.59}$$

(see Section 4.3.2). Due to the uncorrelatedeness of the score vectors $\boldsymbol{T}^{\mathrm{T}}\boldsymbol{T}$ is a diagonal matrix; consequently the inverse is easy and numerically stable to compute. The final regression coefficients for the original model (Equation 4.57) are, see Equation 4.58,

$$\boldsymbol{b}_{\mathrm{PCR}} = \boldsymbol{P}\boldsymbol{g} \tag{4.60}$$

PCR is an alternative method to the much more used regression method PLS (Section 4.7). PCR is a strictly defined method and the model often gives a very similar performance as a PLS model. Usually PCR needs more components than PLS because no information of $\boldsymbol{y}$ is used for the computation of the PCA scores; this is not necessarily a disadvantage because more variance of $\boldsymbol{X}$ is considered and the model may gain stability.

4.6.2 Number of PCA Components

The optimum number of PCA components was already considered in Chapter 3, but with a different goal. In Chapter 3, the goal was to determine the number of components in such a way that as few components as possible explain as much as possible of the total variance—and for this purpose always the first a components are considered. Here, the number of components has to be optimized for best possible prediction of the y-variable; considering the explained variance of the x-data is only of secondary importance. In Section 4.2.2, strategies have been discussed to find the optimum set of the first a components. More general, this goal could be achieved by searching the optimal subset of a certain number of PCs, and not just the "first" PCs with the largest variances, because it could well be that other than the first PCs are best suitable for the prediction of the y-variable. For this purpose, any method of variable selection (Section 4.5, here selection of PCA scores) can be applied. As with variable selection, the prediction quality of the final models has to be checked carefully with the help of CV or bootstrap techniques (see Section 4.2).

Often simple strategies for the selection of a good set of PCA scores (for PCR) are applied: (a) selection of the first PCA scores which cover a certain percentage of the total variance of $\boldsymbol{X}$ (for instance, 99%); (b) selection of the PCA scores with maximum correlation to $\boldsymbol{y}$. Application of PCR within $\mathbb{R}$ is easy for a given number of components. For an example and a comparison of PCR and PLS, see Section 4.9.1.

```
R:  library(pls)
    res <- pcr(y~X,ncomp=5)
```

4.7 PARTIAL LEAST-SQUARES REGRESSION

4.7.1 Overview

PLS stands for *partial least-squares* or/and *projection to latent structures by means of partial least squares* and is a method to relate a matrix $\boldsymbol{X}$ to a vector $\boldsymbol{y}$ or to a matrix $\boldsymbol{Y}$. PLS is the most widely used method in chemometrics for multivariate calibration and finds increasing interest also in other areas. The mathematical concept of PLS is less strictly defined than that of ordinary least-squares (OLS) regression or principal component regression (PCR), actually many versions of PLS algorithms have been described. The original PLS method has been developed around 1975 by the statistician Herman Wold for a treatment of chains of matrices and applications in econometrics. His son, Svante Wold and others introduced the PLS idea into chemometrics; however, PLS was for a long time rather unknown to statisticians. The original ideas of PLS were heuristic and the statistical properties rather a mystery (Frank and Friedman 1993). Later on, theoretical aspects have been developed, and some chemometricians even claim PLS as a philosophy of how to deal with complicated and approximate relationships (Wold et al. 2001; Wold et al. 1998). The history of PLS is closely connected with the history of chemometrics (Geladi and Esbensen 1990).

Essentially, the model structures of PLS and PCR are the same: The x-data are first transformed into a set of a few INTERMEDIATE LINEAR LATENT VARIABLES (COMPONENTS), and these new variables are used for regression (by OLS) with a dependent variable y. PCR uses principal component scores (derived solely from $\boldsymbol{X}$) as components, while PLS uses components that are related to $\boldsymbol{y}$. The criterion for the intermediate latent variables that is mostly applied in PLS is MAXIMUM COVARIANCE between scores and y (or between scores in x-space and scores in y-space). Covariance combines high variance of $\boldsymbol{X}$ (responsible for stability) and high correlation with the interesting property; thus PLS can be considered as a compromise between PCR (maximum variance, modeling $\boldsymbol{X}$) and OLS (maximum correlation, modeling $\boldsymbol{y}$); PLS is a special case of CONTINUUM REGRESSION (Stone and Brooks 1990). Depending on the used algorithm, either the loading vectors (directions of projection) are orthogonal or the scores are uncorrelated (with nonorthogonal loading vectors).

PLS and PCR are linear methods (although nonlinear versions exist) and therefore the final latent variable that predicts the modeled property, y, is a linear combination of the original variables, just as in OLS (Equation 4.1). In general, the resulting regression coefficients are different when applying OLS, PCR, and PLS, and the prediction performances of the models are different.

Some details in the concepts of PLS and the different methods and different software implementations make PLS still a black box or at least a dark gray box to most chemists. Essential facts for the user of PLS can be summarized as follows (Figure 4.24):

- PLS is a powerful linear regression method, insensitive to collinear variables, and accepting a large number of variables.
- Resulting model predicts a property y from the original dependent variables x_1 to x_m.
- Linear model contains regression coefficients b_1 to b_m and an intercept b_0.

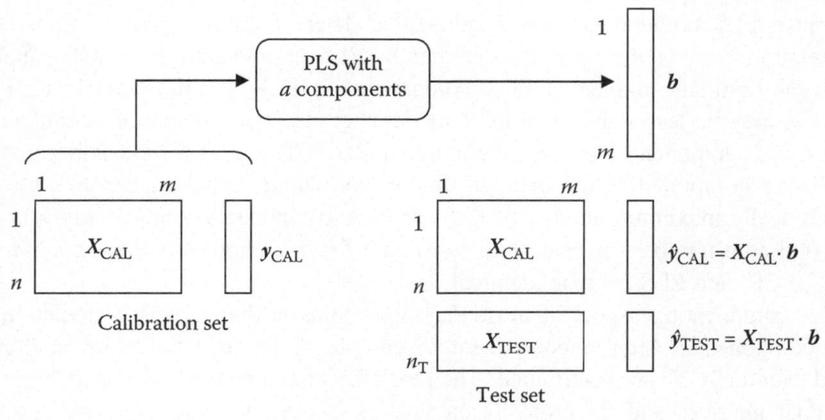

FIGURE 4.24 PLS as a multiple linear regression method for prediction of a property y from variables $x_1, \ldots, x_m$, applying regression coefficients $b_1, \ldots, b_m$ (mean-centered data). From a calibration set, the PLS model is created and applied to the calibration data and to test data.

- During model development, a relatively small number of PLS components (intermediate linear latent variables) are calculated which are internally used for regression.
- The number of PLS components determines the complexity of the model and can be optimized for high prediction performance.

Most often PLS is used for regression of *x*-variables with a single *y*-variable as follows:

1. First PLS-component is calculated as the latent variable which has MAXIMUM COVARIANCE between the scores and modeled property *y*. Note that the criterion "covariance" is a compromise between maximum correlation coefficient (OLS) and maximum variance (PCA).
2. Next, the information (variance) of this component is removed from the *x*-data. This process is called PEELING or DEFLATION; actually it is a projection of the *x*-space on to a (hyper-)plane that is orthogonal to the direction of the found component. The resulting RESIDUAL MATRIX $\boldsymbol{X}_{\mathrm{RES}}$ has the same number of variables as the original $\boldsymbol{X}$-matrix but the intrinsic dimensionality is reduced by one.
3. From the residual matrix, the next PLS component is derived—again with maximum covariance between the scores and $\boldsymbol{y}$.
4. This procedure is continued until no improvement of modeling *y* is achieved. The number of PLS components defines the complexity of the model (see Section 4.2.2).

In the standard versions of PLS, the scores of the PLS components are uncorrelated; the corresponding loading vectors, however, are in general not orthogonal. Each additional PLS component improves the modeling of *y*; the optimum number of components (optimal for prediction performance) is usually estimated by CV. Because PLS components are developed as latent variables possessing a high correlation with *y*, the optimum number of PLS components is usually smaller than the optimum number of PCA components in PCR. On the other hand, PLS models may be less stable than PCR models because less *x*-variance is contained. The more components are used, the more similar PCR and PLS models become. If OLS can be applied to the data (no highly correlating variables and $m < n$), the theoretically maximum number of PLS- or PCA-components is m (usually it is not the optimum number for best prediction), and for m components the models from OLS, PCR, and PLS become identical.

A complicating aspect of most PLS algorithms is the stepwise calculation of the components. After a component is computed, the residual matrices for $\boldsymbol{X}$ (and eventually $\boldsymbol{Y}$) are determined. The next PLS component is calculated from the residual matrices and therefore its parameters (scores, loadings, weights) do not relate to $\boldsymbol{X}$ but to the residual matrices. However, equations exist, that relate the PLS-*x*-loadings and PLS-*x*-scores to the original *x*-data, and that also provide the regression coefficients of the final model for the original x-data. The package "pls" in ℝ provides several versions of PLS as follows (Mevik and Wehrens 2007).

```
ℝ:  library(pls)
    res <- mvr(Y~X,ncomp=5,method="simpls")       # SIMPLS
                      # method="oscorespls") is for O-PLS
                      # method="kernelpls") is for Kernel-PLS
    library(chemometrics)
    res <- pls2_nipals(X,Y,a=5)                   # NIPALS
```

PLS with a matrix $\boldsymbol{Y}$ instead of a vector $\boldsymbol{y}$ is called PLS2. The purpose of data evaluation can still be to create calibration models for a prediction of the y-variables from the x-variables; in PLS2 the models for the various y-variables are connected. In a geometric interpretation (Figure 4.25), the m-dimensional x-space is projected on to a small number of PLS-x-components (summarizing the x-variables), and the q-dimensional y-space is projected on to a small number of PLS-y-components (summarizing the y-variables). The x- and the y-components are related pairwise by maximum covariance of the scores, and represent a part of the relationship between $\boldsymbol{X}$ and $\boldsymbol{Y}$. Scatter plots with the x-scores or the y-scores are projections of

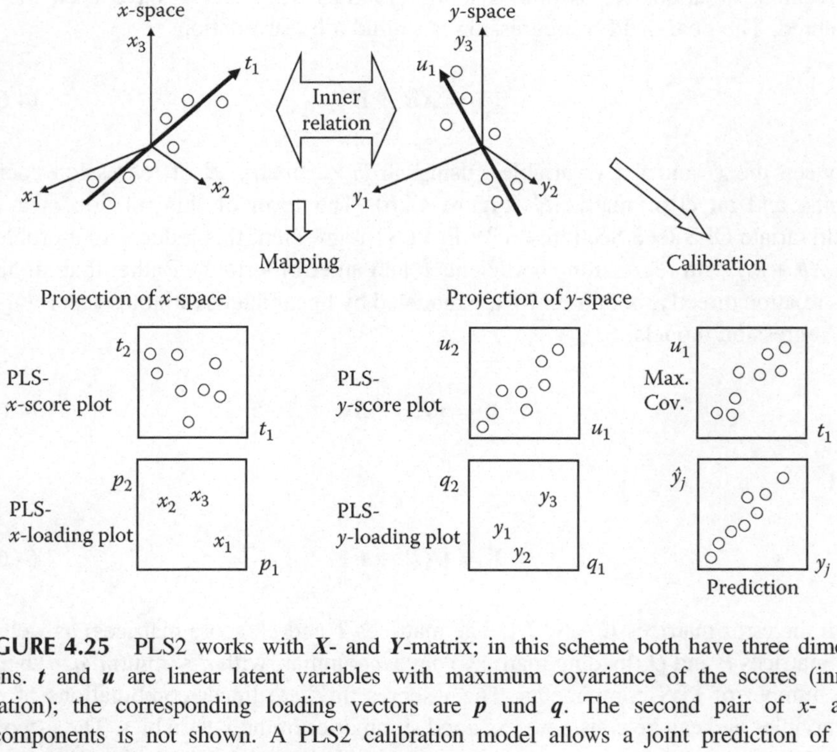

FIGURE 4.25 PLS2 works with $\boldsymbol{X}$- and $\boldsymbol{Y}$-matrix; in this scheme both have three dimensions. $\boldsymbol{t}$ and $\boldsymbol{u}$ are linear latent variables with maximum covariance of the scores (inner relation); the corresponding loading vectors are $\boldsymbol{p}$ und $\boldsymbol{q}$. The second pair of x- and y-components is not shown. A PLS2 calibration model allows a joint prediction of all y-variables from the x-variables via x- and y-scores.

the x-space and the y-space, respectively, with the projection planes influenced by the y-data and the x-data, respectively. Note the difference to score plots from separate PCA of $\boldsymbol{X}$ and $\boldsymbol{Y}$.

4.7.2 Mathematical Aspects

In literature, PLS is often introduced and explained as a numerical algorithm that maximizes an objective function under certain constraints. The objective function is the covariance between x- and y-scores, and the constraint is usually the orthogonality of the scores. Since different algorithms have been proposed so far, a natural question is whether they all maximize the same objective function and whether their results lead to comparable solutions. In this section, we try to answer such questions by making the mathematical concepts behind PLS and its main algorithms more transparent. The main properties of PLS have already been summarized in the previous section.

For PLS2 regression, we assume multivariate x- and y-data given by the matrix $\boldsymbol{X}$ of dimension $n \times m$ and the matrix $\boldsymbol{Y}$ of size $n \times q$. We assume that the data in the rows are obtained from the same n objects, and that $\boldsymbol{X}$ contains the information of m features (predictor variables), and $\boldsymbol{Y}$ describes q properties (response variables). (In case of PLS1 regression, we only have one response variable $\boldsymbol{y}$.) For a more convenient notation, we assume that the columns of $\boldsymbol{X}$ and $\boldsymbol{Y}$ have been mean-centered. The goal of PLS2 regression is to find a linear relation

$$\boldsymbol{Y} = \boldsymbol{X}\boldsymbol{B} + \boldsymbol{E} \tag{4.61}$$

between the x- and the y-variables, using an $m \times q$ matrix $\boldsymbol{B}$ of regression coefficients, and an error matrix $\boldsymbol{E}$ (Figure 4.26). The form of this relation is as in multivariate OLS (see Section 4.3.3). In PLS1 regression, this reduces to a problem $\boldsymbol{y} = \boldsymbol{X}\boldsymbol{b} + \boldsymbol{e}_1$, with regression coefficients $\boldsymbol{b}$ and an error term $\boldsymbol{e}_1$. Rather than finding this relation directly, both $\boldsymbol{X}$ and $\boldsymbol{Y}$ are modeled by linear latent variables according to the regression models

$$\boldsymbol{X} = \boldsymbol{T}\boldsymbol{P}^{\mathrm{T}} + \boldsymbol{E}_X \tag{4.62}$$

and

$$\boldsymbol{Y} = \boldsymbol{U}\boldsymbol{Q}^{\mathrm{T}} + \boldsymbol{E}_Y \tag{4.63}$$

with the error matrices $\boldsymbol{E}_X$ and $\boldsymbol{E}_Y$. The matrices $\boldsymbol{T}$ and $\boldsymbol{U}$ (score matrices) as well as the matrices $\boldsymbol{P}$ and $\boldsymbol{Q}$ (loading matrices) have a columns, with $a \leq \min(m, q, n)$ being the number of PLS components. The x-scores in $\boldsymbol{T}$ are linear combinations of the x-variables and can be considered as good summaries of the x-variables. The y-scores $\boldsymbol{U}$ are linear combinations of the y-variables and can be considered as good summaries of the y-variables. In the following, $\boldsymbol{t}_j$, $\boldsymbol{u}_j$, $\boldsymbol{p}_j$, and $\boldsymbol{q}_j$ denote the jth columns

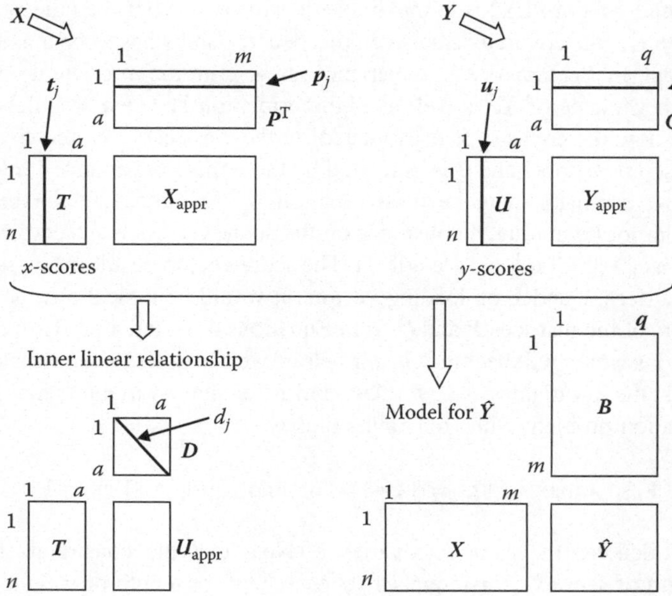

FIGURE 4.26 Matrices in PLS.

of $\boldsymbol{T}$, $\boldsymbol{U}$, $\boldsymbol{P}$, and $\boldsymbol{Q}$, respectively ($j = 1, \ldots, a$). In addition, the x- and y-scores are connected by the INNER LINEAR RELATIONSHIP

$$\boldsymbol{u}_j = d_j \boldsymbol{t}_j + \boldsymbol{h}_j \tag{4.64}$$

with $\boldsymbol{h}_j$ being the residuals and d_j the regression parameters. If for instance the linear relationship between $\boldsymbol{u}_1$ and $\boldsymbol{t}_1$ is strong (if the elements of $\boldsymbol{h}_j$ are small), then the x-score of the first PLS component is good for predicting y-scores and finally for predicting y-data. Usually more than one PLS component is used to model $\boldsymbol{Y}$ by $\boldsymbol{X}$; the optimum number of PLS components can be estimated by CV. The relationship between the scores then becomes

$$\boldsymbol{U} = \boldsymbol{TD} + \boldsymbol{H} \tag{4.65}$$

with $\boldsymbol{D}$ being a diagonal matrix with elements $d_1, d_2, \ldots, d_a$, and $\boldsymbol{H}$ the residual matrix with the columns $\boldsymbol{h}_j$. A schematic overview of the matrix relations is given in Figure 4.26. Equation 4.65 also motivates the name "PLS," because only partial information of the x- and y-data is used for regression, and because the usual classical regression routine to be employed is OLS; in Section 4.7.7, a robust regression technique will be used instead. Since for PLS1 no y-scores are available, Equation 4.65 reduces to

$$\boldsymbol{y} = \boldsymbol{Td} + \boldsymbol{h} \tag{4.66}$$

The fundamental goal of PLS2 is to maximize the covariance between the x- and the y-scores (for PLS1, the covariance between the x-scores and $\boldsymbol{y}$ has to be maximized). As pointed out in Section 4.7.1, covariance as a criterion for latent variables combines high variance of $\boldsymbol{X}$, as well as high correlation between $\boldsymbol{X}$ and $\boldsymbol{Y}$. Again it depends on how the covariance is estimated. In the classical case, the covariance between two score vectors $\boldsymbol{t}$ and $\boldsymbol{u}$ is estimated by the sample covariance $\boldsymbol{t}^{\mathrm{T}}\boldsymbol{u}/(n-1)$, but also robust estimators can be used (Section 4.7.7). Since the maximization problem would not be unique, a constraint on the score vectors is needed, which is usually taken as $\|\boldsymbol{t}\| = \|\boldsymbol{u}\| = 1$ (length 1). The score vectors result from projection of the data matrices $\boldsymbol{X}$ and $\boldsymbol{Y}$ on loading vectors. It would be logical now to use the loading vectors of the matrices $\boldsymbol{P}$ and $\boldsymbol{Q}$ from Equations 4.62 and 4.63. However, for technical reasons which will become clearer below, we use other loading vectors, say a vector $\boldsymbol{w}$ for the x-variables, i.e., $\boldsymbol{t}=\boldsymbol{X}\boldsymbol{w}$, and $\boldsymbol{c}$ for the y-variables, i.e., $\boldsymbol{u}=\boldsymbol{Y}\boldsymbol{c}$. The maximization problem with constraints thus is

$$\mathrm{cov}(\boldsymbol{X}\boldsymbol{w}, \boldsymbol{Y}\boldsymbol{c}) \rightarrow \max \qquad \|\boldsymbol{t}\| = \|\boldsymbol{X}\boldsymbol{w}\| = 1 \quad \text{and} \quad \|\boldsymbol{u}\| = \|\boldsymbol{Y}\boldsymbol{c}\| = 1 \tag{4.67}$$

where "cov" denotes the sample covariance. Note that the constraints have to be either length of score vectors equals 1 (chosen here) or length of weight vectors $\boldsymbol{w}$ and $\boldsymbol{c}$ equals 1. The solutions of the maximization problem are the first score vectors $\boldsymbol{t}_1$ and $\boldsymbol{u}_1$ for the x- and y-space, respectively. For subsequent score vectors, the same criterion (Equation 4.67) is maximized, but additional constraints have to be introduced. Usually these constraints are the ORTHOGONALITY TO PREVIOUS SCORE VECTORS, i.e., $\boldsymbol{t}_j^{\mathrm{T}}\boldsymbol{t}_l = 0$ and $\boldsymbol{u}_j^{\mathrm{T}}\boldsymbol{u}_l = 0$ for $1 \leq j \leq l < a$.

An alternative strategy is to require ORTHOGONALITY OF THE LOADING VECTORS which leads to nonorthogonal and thus not uncorrelated scores. Orthogonal loading vectors are obtained for instance by the eigenvector method (see Section 4.7.6), and this option might be preferable for plotting the scores (mapping). Uncorrelated scores are obtained by most other algorithms (Kernel, NIPALS, SIMPLS, O-PLS) and since each additional score vector covers new variability, this might be preferable for prediction purposes.

The FIRST PLS COMPONENT is found as follows: Since we deal with the sample covariance, the maximization problem (Equation 4.67) can be written as maximization of

$$\boldsymbol{t}^{\mathrm{T}}\boldsymbol{u} = (\boldsymbol{X}\boldsymbol{w})^{\mathrm{T}}\boldsymbol{Y}\boldsymbol{c} = \boldsymbol{w}^{\mathrm{T}}\boldsymbol{X}^{\mathrm{T}}\boldsymbol{Y}\boldsymbol{c} \rightarrow \max \tag{4.68}$$

under the same constraints of length 1 vectors, see Equation 4.67. The solutions for $\boldsymbol{w}$ and $\boldsymbol{c}$ are easily found by singular value decomposition (SVD) of $\boldsymbol{X}^{\mathrm{T}}\boldsymbol{Y}$ (see Section 3.6.3). According to that, among all possible directions $\boldsymbol{w}$ and $\boldsymbol{c}$, the maximum of Equation 4.68 is attained for the vectors $\boldsymbol{w}=\boldsymbol{w}_1$ and $\boldsymbol{c}=\boldsymbol{c}_1$ corresponding to the largest singular value of $\boldsymbol{X}^{\mathrm{T}}\boldsymbol{Y}$ (Hoeskuldsson 1988).

Different algorithms have been proposed for finding the solution. Especially for subsequent directions, the algorithms differ substantially. In the following, we will describe the most used algorithms for PLS.

4.7.3 Kernel Algorithm for PLS

Lindgren et al. (1993) introduced the Kernel algorithm; the name results from using eigen-decompositions of so-called kernel matrices, being products of $\boldsymbol{X}$ and $\boldsymbol{Y}$. Recall the maximization problem (Equation 4.68) with the solutions $\boldsymbol{w}_1$ and $\boldsymbol{c}_1$ as the left and right eigenvectors from an SVD of $\boldsymbol{X}^{\mathrm{T}}\boldsymbol{Y}$. Using the properties of SVD, the solutions can also be found by (Hoeskuldsson 1988)

$$\boldsymbol{w}_1 \text{ is the eigenvector to the largest eigenvalue of } \boldsymbol{X}^{\mathrm{T}}\boldsymbol{Y}\boldsymbol{Y}^{\mathrm{T}}\boldsymbol{X} \tag{4.69}$$

$$\boldsymbol{c}_1 \text{ is the eigenvector to the largest eigenvalue of } \boldsymbol{Y}^{\mathrm{T}}\boldsymbol{X}\boldsymbol{X}^{\mathrm{T}}\boldsymbol{Y} \tag{4.70}$$

According to Equation 4.67, both vectors have to be normalized such that $\|\boldsymbol{X}\boldsymbol{w}_1\| = \|\boldsymbol{Y}\boldsymbol{c}_1\| = 1$. The scores to the found directions are the projections $\boldsymbol{t}_1 = \boldsymbol{X}\boldsymbol{w}_1$ and $\boldsymbol{u}_1 = \boldsymbol{Y}\boldsymbol{c}_1$, and they are already normalized to length 1. The latent variable $\boldsymbol{p}_1$ is found by OLS regression according to the model (Equation 4.62) by

$$\boldsymbol{p}_1^{\mathrm{T}} = \left(\boldsymbol{t}_1^{\mathrm{T}}\boldsymbol{t}_1\right)^{-1}\boldsymbol{t}_1^{\mathrm{T}}\boldsymbol{X} = \boldsymbol{t}_1^{\mathrm{T}}\boldsymbol{X} = \boldsymbol{w}_1^{\mathrm{T}}\boldsymbol{X}^{\mathrm{T}}\boldsymbol{X} \tag{4.71}$$

which also makes the association to the vector $\boldsymbol{w}_1$ visible.

We continue with deriving the next set of components by maximizing the initial problem (Equation 4.67). This maximum is searched in a direction orthogonal to $\boldsymbol{t}_1$, and searching in the orthogonal complement is conveniently done by DEFLATION OF $\boldsymbol{X}$. The deflated matrix $\boldsymbol{X}_1$ is

$$\boldsymbol{X}_1 = \boldsymbol{X} - \boldsymbol{t}_1\boldsymbol{p}_1^{\mathrm{T}} = \boldsymbol{X} - \boldsymbol{t}_1\boldsymbol{t}_1^{\mathrm{T}}\boldsymbol{X} = \left(\boldsymbol{I} - \boldsymbol{t}_1\boldsymbol{t}_1^{\mathrm{T}}\right)\boldsymbol{X} \tag{4.72}$$

where we used the relation in Equation 4.71.

A DEFLATION OF $\boldsymbol{Y}$ is not necessary because when using the inner relationship (Equation 4.64) it turns out that the deflation would be done by multiplication of $\boldsymbol{Y}$ with the same matrix $\boldsymbol{G}_1 = (\boldsymbol{I} - \boldsymbol{t}_1\boldsymbol{t}_1^{\mathrm{T}})$ as for the $\boldsymbol{X}$ matrix. Since $\boldsymbol{G}_1$ is symmetric, $\boldsymbol{G}_1^{\mathrm{T}} = \boldsymbol{G}_1$, and idempotent, $\boldsymbol{G}_1\boldsymbol{G}_1 = \boldsymbol{G}_1$, the matrix products (Equation 4.69 and 4.70) for the eigen-decompositions to obtain $\boldsymbol{w}_2$ and $\boldsymbol{c}_2$ deliver the same result for $\boldsymbol{Y}$ being deflated or not. In more detail, $\boldsymbol{w}_2$ is the eigenvector to the largest eigenvalue of $\boldsymbol{X}_1^{\mathrm{T}}\boldsymbol{Y}\boldsymbol{Y}^{\mathrm{T}}\boldsymbol{X}_1 = \boldsymbol{X}^{\mathrm{T}}\boldsymbol{G}_1^{\mathrm{T}}\boldsymbol{Y}\boldsymbol{Y}^{\mathrm{T}}\boldsymbol{G}_1\boldsymbol{X} = \boldsymbol{X}^{\mathrm{T}}\boldsymbol{G}_1^{\mathrm{T}}(\boldsymbol{G}_1\boldsymbol{Y})(\boldsymbol{Y}^{\mathrm{T}}\boldsymbol{G}_1^{\mathrm{T}})\boldsymbol{G}_1\boldsymbol{X}$. The matrices in brackets would be the deflated matrices of the y-part, but the eigenvalue is the same without the deflation. The argument for $\boldsymbol{c}_2$ is similar.

Further PLS components ($\boldsymbol{t}_2$, $\boldsymbol{p}_2$, and so on) are obtained by the same algorithm as the first components using the deflated $\boldsymbol{X}$ matrix obtained after calculation of the previous component. The procedure is continued until a components have been extracted.

The y-SCORES $\boldsymbol{u}_j$ for components $j = 1, \ldots, a$ are derived from the x-scores by

$$\boldsymbol{u}_j = \boldsymbol{Y}\boldsymbol{c}_j \tag{4.73}$$

The y-LOADINGS $\boldsymbol{q}_j$ for components 1 to a are computed through the regression model (Equation 4.63).

$$\boldsymbol{q}_j^{\mathrm{T}} = \left(\boldsymbol{u}_j^{\mathrm{T}}\boldsymbol{u}_j\right)^{-1}\boldsymbol{u}_j^{\mathrm{T}}\boldsymbol{Y} \tag{4.74}$$

However, for estimating the final regression coefficients $\boldsymbol{B}$ for the model (Equation 4.61), they are not needed. It can be shown (Manne 1987) that the regression coefficients are estimated by

$$\boldsymbol{B} = \boldsymbol{W}(\boldsymbol{P}^{\mathrm{T}}\boldsymbol{W})^{-1}\boldsymbol{C}^{\mathrm{T}} \tag{4.75}$$

and they finally link the y-data with the x-data.

The kernel algorithm also works for UNIVARIATE Y-DATA (PLS1). Like for PLS2, the deflation is carried out only for the matrix $\boldsymbol{X}$. Now there exists only one positive eigenvalue for Equation 4.69, and the corresponding eigenvector is the vector $\boldsymbol{w}_1$. In this case, the eigenvectors for Equation 4.70 are not needed.

This version of the kernel algorithm is especially designed for data matrices with a large number of objects n. In this case—and if the dimensions of the data matrices are not too large—the kernel matrices have dimension much smaller than n and the eigen-decomposition is fast to compute. There is also an alternative version for the kernel algorithm for a high number of variables for the x- and y-data, and for a moderate number of objects, also called WIDE KERNEL METHOD (Rännar et al. 1994). The idea is to multiply Equation 4.69 with $\boldsymbol{X}$ from the left and Equation 4.70 with $\boldsymbol{Y}$ from the left. With the relations $\boldsymbol{t}_1 = \boldsymbol{X}\boldsymbol{w}_1$ and $\boldsymbol{u}_1 = \boldsymbol{Y}\boldsymbol{c}_1$, the new kernel matrices are $\boldsymbol{X}\boldsymbol{X}^{\mathrm{T}}\boldsymbol{Y}\boldsymbol{Y}^{\mathrm{T}}$ and $\boldsymbol{Y}\boldsymbol{Y}^{\mathrm{T}}\boldsymbol{X}\boldsymbol{X}^{\mathrm{T}}$ and the eigenvectors to the largest eigenvalues are $\boldsymbol{t}_1$ and $\boldsymbol{u}_1$, respectively. The rest of the procedure is similar.

4.7.4 NIPALS Algorithm for PLS

The NIPALS algorithm was the first algorithm for solving the PLS problem. Although the results turned out to be useful, there was confusion about what the algorithm is actually doing. The proposal of several slightly different versions of the algorithm was also not helpful with this respect. Note that the NIPALS algorithm gives the same results as the kernel algorithm because the same deflation is used; only the components are calculated differently (by the NIPALS iteration or as eigenvectors, respectively) but with the same result up to numerical precision.

We describe the most used version with the notation used in the previous section. The main steps of the NIPALS algorithm are as follows. Suppose we want to find the first PLS component, then the pseudocode is

(1) initialize $\boldsymbol{u}_1$ for instance by the first column of $\boldsymbol{Y}$
(2) $\boldsymbol{w}_1 = \boldsymbol{X}^{\mathrm{T}}\boldsymbol{u}_1/(\boldsymbol{u}_1^{\mathrm{T}}\boldsymbol{u}_1)$
(3) $\boldsymbol{w}_1 = \boldsymbol{w}_1/\|\boldsymbol{w}_1\|$
(4) $\boldsymbol{t}_1 = \boldsymbol{X}\boldsymbol{w}_1$
(5) $\boldsymbol{c}_1 = \boldsymbol{Y}^{\mathrm{T}}\boldsymbol{t}_1/(\boldsymbol{t}_1^{\mathrm{T}}\boldsymbol{t}_1)$
(6) $\boldsymbol{c}_1 = \boldsymbol{c}_1/\|\boldsymbol{c}_1\|$
(7) $\boldsymbol{u}_1^* = \boldsymbol{Y}\boldsymbol{c}_1$

(8) $\boldsymbol{u}_\Delta = \boldsymbol{u}_1^* - \boldsymbol{u}_1$
(9) $\Delta u = \boldsymbol{u}_\Delta^{\mathrm{T}} \boldsymbol{u}_\Delta$
(10) stop if $\Delta u < \varepsilon$ (with ε for instance set to 10^{-6}); otherwise $\boldsymbol{u}_1 = \boldsymbol{u}_1^*$ and go to step 2

Considering the latent variable models (Equations 4.62 and 4.63), it can be seen that step 2 finds the OLS regression coefficients $\boldsymbol{w}_1^j$ in the regression model $\boldsymbol{X} = \boldsymbol{u}_1^{j-1}(\boldsymbol{w}_1^j)^{\mathrm{T}} + \boldsymbol{e}$ with an error term $\boldsymbol{e}$. So, this is a regression of the x-data on the first "potential" score vector of the y-data, and it can be viewed as an approximation of the x-information by only one component. The denominator $\boldsymbol{u}_1^{\mathrm{T}}\boldsymbol{u}_1$ in step 2 is shown to make the regression evident. After normalizing $\boldsymbol{w}$ in step 3, the x-data are projected in step 4 to find the first score vector of the x-data. Then in step 5, there is an OLS regression of the y-data on the first score vector of the x-data. Again the denominator is shown to make the regression evident. Normalization in step 6 and projection in step 7 yield an update of the first y-score vector. This alternating regression scheme from x-information to y-information and vice versa usually converges quickly.

It is not immediately visible why this algorithm solves the initial problem (Equation 4.67) of maximizing the covariance between x-scores and y-scores. This can be shown by relating the equations in the above pseudocode. For example, when starting with the equation in step 2 for iteration $j+1$, we obtain $\boldsymbol{w}_1^{j+1} = \boldsymbol{X}^{\mathrm{T}}\boldsymbol{u}_1^j / \left((\boldsymbol{u}_1^j)^{\mathrm{T}}\boldsymbol{u}_1^j\right)$. Now we can plug in the formula for $\boldsymbol{u}_1^j$ from step 7; can again replace $\boldsymbol{c}_1^j$ by step 6 and step 5, and so on. Finally we obtain

$$\boldsymbol{w}_1^{j+1} = \boldsymbol{X}^{\mathrm{T}}\boldsymbol{Y}\boldsymbol{Y}^{\mathrm{T}}\boldsymbol{X}\boldsymbol{w}_1^j \cdot \text{constant} \tag{4.76}$$

where the constant depends on the norms of the different vectors. This shows that after convergence to $\boldsymbol{w}_1$, Equation 4.76 is an eigenvalue problem where $\boldsymbol{w}_1$ is the eigenvector of $\boldsymbol{X}^{\mathrm{T}}\boldsymbol{Y}\,\boldsymbol{Y}^{\mathrm{T}}\boldsymbol{X}$ to the largest eigenvalue. This was the task in Equation 4.69 for the kernel method where we have shown that this solves the initial problem (Equation 4.67). Similarly, it can be shown that $\boldsymbol{c}_1^{j+1} = \boldsymbol{Y}^{\mathrm{T}}\boldsymbol{X}\boldsymbol{X}^{\mathrm{T}}\boldsymbol{Y}\boldsymbol{c}_1^j \cdot$ constant, which results in the same problem (Equation 4.70) as in the kernel method.

For subsequent PLS components, the NIPALS algorithm works differently than the kernel method; however, the results are identical. NIPALS requires a deflation of $\boldsymbol{X}$ and of $\boldsymbol{Y}$ and the above pseudocode is continued by

(11) $\boldsymbol{p}_1 = \boldsymbol{X}^{\mathrm{T}}\boldsymbol{t}_1 / \left(\boldsymbol{t}_1^{\mathrm{T}}\boldsymbol{t}_1\right)$
(12) $\boldsymbol{q}_1 = \boldsymbol{Y}^{\mathrm{T}}\boldsymbol{u}_1 / \left(\boldsymbol{u}_1^{\mathrm{T}}\boldsymbol{u}_1\right)$
(13) $d_1 = \boldsymbol{u}_1^{\mathrm{T}}\boldsymbol{t}_1 / \left(\boldsymbol{t}_1^{\mathrm{T}}\boldsymbol{t}_1\right)$
(14) $\boldsymbol{X}_1 = \boldsymbol{X} - \boldsymbol{t}_1\boldsymbol{p}_1^{\mathrm{T}}$ and $\boldsymbol{Y}_1 = \boldsymbol{Y} - d_1\boldsymbol{t}_1\boldsymbol{c}_1^{\mathrm{T}}$

Steps 11–13 are the OLS estimates using the regression models (Equations 4.62 through 4.64). Step 14 performs a deflation of the $\boldsymbol{X}$ and of the $\boldsymbol{Y}$ matrix. The residual matrices $\boldsymbol{X}_1$ and $\boldsymbol{Y}_1$ are then used to derive the next PLS components, following the scheme of steps 1–10. Finally, the regression coefficients $\boldsymbol{B}$ from Equation 4.61 linking the y-data with the x-data are obtained by $\boldsymbol{B} = \boldsymbol{W}(\boldsymbol{P}^{\mathrm{T}}\boldsymbol{W})^{-1}\boldsymbol{C}^{\mathrm{T}}$, see Equation 4.75. Here, the matrices $\boldsymbol{W}$, $\boldsymbol{P}$, and $\boldsymbol{C}$ collect the vectors $\boldsymbol{w}_j$, $\boldsymbol{p}_j$, and $\boldsymbol{c}_j$ ($j = 1, \ldots, a$) in their columns, with a being the number of PLS components.

For PLS1 regression, the NIPALS algorithm simplifies. It is no longer necessary to use iterations for deriving one PLS component. Thus the complete pseudocode for extracting a components is as follows:

(1) initialize $\boldsymbol{X}_1 = \boldsymbol{X}$ and $\boldsymbol{y}_1 = \boldsymbol{y}$ and iterate steps 2 to 7 for $j = 1, \ldots, a$
(2) $\boldsymbol{w}_j = \boldsymbol{X}_j^{\mathrm{T}} \boldsymbol{y}_j / (\boldsymbol{y}_j^{\mathrm{T}} \boldsymbol{y}_j)$
(3) $\boldsymbol{w}_j = \boldsymbol{w}_j / \|\boldsymbol{w}_j\|$
(4) $\boldsymbol{t}_j = \boldsymbol{X}\boldsymbol{w}_j$
(5) $c_j = \boldsymbol{y}_j^{\mathrm{T}} \boldsymbol{t}_j / (\boldsymbol{t}_j^{\mathrm{T}} \boldsymbol{t}_j)$
(6) $\boldsymbol{p}_j = \boldsymbol{X}_j^{\mathrm{T}} \boldsymbol{t}_j / (\boldsymbol{t}_j^{\mathrm{T}} \boldsymbol{t}_j)$
(7) $\boldsymbol{X}_{j+1} = \boldsymbol{X}_{j+1} - \boldsymbol{t}_j \boldsymbol{p}_j^T$

The final regression coefficients in the model $\boldsymbol{y} = \boldsymbol{X}\boldsymbol{b} + \boldsymbol{e}_1$ are then estimated by $\boldsymbol{b} = \boldsymbol{W}(\boldsymbol{P}^{\mathrm{T}}\boldsymbol{W})^{-1}\boldsymbol{c}$, where $\boldsymbol{W}$ and $\boldsymbol{P}$ collect the vectors $\boldsymbol{w}_j$ and $\boldsymbol{p}_j$ in the columns, and $\boldsymbol{c}$ is the vector with the elements c_j.

4.7.5 SIMPLS Algorithm for PLS

The name for this algorithm proposed by de Jong (1993) originates from "straightforward implementation of a statistically inspired modification of the PLS method according to a simple concept." This algorithm directly maximizes the initial problem (Equation 4.67) under the constraint of orthogonality of the $\boldsymbol{t}$-scores for different components. The first PLS component from SIMPLS is identical with the result from NIPALS or the kernel algorithm. Subsequent components are in general slightly different. The main difference to NIPALS and the kernel algorithm is the kind of deflation. In SIMPLS, no deflation of the centered data matrices $\boldsymbol{X}$ and $\boldsymbol{Y}$ is made, but the deflation is carried out for the covariance matrix, or more precisely, the cross-product matrix $\boldsymbol{S} = \boldsymbol{X}^{\mathrm{T}}\boldsymbol{Y}$ between the x- and y-data. The pseudocode for the SIMPLS algorithm is as follows:

(1) initialize $\boldsymbol{S}_0 = \boldsymbol{X}^{\mathrm{T}}\boldsymbol{Y}$ and iterate steps 2 to 6 for $j = 1, \ldots, a$
(2) if $j = 1$, $\boldsymbol{S}_j = \boldsymbol{S}_0$; if $j > 1$, $\boldsymbol{S}_j = \boldsymbol{S}_{j-1} - \boldsymbol{P}_{j-1}(\boldsymbol{P}_{j-1}^{\mathrm{T}}\boldsymbol{P}_{j-1})^{-1}\boldsymbol{P}_{j-1}^{\mathrm{T}}\boldsymbol{S}_{j-1}$
(3) compute $\boldsymbol{w}_j$ as the first (left) singular vector of $\boldsymbol{S}_j$
(4) $\boldsymbol{w}_j = \boldsymbol{w}_j / \|\boldsymbol{w}_j\|$
(5) $\boldsymbol{t}_j = \boldsymbol{X}\boldsymbol{w}_j$
(6) $\boldsymbol{t}_j = \boldsymbol{t}_j / \|\boldsymbol{t}_j\|$
(7) $\boldsymbol{p}_j = \boldsymbol{X}_j^{\mathrm{T}} \boldsymbol{t}_j$
(8) $\boldsymbol{P}_j = [\boldsymbol{p}_1, \boldsymbol{p}_2, \ldots, \boldsymbol{p}_{j-1}]$

The resulting weights $\boldsymbol{w}_j$ and scores $\boldsymbol{t}_j$ are stored as columns in the matrices $\boldsymbol{W}$ and $\boldsymbol{T}$, respectively. Note that the matrix $\boldsymbol{W}$ differs now from the previous algorithms because the weights are directly related to $\boldsymbol{X}$ and not to the deflated matrices. Step 2 accounts for the orthogonality constraint of the scores $\boldsymbol{t}_j$ to all previous score vectors, because the search is done in the orthogonal complement of S_{j-1}. Step 3 directly maximizes the initial problem (Equation 4.67), compare to Equation 4.68. The scores in step 4 are obtained directly projecting X on the optimal direction, and the loadings in step 5 are obtained by OLS regression for the model (Equation 4.62).

The final regression coefficients for Equation 4.61 are now

$$B = WT^{\mathrm{T}}Y \tag{4.77}$$

where no matrix inversion is needed, compare to Equation 4.75.

The algorithm simplifies somewhat for PLS1. The orthogonality in step 2 is already fulfilled by using the projection $S_j = S_{j-1} - p_{j-1}(p_{j-1}^{\mathrm{T}}p_{j-1})^{-1}p_{j-1}^{\mathrm{T}}S_{j-1}$ if $j > 1$.

4.7.6 Other Algorithms for PLS

Orthogonal Projections to Latent Structures (O-PLS) was introduced by Trygg and Wold (2002). This method aims at removing variation from X that is orthogonal to the y-data. This orthogonal variation is modeled by extra components for the x-data and results in the decomposition $X = TP^{\mathrm{T}} + T_{\mathrm{o}}P_{\mathrm{o}}^{\mathrm{T}} + E$; compare with Equation 4.62, where T_{o} represent the scores and P_{o} the loadings of the orthogonal variation. This decomposition could also be viewed as a PLS model for the "filtered" data matrix $X - T_{\mathrm{o}}P_{\mathrm{o}}^{\mathrm{T}}$. By removing Y-orthogonal variation, O-PLS intends to maximize both correlation and covariance between the x- and y-scores to achieve both good prediction and interpretation. Remarkably, the O-PLS method has been patented (Trugg and Wold 2000).

The eigenvector algorithm (Hoeskuldsson 1988) is a similar approach as the kernel algorithm (Section 4.7.3) but works much easier. The idea is to compute not just the eigenvector to the largest eigenvalue in Equations 4.69 and 4.70 but to compute all eigenvectors to the largest a eigenvalues, where a is the desired number of PLS components. Thus, $p_1, \ldots, p_a$ are orthogonal PLS loading vectors in x-space given by the eigenvectors to the a largest eigenvalues of $X^{\mathrm{T}}Y\,Y^{\mathrm{T}}X$. Orthogonal PLS loading vectors in y-space $q_1, \ldots, q_a$ are the eigenvectors to the a largest eigenvalues of $Y^{\mathrm{T}}X\,X^{\mathrm{T}}Y$. The x- and y-scores are found by projecting the data on the loading vectors, i.e., x-scores $t_j = X\,p_j$ and y-scores $u_j = Y\,q_j$, for $j = 1, \ldots, a$; see Equations 4.62 and 4.63. No deflation is applied, and therefore the score vectors are not uncorrelated. This approach therefore does not solve the maximization problem (Equation 4.67). However, by definition of eigenvectors, the loading vectors are orthogonal, and this can be preferable especially for mapping. Using the score vectors, the x-data and the y-data can be mapped in lower-dimensional orthogonal coordinate systems, and interesting data structures can be revealed. For instance an application to combined mass spectrometric data (X) and chemical structure data (Y) has been published (Varmuza 2005). For use of the eigenvector PLS method within ℝ a function has been provided as follows.

```
ℝ:  library(chemometrics)
    res <- pls_eigen(X,Y,a=5)  # 5 PLS components
               # returned are the results for T, Q, U, and P
```

All algorithms described so far solve the linear PLS regression models (Equations 4.61 through 4.65). For NONLINEAR PLS we assume nonlinear relations, and two major approaches are mentioned here (Rosipal and Krämer 2006).

- Linear inner relation (Equation 4.65) is changed to a nonlinear inner relation, i.e., the y-scores have no longer a linear relation to the x-scores but a nonlinear one. Several approaches for modeling this nonlinearity have been introduced, like the use of polynomial functions, splines, ANNs, or RBF networks (Wold 1992; Wold et al. 1989).
- Original x- and y-data are mapped to a new representation using a nonlinear function. For this purpose the theory of kernel-based learning has been adapted to PLS. In the new data space linear PLS can be applied (Rosipal and Trejo 2001).

With the first approach the results are still interpretable in terms of the original variables. However, in some situations the second approach might be more adequate for a better description of the relations.

4.7.7 ROBUST PLS

All previous algorithms solved the problem stated in Equation 4.67, and the estimation of the covariance "cov" between the x-scores and the y-scores was done by the classical sample covariance. A robust covariance estimation for this purpose has been suggested (Gil and Romera 1998). Other approaches are based on replacing the OLS regressions by robust regressions (Cummins and Andrews 1995; Wakeling and Macfie 1992) or on robustifying SIMPLS (Hubert and Vanden Branden 2003).

Here we want to refer to a method that is directly following the idea of "partial least-squares," but uses robust M-regression (see Section 4.4) instead of "least-squares" to result in a method called "partial robust M-regression" (Serneels et al. 2005). We do not directly solve the original regression problem (Equation 4.61), but we regress the y-data only on partial information of the x-data, given by the latent variable model (Equation 4.62). Joining both formulas gives

$$\boldsymbol{y} = \boldsymbol{X}\boldsymbol{b} + \boldsymbol{e}_1 = \boldsymbol{T}\boldsymbol{P}^{\mathrm{T}}\boldsymbol{b} + \boldsymbol{e}_2 \tag{4.78}$$

We explain the ideas with univariate y-data, but they can be extended to multivariate y. The task is to robustly estimate the new regression coefficients $\boldsymbol{g} = \boldsymbol{P}^{\mathrm{T}}\boldsymbol{b}$ in Equation 4.78, where $\boldsymbol{T}$ is still unknown. Like in M-regression (Section 4.4), the idea is to apply a function ρ to the residuals $r_i = y_i - \boldsymbol{t}_i^{\mathrm{T}}\boldsymbol{g}$, where $\boldsymbol{t}_i$ form the rows of $\boldsymbol{T}$, for $i = 1, \ldots, n$. The function ρ downweights large absolute residuals, and thus we minimize $\sum \rho(y_i - \boldsymbol{t}_i^{\mathrm{T}}\boldsymbol{g})$. Alternatively, this can be written as minimization of $\sum w_i^r (y_i - \boldsymbol{t}_i^{\mathrm{T}}\boldsymbol{g})^2$ with appropriate residual weights $w_i^r = \rho(r_i)/r_i^2$. Not only large residuals but also leverage points can spoil the estimation of the regression coefficients (see Section 4.4), and thus we need to introduce additional weights for downweighting leverage points. These are outlying objects in the space of the regressor variables $\boldsymbol{T}$, and the resulting weights assigned to each object $\boldsymbol{t}_i$ are denoted by w_i^t (for a possible choice of the weight function see Serneels et al. 2005). Both types of weights can be combined by $w_i = w_i^r w_i^t$, and the regression coefficients $\boldsymbol{g}$ are resulting from minimizing

$$\sum_{i=1}^{n} w_i \left(y_i - \boldsymbol{t}_i^{\mathrm{T}} \boldsymbol{g}\right)^2 = \sum_{i=1}^{n} \left(\sqrt{w_i} y_i - (\sqrt{w_i} \boldsymbol{t}_i)^{\mathrm{T}} \boldsymbol{g}\right)^2 \tag{4.79}$$

This, however, means that both the y-data and the scores have to be multiplied by the appropriate weights $\sqrt{w_i}$ and then the classical OLS-based procedure can be applied. Practically, starting values for the weights have to be determined, and they are updated using an iterative algorithm.

The remaining task is to robustly estimate the score vectors $\boldsymbol{T}$ that are needed in the above regression. According to the latent variable model (Equation 4.62) for the x-data, the jth score vector is given by $\boldsymbol{t}_j = \boldsymbol{X}\boldsymbol{p}_j$, for $j = 1, \ldots, a$. Here, $\boldsymbol{t}_j$ denotes the jth column of $\boldsymbol{T}$, and $\boldsymbol{p}_j$ the jth loading vector (jth column of $\boldsymbol{P}$). According to Equation 4.67, the loading vectors $\boldsymbol{p}_j$ are obtained in a sequential manner via the maximization problem

$$\text{Maximize } \operatorname{cov_w}(\boldsymbol{X}\boldsymbol{p}, \boldsymbol{y}) \tag{4.80}$$

under the constraints $\|\boldsymbol{p}\| = 1$ and $\operatorname{cov_w}(\boldsymbol{X}\boldsymbol{p}, \boldsymbol{X}\boldsymbol{p}_l) = 0$ for $1 \leq l < j$. The notation $\operatorname{cov_w}(\boldsymbol{u}, \boldsymbol{y})$, with a vector $\boldsymbol{u}$ of length n, stands for the weighted covariance, and is defined with the above weights $\operatorname{cov_w}(\boldsymbol{u}, \boldsymbol{y}) = 1/n \, \Sigma w_i y_i u_i$. Thus, the constraints ensure loading vectors of length 1 that are uncorrelated to all previously extracted loading vectors. Once all loading vectors have been determined, the scores are computed by $\boldsymbol{T} = \boldsymbol{X}\boldsymbol{P}$. Solving the robust regression problem (Equation 4.79) yields the coefficients $\boldsymbol{g} = \boldsymbol{P}^{\mathrm{T}}\boldsymbol{b}$. The final regression parameters for Equation 4.78 are then $\boldsymbol{b} = \boldsymbol{P}\boldsymbol{g}$.

```
R:  library(chemometrics)
    res <- prm(X,y,a=5)   # 5 PLS components
```

4.8 RELATED METHODS

4.8.1 CANONICAL CORRELATION ANALYSIS

PLS regression as described in Section 4.7 allows finding (linear) relations between two data matrices $\boldsymbol{X}$ and $\boldsymbol{Y}$ that were measured on the same objects. This is also the goal of CCA, but the linear relations are determined by using a different objective function. While the objective in the related method PLS2 is to MAXIMIZE the COVARIANCE between the scores of the x- and y-data, the objective of CCA is to MAXIMIZE their CORRELATION. In CCA, it is usually assumed that the number n of objects is larger than the rank of $\boldsymbol{X}$ and of $\boldsymbol{Y}$. The reason is that the inverse of the covariance matrices are needed which would otherwise not be computable; applicability of CCA to typical chemistry data is therefore limited.

Using the same notation as in Section 4.7, the data matrices $\boldsymbol{X}(n \times m_X)$ and $\boldsymbol{Y}(n \times m_Y)$ are decomposed into loadings and scores, see Equations 4.62 and 4.63.

The score vectors $\boldsymbol{t}_j$ and $\boldsymbol{u}_j$ are linear projections of the data onto the corresponding loading vectors $\boldsymbol{p}_j$ and $\boldsymbol{q}_j$, i.e., $\boldsymbol{t}_j = \boldsymbol{X}\boldsymbol{p}_j$ and $\boldsymbol{u}_j = \boldsymbol{Y}\boldsymbol{q}_j$, for $j = 1, \ldots, a$ components. The goal of CCA is to find directions $\boldsymbol{p}$ and $\boldsymbol{q}$ in the x- and y-space which maximize the correlation

$$\text{cor}(\boldsymbol{X}\boldsymbol{p}, \boldsymbol{Y}\boldsymbol{q}) \rightarrow \max \tag{4.81}$$

under the constraints $\|\boldsymbol{X}\boldsymbol{p}\| = 1$ and $\|\boldsymbol{Y}\boldsymbol{q}\| = 1$. "cor" denotes the Pearson correlation coefficient (but also other correlation measures could be considered, see Section 2.3.2). Similar to PLS there is a subspace of solutions, and the dimension of the subspace is $a = \min(m_X, m_Y)$, the minimum of the dimensions of the x- and y-spaces. The solutions of the maximization problem (Equation 4.81) are the loadings vectors $\boldsymbol{p}_j$ and $\boldsymbol{q}_j$, for $j = 1, \ldots, a$; hereby assuming that the corresponding score vectors are uncorrelated, i.e., $\text{cor}(\boldsymbol{t}_j, \boldsymbol{t}_k) = 0$ and $\text{cor}(\boldsymbol{u}_j, \boldsymbol{u}_k) = 0$ for $j \neq k$. The resulting maximal correlations $r_j = \text{cor}(\boldsymbol{t}_j, \boldsymbol{u}_j)$ are called jth CANONICAL CORRELATION COEFFICIENTS. In general, the x-loading vectors $\boldsymbol{p}_j$ and the y-loading vectors $\boldsymbol{q}_j$ are not orthogonal. Figure 4.27 shows a schematic overview of the resulting matrices and vectors.

The solutions for the loading vectors $\boldsymbol{p}_j$ and $\boldsymbol{q}_j$ are found by solving two eigenvector/eigenvalue problems. Let $\boldsymbol{S}_X = \text{cov}(\boldsymbol{X})$, $\boldsymbol{S}_Y = \text{cov}(\boldsymbol{Y})$, and $\boldsymbol{S}_{XY} = \text{cov}(\boldsymbol{X}, \boldsymbol{Y})$ be the sample covariance matrices of $\boldsymbol{X}$ and $\boldsymbol{Y}$, and the sample covariance matrix between $\boldsymbol{X}$ and $\boldsymbol{Y}$ (a matrix $m_X \times m_Y$ containing the covariances between all x- and all y-variables), respectively. Also other covariance measures could be considered, see Section 2.3.2. Then the solutions are (Johnson and Wichern 2002)

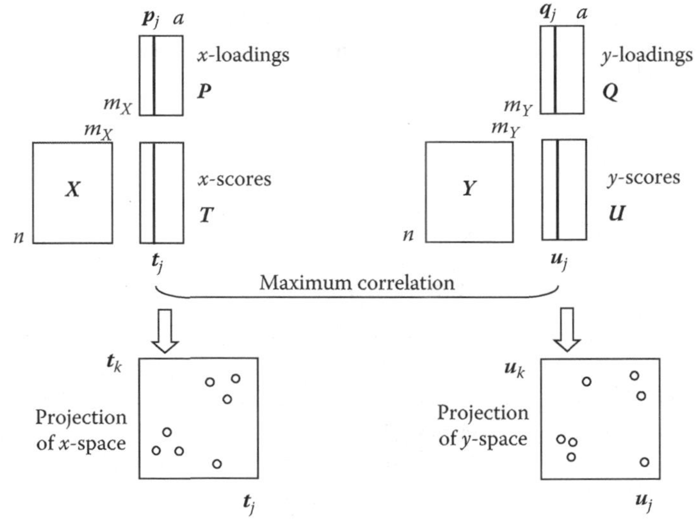

FIGURE 4.27 Canonical correlation analysis (CCA). x-scores are uncorrelated; y-scores are uncorrelated; pairs of x- and y-sores (for instance $\boldsymbol{t}_1$ and $\boldsymbol{u}_1$) have maximum correlation; loading vectors are in general not orthogonal. Score plots are connected projections of x- and y-space.

$$r_j^2 \text{ is eigenvalue of } \boldsymbol{S}_X^{-1}\boldsymbol{S}_{XY}\boldsymbol{S}_Y^{-1}\boldsymbol{S}_{XY}^T \text{ to the eigenvector } \boldsymbol{p}_j \tag{4.82}$$

$$r_j^2 \text{ is eigenvalue of } \boldsymbol{S}_Y^{-1}\boldsymbol{S}_{XY}^{\mathrm{T}}\boldsymbol{S}_X^{-1}\boldsymbol{S}_{XY} \text{ to the eigenvector } \boldsymbol{q}_j \tag{4.83}$$

for $j = 1, \ldots, a$. Note that the eigenvalues in Equations 4.82 and 4.83 are the same, and they are the squares of the canonical correlation coefficients. The canonical correlation coefficients are in the interval [0, 1], where 1 indicates a direction in the x-space and a direction in the y-space with perfect linear relation. Usually the eigenvectors are sorted according to decreasing eigenvalues, and so the first canonical correlation coefficient measures the maximal linear relation between the x- and y-data, the second canonical correlation coefficient measures the maximum linear relation but only among directions that lead to uncorrelated scores, and so on.

From the definition of the canonical correlation it follows that a high canonical correlation coefficient can already be obtained if a single x-variable is highly correlated with a single y-variable. So, the canonical correlation coefficient is not a measure for "overall correspondence" of the x- and y-data. If this is desired then REDUNDANCY ANALYSIS is the correct method (van den Wollenberg 1977). For this reason, CCA will in general not be useful for prediction purposes. This means that the score and loading vectors of the x-data will in general not be able to predict the y-data with sufficient precision, and vice versa. Thus CCA is mainly used as mapping method, where the pairs of score vectors $\boldsymbol{t}_j$ and $\boldsymbol{u}_j$ are plotted. Since these pairs reflect the maximal correlation of the x- and y-data, the resulting structure in the plot can be useful. The scatter plot with x-scores is a projection of the x-space; the scatter plot with y-scores is a projection of the y-space; both plots are related by maximum correlation between corresponding pairs of x- and y-scores. Note that PLS2 has some advantages compared to CCA: PLS2 is insensitive to highly correlating variables, accepts a large number of variables, and can be used for prediction purposes.

The canonical correlation coefficients can also be used for hypothesis testing. The most important test is a TEST FOR UNCORRELATEDNESS of the x- and y-variables. This corresponds to testing the null hypothesis that the theoretical covariance matrix between the x- and y-variables is a zero matrix (of dimension $m_X \times m_Y$). Under the assumption of multivariate normal distribution, the test statistic

$$T = \left(1 - r_1^2\right) \cdot \left(1 - r_2^2\right) \cdots \left(1 - r_a^2\right) \tag{4.84}$$

follows a Wilk's lambda distribution which, in case of a reasonably large number n of samples, can be approximated using the modified test statistic

$$T^* = -\left(n - \frac{m_X + m_Y + 3}{2}\right) \log T \tag{4.85}$$

by a chi-square distribution with $m_X \cdot m_Y$ degrees of freedom. The null hypothesis is rejected if the value of T^* computed for the data is larger than e.g., the quantile 0.95 of the above chi-square distribution. Note, however, that rejecting the null hypothesis of uncorrelatedness does not imply that the x-variables are highly correlated with the y-variables.

4.8.2 Ridge and Lasso Regression

Ridge and Lasso regression are alternatives to PCR or variable selection. While Ridge regression uses all x-variables in the final regression model, Lasso regression only uses a subset of the x-variables. Both methods depend on the selection of a parameter to find the best model for prediction (Hoerl and Kennard 1970; Tibshirani 1996). The term "Lasso" comes from the abbreviation *least absolute shrinkage and selection operator*, but is also motivated by the picture of catching x-variables with a lasso.

Rigde and Lasso regression are so-called SHRINKAGE METHODS, because they shrink the regression coefficients in order to stabilize their estimation. Shrinkage means that the allowed range of the absolute regression coefficients is limited; otherwise highly correlating x-variables would give instable OLS models with varying large regression coefficients. These methods are typically used for regression of a single y-variable on a high-dimensional $\boldsymbol{X}$ matrix with collinear variables. Instead of minimizing the sum of squared residuals like in OLS regression,

$$\sum_{i=1}^{n}\left(y_i - b_0 - \sum_{j=1}^{m} x_{ij}b_j\right)^2 \rightarrow \min \tag{4.86}$$

a penalized sum of squared residuals in minimized. Ridge and Lasso regression use a different penalization (Hastie et al. 2001).

More specifically, let us consider the multiple linear regression model $\boldsymbol{y} = \boldsymbol{X}\boldsymbol{b} + \boldsymbol{e}$, see Equation 4.36, which can be denoted for each object as

$$y_i = b_0 + \sum_{j=1}^{m} x_{ij}b_j + e_i \tag{4.87}$$

Then the objective of Ridge regression is

$$\sum_{i=1}^{n}\left(y_i - b_0 - \sum_{j=1}^{m} x_{ij}b_j\right)^2 + \lambda_{\mathrm{R}}\sum_{j=1}^{m} b_j^2 \rightarrow \min \tag{4.88}$$

and the objective function of Lasso regression is

$$\sum_{i=1}^{n}\left(y_i - b_0 - \sum_{j=1}^{m} x_{ij}b_j\right)^2 + \lambda_{\mathrm{L}}\sum_{j=1}^{m} |b_j| \rightarrow \min \tag{4.89}$$

Solving these minimization problems results in the estimated regression coefficients $\boldsymbol{b}_{\mathrm{RIDGE}}$ and $\boldsymbol{b}_{\mathrm{LASSO}}$, respectively. The parameters λ_{R} and λ_{L} are complexity parameters and they control the amount of shrinkage. They have to be chosen to be ≥ 0, the larger the complexity parameter the more penalty is on the regression coefficients and the more they are shrunk towards zero. If the complexity parameter is zero, the result of both Ridge and Lasso regression is the same as for OLS regression. Practically, the choice of the complexity parameter is made by CV or bootstrap. It should be chosen such that the prediction error is minimized, see Section 4.2.

From the definition of the objective functions (Equations 4.88 and 4.89), it can be seen that different scalings of the x-variables would result in different penalization, because only the coefficients themselves but no information about the scale of the x-variables is included in the term for penalization. Therefore the x-variables are usually autoscaled. Note that the intercept b_0 is not included in the penalization term in order to make the result not be depending on the origin of the y-variable.

The only difference in the objective functions for Ridge and Lasso regression is the different way to penalize the regression coefficients. While Ridge regression penalized the so-called L_2-norm (sum of squares of the regression coefficients), Lasso regression penalized the L_1-norm (sum of absolute regression coefficients). This seems to be a marginal difference, but it has big consequences. The use of the L_2-norm has the nice effect that the estimated Ridge regression parameters $\boldsymbol{b}_{\text{RIDGE}}$ are a linear function of $\boldsymbol{y}$. This can be seen by writing down Equation 4.88 in matrix form

$$(\boldsymbol{y} - \boldsymbol{X}\boldsymbol{b})^{\mathrm{T}}(\boldsymbol{y} - \boldsymbol{X}\boldsymbol{b}) + \lambda_{\mathrm{R}}\boldsymbol{b}^{\mathrm{T}}\boldsymbol{b} \rightarrow \min \tag{4.90}$$

where
$\boldsymbol{X}$ has been mean-centered
$\boldsymbol{b}$ includes no intercept term

The solution to this problem is

$$\boldsymbol{b}_{\text{RIDGE}} = \left(\boldsymbol{X}^{\mathrm{T}}\boldsymbol{X} + \lambda_{\mathrm{R}}\boldsymbol{I}\right)^{-1}\boldsymbol{X}^{\mathrm{T}}\boldsymbol{y} \tag{4.91}$$

which is indeed a linear function in $\boldsymbol{y}$. This solution is similar to the OLS solution (Equation 4.38), but the inverse is stabilized by a constant, the Ridge parameter λ_{R}; in other words the diagonal elements of $\boldsymbol{X}^{\mathrm{T}}\boldsymbol{X}$ (considered as a mountain ridge) are enlarged.

The Lasso regression coefficients can no longer be written as a function of $\boldsymbol{y}$, and the solution has to be found by an optimization through quadratic programming. It can, however, be shown that the L_1 penalty causes some regression coefficients to be exactly zero, depending on the choice of the Lasso parameter λ_{L}. Thus this method comes down to a VARIABLE SELECTION algorithm, where variables with coefficients being exactly zero are not in the model (so to say, hopefully relevant variables are caught by a lasso). This was different in Ridge regression, because a larger value of λ_{R} will cause that all regression parameters are shrunk towards zero, but they are in general not exactly zero.

There is also a link between Ridge regression and PCR (Section 4.6). PCR finds new regressor variables, the principal components of the x-variables, and they can be ordered according to decreasing variance. Since the first few PCs cover the most important information of the x-variables, they are often considered to be most useful for the prediction of the y-variable (although this might not be necessarily true—see

Section 4.6). It can be shown that Ridge regression gives most weight along the directions of the first PCs of the x-variables, and downweights directions related to PCs with small variance (Hastie et al. 2001). Thus, the shrinkage in Ridge regression is proportional to the variance of the PCs. However, in contrast to PCR where only a subset of the PCs is used for regression, here all PCs are used, but with varying weight (shrinkage).

Ridge and Lasso regression can be applied within $\mathbb{R}$ as follows:

```
R:  library(MASS)
    Ridge <- lm.ridge(y~X,lambda=seq(0,10,by=0.1))
                       # Ridge regression for values 0 to 10 with
                       # step 0.1 of lambda
    select(Ridge)  # gives the information of the optimal lambda
    plot(Ridge$lambda,Ridge$GCV)
            # plots the lambda values versus the evaluation by
            # generalized cross validation (Section 4.3.2)
```

```
R:  library(lars)
    Lasso <- lars(X,y)    # Lasso regression
    plot(Lasso)           # Output plot for Lasso regression
    cv.lars(X,y)          # cross validation and result plot
```

4.8.3 Nonlinear Regression

Linear regression methods are convenient because they usually allow an interpretation, and methods like PLS or PCR avoid the problem of overfitting, especially if many parameters have to be estimated with only a few objects. However, the relation between a response variable and one or several predictor variables can in some situations be better described by nonlinear functions, like by the functional relation

$$y = b_0 + b_1 f_1(x_1) + b_2 f_2(x_2) + \cdots + b_m f_m(x_m) + e \tag{4.92}$$

where $f_1, f_2, \ldots, f_m$ are nonlinear functions. In this case, the functions are applied to the whole range of the x-variables. There exist also different approaches for nonlinear regression that work in a local neighborhood around a considered object. The main methods for nonlinear regression are briefly described here; the interested reader is referred to Bates and Watts (1998), Hastie et al. (2001), Huet et al. (2003), Seber and Wild (2003), nonlinear PLS is described in Section 4.7.6.

4.8.3.1 Basis Expansions

The main idea is to replace the x-variables with new variables which are transformations of the x-variables. The transformed variables can be used instead of the original ones or can be added (augmented variable set). The relation between the y-variable and the derived x-variables is established by linear models of the form

$$y = b_0 + b_1 h_1(x_1) + b_2 h_2(x_2) + \cdots + b_r h_r(x_r) + e \tag{4.93}$$

where h_l are called BASIS FUNCTIONS. Usually, the number r of basis functions is larger than the number of x-variables. In the simplest case, $h_l(x_l)=x_l$, for $l=1,\ldots, m$, which results in the usual linear model directly for the x-variables. Another possibility is to use transformations of the x-variables, like $h_l(x_l)=x_l^2$, or $h_l(x_l)=x_jx_k$. All these transformation are often defined for the whole data range of the x-variables, but one could also restrict basis functions to a certain range. A prominent method for this purpose is to use PIECEWISE POLYNOMIAL SPLINES which are polynomial functions of a certain degree (e.g., cubic splines for degree 3) that are defined in certain data ranges of the x-variables. Thus each x-variable will be described by several (e.g., cubic) functions, and each of the functions covers a part of the x-variable. A different concept is used with WAVELET TRANSFORMATIONS where a complete orthonormal basis is used to represent the functional relations (Chau et al. 2004). The use of more basis functions than x-variables implies that more than m parameters have to be estimated. Therefore, not the RSS is minimized but a penalized RSS, where the penalization is done for a high number of basis functions used in the model.

```
R: library(mgcv)
   res <- gam(y~s(x1)+s(x2)+s(x3),subset=train,data=dat)
       # for each regressor variable cubic splines are used
       # which are generated by the function "s"
       # "train" includes object numbers of training data
   predict(res,dat[-train,])  # prediction for test data
```

4.8.3.2 Kernel Methods

The regression functions f_j in Equation 4.92 are estimated by using different but simple models at each query point x_0. This local fitting is done by using a kernel function $K(x_0, x_i)$ which assigns a weight to an object x_i, depending on its distance to x_0. The larger the distance the smaller is the weight for x_i. The exact weighting scheme depends on the actual type of the kernel function and on parameters that determine the width of the neighborhood. RADIAL BASIS FUNCTIONS (RBF) combine the idea of basis functions and kernel methods. Each basis function h_j in Equation 4.93 is represented by an own kernel function, with an own choice of the weighting scheme (type, location, and width of the kernel). A popular choice for the kernel function is the standard Gaussian density function.

For any point $\boldsymbol{x}$ in the data space one obtains an estimation $f(\boldsymbol{x})$ by the equation

$$f(\boldsymbol{x}) = b_0 + b_1K(\boldsymbol{x}, \boldsymbol{m}_1, s_1) + b_2K(\boldsymbol{x}, \boldsymbol{m}_2, s_2) + \cdots + b_rK(\boldsymbol{x}, \boldsymbol{m}_r, s_r) \qquad (4.94)$$

Here, $K(\boldsymbol{x}, \boldsymbol{m}_j, s_j)$ are the kernel functions with prototypes $\boldsymbol{m}_j$ and scale parameters s_j. For example, if the kernel function is the standard normal density function φ, Equation 4.94 can be formulated as

$$\begin{aligned} f(\boldsymbol{x}) &= b_0 + b_1\varphi(||\boldsymbol{x}-\boldsymbol{m}_1||/s_1) + b_2\varphi(||\boldsymbol{x}-\boldsymbol{m}_2||/s_2) \\ &\quad + \cdots + b_r\varphi(||\boldsymbol{x}-\boldsymbol{m}_r||/s_r) \end{aligned} \qquad (4.95)$$

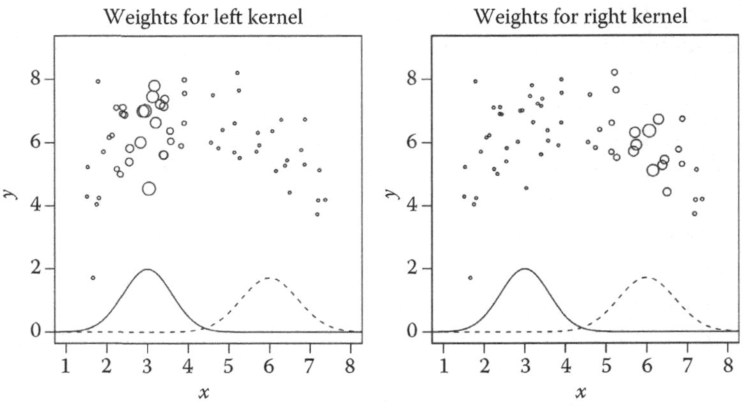

FIGURE 4.28 Visualization of kernel regression with two Gaussian kernels. The point sizes reflect the influence on the regression model. The point sizes in the left plot are for the solid kernel function, those in the right plot are for the dashed kernel function.

There are various procedures to estimate the unknown regression parameters and the parameters for the kernel functions. One approach is to estimate prototypes $\boldsymbol{m}_j$ and scale parameters s_j separately by clustering methods, and then to estimate the regression parameters, however, this approach does not incorporate information of the y-variable. Another approach is to use optimization techniques to minimize the RSS for the residuals $y_i - f(\boldsymbol{x}_i)$ obtained via Equation 4.95, for $i = 1, \ldots, n$.

Figure 4.28 visualizes the idea of kernel based regression methods in the case of univariate x and y. Two data groups are visible, and thus two Gaussian kernel functions are chosen with certain prototypes (3 and 6) and scale parameters (0.6 and 0.7). The size of the symbols represents the weights of the objects for the regression Equation 4.95. Figure 4.28 (left) shows the weights for the left kernel function (solid line) and Figure 4.28 (right) shows the weights for the right kernel function (dashed line). Using these weights, the regression parameters can be estimated, resulting in a fitted y-value for each object.

```
R: library(neural)  # neural networks, includes RBF
   dat <- rbftrain(X,neurons=8,y)  # select 8 basis functions
   res <- rbf(X,dat$weight,dat$dist, dat$neurons, dat$sigma)
```

4.8.3.3 Regression Trees

Here, the space of the x-variables is partitioned into rectangular regions, and rather simple models are then fit to each region for predicting the y-variable. Partitioning is usually done directly along the x-coordinates, e.g., by splitting variable x_j at a value v_j into two regions. The task is to find the optimal split variables and the optimal split points, where "optimal" refers to a fit criterion, like the RSS that is to be minimized. Regression trees are grown to a certain size, followed by "pruning," which reduces the tree complexity for increasing the prediction performance. For a more detailed discussion see Section 5.4 about classification trees (CART method).

```
ℝ: library(rpart)  # regression trees
   tree1 <- rpart(y~.,data=dat,subset=train)
          # use all remaining variables in "dat" for the tree
   plot(tree1)           # plots the regression tree
   text(tree1)           # adds text labels to the plot
   predict(tree1,dat[-train,]  # prediction for test data
   printcp(tree1)  # plot results of cross validation for
                   # finding the optimal complexity
                   # parameter cp
   tree2 <- prune(tree1,cp=0.01)  # pruning of the above
                  # tree using the optimal cp
                  # plot, text, predict can be applied to
                  # tree2
```

4.8.3.4 Artificial Neural Networks

ANNs are very popular among some scientists (usually not "puritanical" chemometricians); ANNs include a large class of learning methods that were developed separately in statistics and artificial intelligence. The most widely used method is the single layer PERCEPTRON, also called single hidden layer BACK-PROPAGATION NETWORK. Behind this complicated name and a biology-related terminology is just a nonlinear statistical model. Only a very brief description is given here; the interested reader is referred for instance to Cheng and Titterington (1994), Jansson (1991), Looney (1997), Ripley (1996), Schalkoff (1997), a book is dedicated to Kohonen maps in chemoinformatics (Zupan and Gasteiger 1999), and a summary of ANN strategies from a chemometricians point of view is in Otto (2007).

First, r different linear combinations of the x-variables are built

$$v_j = a_{0j} + a_{1j}x_1 + \cdots + a_{mj}x_m \quad \text{for } j = 1, \ldots, r \tag{4.96}$$

and then a nonlinear function σ—often the SIGMOID FUNCTION—is applied

$$z_j = \sigma(v_j) = \frac{1}{1 + \exp(-v_j)} \quad \text{for } j = 1, \ldots, r \tag{4.97}$$

Equations 4.96 and 4.97 constitute a neuron with several inputs x and one output z.

The new variables z_j can be used in different ways to produce the final output y: (a) as inputs of a neuron with output y, (b) in a linear regression model,

$$y = b_0 + b_1z_1 + b_2z_2 + \cdots + b_rz_r + e \tag{4.98}$$

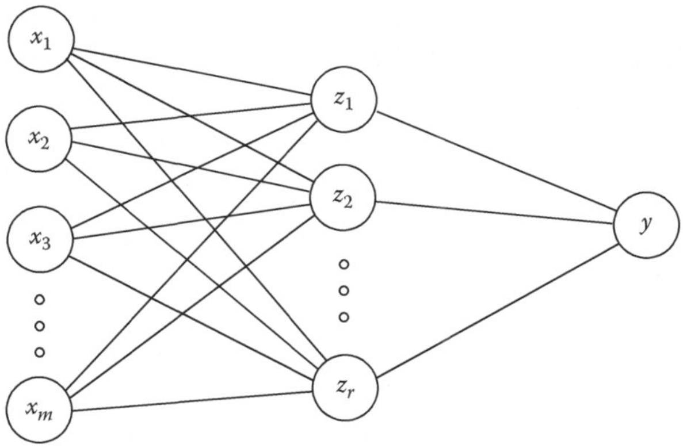

FIGURE 4.29 Schematic of a neural network with a single hidden layer.

and (c) in a nonlinear regression model

$$y = b_0 + b_1 f_1(z_1) + b_2 f_2(z_2) + \cdots + b_r f_r(z_r) + e \tag{4.99}$$

Nonlinear models often tend to overfitting and should thus be used very carefully and only after having shown that linear models are not satisfactory.

Since the variables z_j are not directly observed, they are called HIDDEN UNITS, and often they are arranged in a graphical presentation as a HIDDEN LAYER. Figure 4.29 shows such a schematic for neural networks with one hidden layer. The x-variables are the inputs to the network. Linear combinations of the x-variables and functions thereof lead to the z-variables. Finally, the z-variables are used for (linear or nonlinear) regression, resulting in an output variable y. In the more general case there can be several y-variables. Also more than one hidden layer can be used, however, with a great tendency to overfitting.

ANNs can not only be used in the regression case but also for classification tasks, see Section 5.5.

```
ℝ: library(nnet)  # neural networks
   resNN <- nnet(y~.,size=3,data=dat,subset=train)
   # fits neural network with size=3 units in the hidden layer
   predict(resNN,dat[-train,])
```

4.9 EXAMPLES

4.9.1 GC RETENTION INDICES OF POLYCYCLIC AROMATIC COMPOUNDS

This example belongs to the area QUANTITATIVE STRUCTURE–PROPERTY RELATIONSHIPS (QSPR) in which chemical–physical properties of chemical compounds are modeled by chemical structure data—mostly built by multivariate calibration methods as described in this chapter und using molecular descriptors (Todeschini and Consonni

2000) as variables. Results from many QSPR models—for many different properties and using different methods—have been published, however, mostly without providing the used data or the parameters of the created models, and unfortunately many papers do not contain a sufficient evaluation of the prediction performance of the models.

Gas chromatographic (GC) retention indices for several substance classes have been modeled by molecular descriptors. Typical works deal with 50–200 organic compounds, often belonging only to a single substance class. Starting from 20 to 300 molecular descriptors, subsets with less than 10 descriptors are selected and the resulting regression coefficients are discussed in terms of the physical-chemical background. Recently published studies belong for instance to alkanes (Xu et al. 2003), alkenes (Du et al. 2002), alkylbenzenes (Yan et al. 2000), polycyclic aromatic hydrocarbons (Liu et al. 2002), esters, alcohols, aldehydes and ketones (Junkes et al. 2003; Körtvelyesi et al. 2001), terpenes (Jalali-Heravi and Fatemi 2001), but also to chemical warfare agents (Woloszyn and Jurs 1992), about 400 diverse organic compounds (Lucic et al. 1999; Pompe and Novic 1999), and about 800 compounds relevant in forensic chemistry (Garkani-Nejad et al. 2004).

A set of $n = 209$ polycyclic aromatic compounds (PAC) was used in this example. The chemical structures have been drawn manually by a structure editor software; approximate 3D-structures including all H-atoms have been made by software CORINA (Corina 2004), and software DRAGON, version 5.3 (Dragon 2004), has been applied to compute 1630 molecular descriptors. These descriptors cover a great diversity of chemical structures and therefore many descriptors are irrelevant for a selected class of compounds as the PACs in this example. By a simple variable selection, descriptors which are constant or almost constant (all but a maximum of five values constant), and descriptors with a correlation coefficient >0.95 to another descriptor have been eliminated. The resulting $m = 467$ descriptors have been used as x-variables. The y-variable to be modeled is the Lee retention index (Lee et al. 1979) which is based on the reference values 200, 300, 400, and 500 for the compounds naphthalene, phenanthrene, chrysene, and picene, respectively.

QSPR models have been developed by six multivariate calibration methods as described in the previous sections. We focus on demonstration of the use of these methods but not on GC aspects. Since the number of variables is much larger than the number of observations, OLS and robust regression cannot be applied directly to the original data set. These methods could only be applied to selected variables or to linear combinations of the variables.

4.9.1.1 Principal Component Regression

The crucial point for building a prediction model with PCR (Section 4.6) is to determine the number of PCs to be used for prediction. In principle we could perform variable selection on the PCs, but for simplicity we limit ourselves to finding the appropriate number of PCs with the largest variances that allows the probably best prediction. In other words, the PCs are sorted in decreasing order according to their variance, and the prediction error for a regression model with the first a components will tell us which number of components is optimal. As discussed in Section 4.2

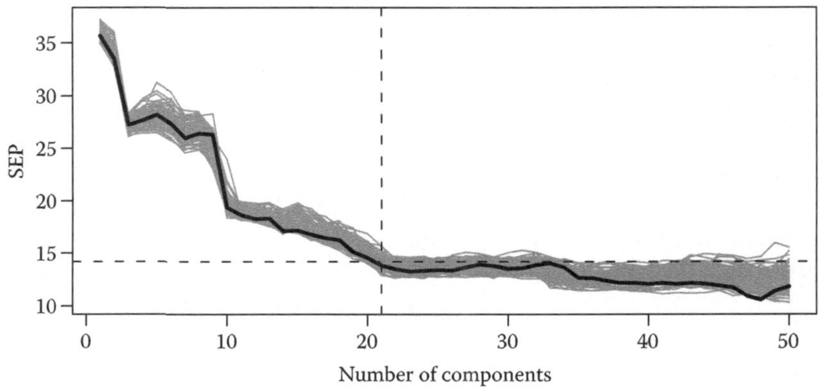

FIGURE 4.30 Results of PCR for the PAC data set. The black line results from a single 10-fold CV, the gray lines from repeating the 10-fold CV 100 times. Indicated is the optimal number of PCs as obtained by repeated double CV, see text.

there are strategies and measures for estimating the prediction errors and determining the optimum number of components. For Figure 4.30 we used the SEP_{CV} (see Section 4.2.3) as performance measure. The thick black line shows the resulting SEP_{CV} values when using a single CV with 10 segments into which the objects have been assigned randomly. The gray lines result from repeating this procedure 100 times. This makes the "randomness" of CV clearly visible.

The choice of the number of PCs was based on REPEATED DOUBLE CV with 100 repetitions, also allowing us to estimate the distribution of prediction errors for new cases. The inner CV was done with 10 segments, and the outer CV with four segments (compare Figure 4.6). From the four segments three were used as calibration set, and the optimum number of components was determined from the calibration set by 10-fold CV (details on how the optimal number of components is determined are given below). The outer CV yields four values for the optimum number of components which may differ. Repeating the procedure 100 times (repeated double CV) thus results in 400 values for the optimum number of components. The relative frequencies of these 400 numbers are shown in Figure 4.31. Models with 21 components were optimal most frequently, and thus this number can be taken as the resulting optimal number of PCs.

The choice of the optimal number of PCs was based on the ideas presented in Figure 4.4 (right). The procedure is performed in the inner CV, carried out within one run of the outer CV (four segments) and within one (out of 100) repititions. Therefore this procedure has to be applied 400 times and gives 400 estimates for the optimum number of PCs. The prediction error was measured by the MSE_{CV}, which was computed separately for each of the 10 segments of the inner CV. For the resulting 10 numbers, the mean and the standard error around the mean (standard deviation divided by $\sqrt{10}$) are computed. The maximum number of PCs to be considered is where the minimum of the mean MSE_{CV} values occurs. However, we opt for the most parsimonious model where the mean MSE_{CV} is not larger than

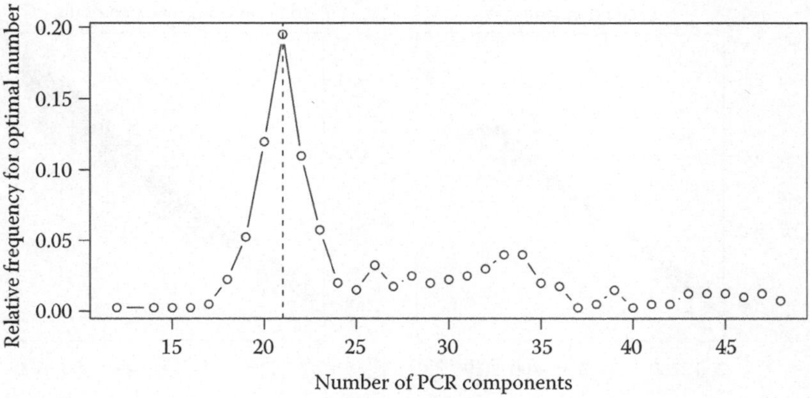

FIGURE 4.31 Results of repeated double CV for the optimal number of PCs for PCR applied to the PAC data set. Models with 21 components are most frequent.

this minimum plus two times the standard error corresponding to that minimum (see Figure 4.4, right). We choose a two-standard-error rule here in order to give preference to smaller models avoiding overfitting.

As mentioned above, models with 21 components resulted most frequently, and this is also visualized by the dashed vertical line in Figure 4.30. The dashed horizontal line is drawn at the final SEP_{TEST} value 14.2. This performance measure is obtained from 20,900 test set predicted $\hat{y}$ values (number of objects times number of repetitions), calculated by models with 21 components.

```
R: library(chemometrics)
   data(PAC)                # load PAC data
   pcr_dcv<-mvr_dcv(y~X,ncomp=50,data=PAC,method="svdpc")
              # PCR with repeated double cross validation
              # Default are 100 repetitions
   pcr_plot2<-plotcompmvr(pcr_dcv)
                    # generates plot in Figure 4.31
   pcr_plot1<-plotSEPmvr(pcr_dcv,opt=pcr_plot2$opt,PAC$y,
             PAC$X, method="svdpc")
                        # generates plot in Figure 4.30
```

Although the SEP_{TEST} value allows drawing conclusions about the ability of the model for prediction, additional plots can be instructive. The plot of the measured *y*-variable versus the predicted *y*-values is shown in Figure 4.32. For the prediction we used the models with the first 21 PCs. The left picture is the prediction as obtained from a single CV procedure (all 209 objects, 10 segments), whereas the right picture visualizes the results from repeated double CV as described above. Since repeated double CV yielded for each *y*-value 100 predicted values, we plot (in gray) all 100 results. Additionally, the average over the 100 predicted values to each *y*-value is plotted in black. The results from repeated double CV allow a better insight into the behavior of the prediction error, because it makes the error distribution visible.

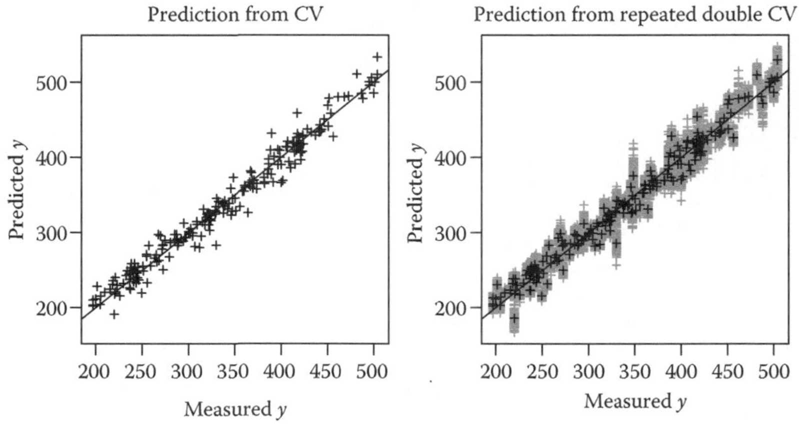

FIGURE 4.32 Comparison of the measured y-values of the PAC data set with the predictions using PCR with 21 components. The 100 predictions to each y-value from repeated double CV (right) give a better insight into the distribution of the prediction errors than the results from a single CV (left).

```
R: plotpredmvr(pcr_dcv,opt=21,PAC$y,PAC$X,method="svdpc")
                        # generates plot in Figure 4.32
```

A plot of the predicted y-values versus the residuals (= measured y minus predicted y) is shown in Figure 4.33. Again, the results from a single CV procedure are shown on the left, and those from repeated double CV with 100 replications are presented on the right hand side. Similar to Figure 4.32 we show in the right figure all resulting residuals using the 100 predictions to each y-value on the vertical axis (in gray), and additionally the average of these values (in black). On the horizontal axis

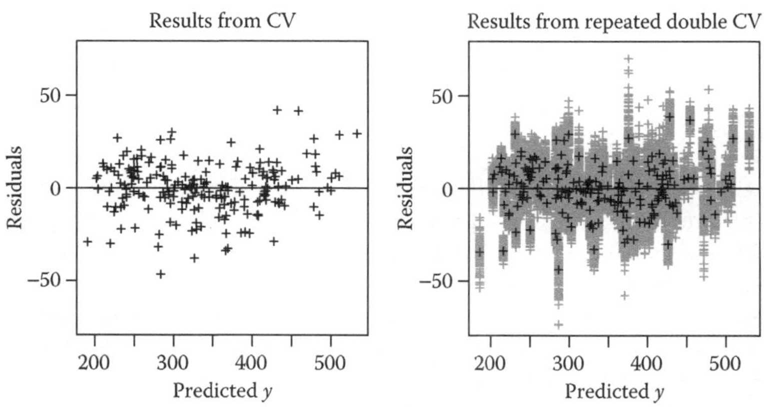

FIGURE 4.33 Comparison of the predicted y-values of the PAC data set with the residuals using PCR with 21 components. The residuals from repeated double CV (right) allow a better view of the error distributions than those from a single CV (left).

the averages of the 100 predictions to each y-value are shown. From the plot it is evident that some objects have a higher variability of the prediction errors than others.

```
R: plotresmvr(pcr_dcv,opt=21,PAC$y,PAC$X,method="svdpc")
                  # generates plot in Figure 4.33
```

4.9.1.2 Partial Least-Squares Regression

The results from PCR for the PAC data set will now be compared with PLS regression (Section 4.7). We apply the algorithm SIMPLS, other PLS algorithms give very similar or even identical results. Firstly, a decision on the optimum number of PLS components has to be made. Similar to PCR, Figure 4.34 shows the SEP_{CV} values from a single CV with 10 segments (black line) while the gray lines result from repeating this procedure 100 times. We used the same procedure based on repeated double CV as described above for determining the number of PLS components, and a model with 11 components turned out to be optimal, resulting in a SEP_{TEST} value of 12.2. Thus, for this data set PLS requires considerably less components than PCR, and gives a smaller SEP_{TEST} value. This underlines the advantage of using both x- and y-information for intermediate latent variables which are then used for OLS.

The plots with the predicted values and residuals obtained from PLS models are visually very similar to the plots for the PCR results. The ℝ code for PLS is almost identical to the code for PCR shown above, only the argument `method` has to be changed.

```
R: pls_dcv=mvr_dcv(y~X,ncomp=50,data=PAC,method="simpls")
                  # PLS with repeated double cross validation
                  # Default are 100 repetitions
   pls_plot1=plotSEPmvr(pls_dcv,opt=11,PAC$y,PAC$X,
             method="simpls")
                  # generates plot in Figure 4.34
```

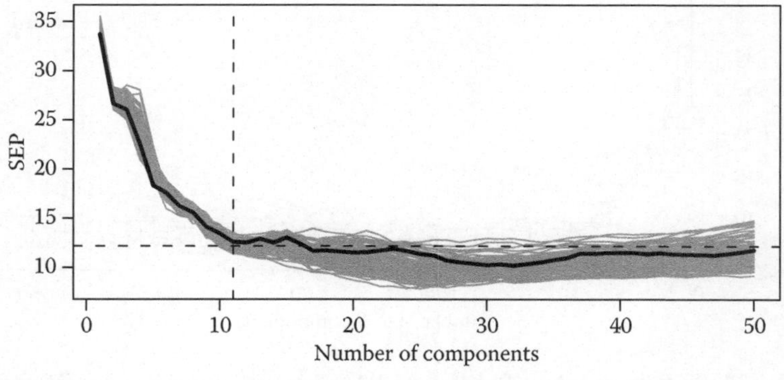

FIGURE 4.34 Results of PLS for the PAC data set. The black line results from a single 10-fold CV, the gray lines from repeating the 10-fold CV 100 times. The choice of the optimal number of PCs is based on repeated double CV.

4.9.1.3 Robust PLS

In Figure 4.33 we could observe somewhat inflated residuals for some objects. This might be data points that are less reliable, and a model fit with PCR or PLS can be unduly influenced by such outliers. The idea of robust PLS (Section 4.7.7) is to downweight atypical objects that cause deviations from the assumptions used for the classical methods.

Although the algorithm for the robust PLS procedure mentioned in Section 4.7.7 is relatively fast, an exhaustive evaluation by repeated double CV would require a lot of time. Therefore, we estimate the prediction error for different numbers of components with a single 10-fold CV using all 209 objects. Figure 4.35 shows the resulting SEP_{CV} values (dashed line—using all 209 objects). However, the SEP_{CV} values using *all* residuals are not very informative because they will be biased upwards by the outliers. Therefore, a trimmed SEP_{CV} value where a certain percentage of the largest absolute residuals (coming from potential outliers) has been eliminated is shown for comparison (we used 20% trimming). Note that the trimmed SEP_{CV} is used for the determination of the optimum number of components, but it can also serve as a measure of the prediction performance. However, a comparison with other methods is only possible if the performance measure is always calculated with the same trimming. In Figure 4.35 we additionally show intervals that correspond to mean plus/minus 2 standard errors of the trimmed SEP_{CV} values. The horizontal line corresponds to the minimum of these means (at 24 components) plus two times the standard error. As suggested above, the optimal number of components for robust PLS is the smallest model where the mean is still below the horizontal line. A model with 21 components is thus optimal, resulting in a 20%-trimmed SEP_{CV} value of 6.2 (the SEP_{CV} with all values would be 14.7). However, a reduction to 10 components would result only in a marginal increase to 7.0 of the SEP_{CV} value (without trimming the SEP_{CV} is 13.4).

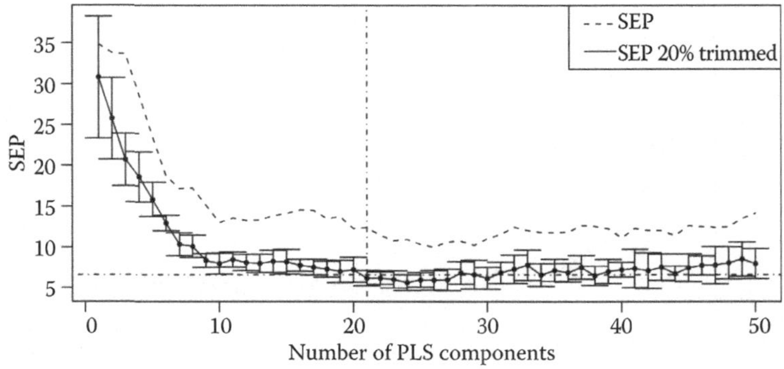

FIGURE 4.35 Results of robust PLS for the PAC data set. Using 10-fold CV, the SEP_{CV} values for all objects are compared to the SEP_{CV} values where the largest 20% of absolute residuals have been eliminated. A model with 21 components is optimal.

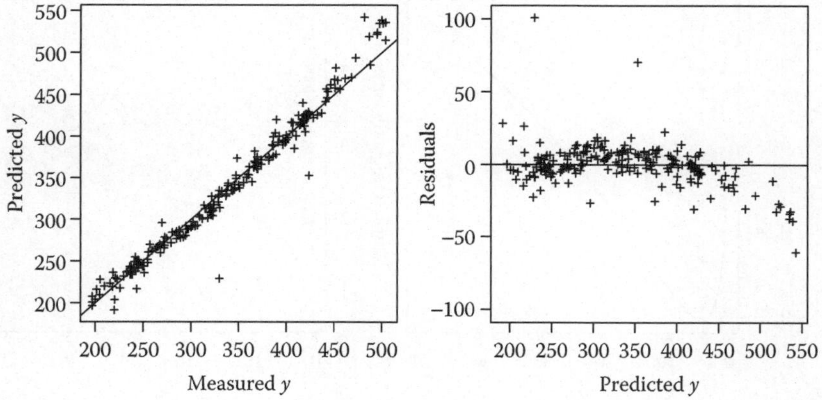

FIGURE 4.36 Robust PLS for the PAC data set using 21 PLS components. The plots show measured versus predicted values (left) and predicted values versus residuals (right) using a single 10-fold CV.

```
R:  library(chemometrics)
    data(PAC)
    rpls=prm_cv(PAC$X,PAC$y,a=50,trim=0.2,plot.opt=TRUE)
                # generates the plot in Figure 4.35
```

Figure 4.36 shows the measured versus predicted values (left) and predicted values versus residuals (right). Some of the 209 objects were strongly downweighted for building the robust PLS model, and therefore they appear with large residuals. The downweighting scheme also resulted in a certain curvature, because for the fit the lower and higher values of *y* received lower weight.

```
R:  plotprm(rpls,PAC$y)  # generates the plot in Figure 4.36
```

4.9.1.4 Ridge Regression

The singularity problem that would appear with OLS for this data set is avoided by using a Ridge parameter λ_R that constrains the size of the regression coefficients (Section 4.8.2). The first task is thus to find the optimal ridge parameter that results in the smallest prediction error. This is numerically efficiently done by GENERALIZED CV (GCV) (see Section 4.3.2.2) which approximates the MSEP. Figure 4.37 (left) shows the dependency of the MSEP on the Ridge parameter λ_R. The optimal choice leading to the smallest prediction error is $\lambda_R = 4.3$. An instructive plot is Figure 4.37 (right) where the size of the regression coefficients as a function of λ_R becomes visible. Each curve represents the change with respect to λ_R of the regression coefficient of one particular variable. The larger the Ridge parameter the more shrinkage is on the coefficients, and the curves move closer towards zero. The optimal choice $\lambda_R = 4.3$ is shown by a vertical line, and the intersections with the curves result in the optimized regression coefficients that can be used for prediction.

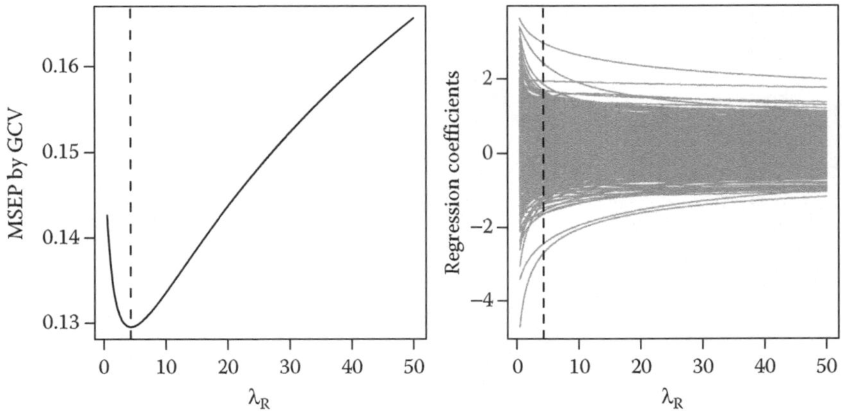

FIGURE 4.37 Ridge regression for the PAC data set. The optimal ridge parameter $\lambda_R = 4.3$ is found by GCV (left), and the resulting regression coefficients are the intersections of the curves representing the size of the regression coefficients with the vertical line at 4.3 (right).

```
ℝ:  ridge_res=plotRidge(y~X,data=PAC,lambda=seq(0.5,50, by=0.05))
    # generates the plot in Figure 4.37
```

Although GCV is fast to compute, it may result in a too optimistic estimation of the prediction error. Therefore, we will carefully evaluate the prediction error for the choice of the Ridge parameter $\lambda_R = 4.3$ by repeated 10-fold CV. In more detail, the regression parameters are estimated on 9 parts of the data using the Ridge parameter $\lambda_R = 4.3$, and the prediction is done for the 10th part. Each part is once left out for estimation and used for prediction. The whole procedure is repeated 100 times, resulting in 100 predicted values for each y-value. Figure 4.38 (left)

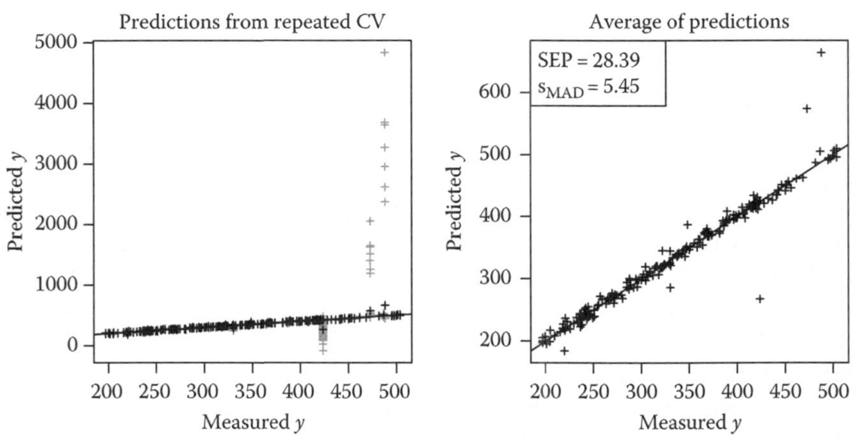

FIGURE 4.38 Ridge regression for the PAC data set. The optimal ridge parameter $\lambda_R = 4.3$ is evaluated using repeated 10-fold CV. The resulting (average) predictions (in black) versus the measured y-values are shown with two different scales because of severe prediction errors for two objects.

shows all these predicted values (in gray) versus the measured values, and additionally the averages of the 100 predictions (in black). Some objects appear as severe outliers, and it could be advisable from this result to remove them from the data set. Actually, the structures of the two compounds giving much too high predicted retention indices do not contain condensed benzene rings but a chain of four benzene rings connected by single bonds (quarter-phenyl compounds, $C_{24}H_{18}$, CAS registry numbers 1166–18–3 and 135–70–6). The averages (which are less extreme) are again shown in the right plot of Figure 4.38, and measures for the prediction error are provided. Here we show the SEP_{CV} which is the average of the standard deviations of the residuals. A more robust measure is the average of the median absolute deviations (MAD) of the residuals, resulting in the measure s_{MAD} (see Section 1.6.4). Due to some severe outliers for the predictions, the robust measure is much more reliable. Thus, Ridge regression gives a very good prediction for most objects, but fails dramatically for some few objects that deviate slightly from the linear trend.

```
ℝ:  res_CV=ridgeCV(y~X,data=PAC,repl=100,
    lambda=ridge_res$lambdaopt)
        # generates the plot in Figure 4.38
```

4.9.1.5 Lasso Regression

Although the concepts of Ridge and Lasso regression (Section 4.8.2) are very similar, Lasso regression can be viewed as a variable selection method because—depending on the Lasso parameter λ_L—some regression coefficients will become exactly zero. The optimal choice for the Lasso parameter can be made by CV. Since Lasso regression is computationally more expensive than Ridge regression, we only use a single 10-fold CV, but additionally visualize the range of plus/minus two standard errors around the mean prediction errors, similar to Figure 4.35. The result is shown in Figure 4.39, where the horizontal axis is the fraction

$$\beta = \frac{\sum_{j=1}^{m} |b_j|}{\sum_{j=1}^{m} |b_j^{\max}|} \tag{4.100}$$

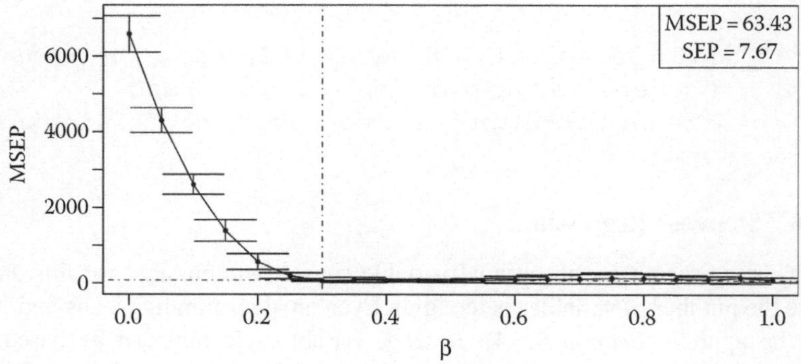

FIGURE 4.39 Lasso regression for the PAC data set. The optimal Lasso parameter is for a fraction 0.3 leading to a SEP value 7.67.

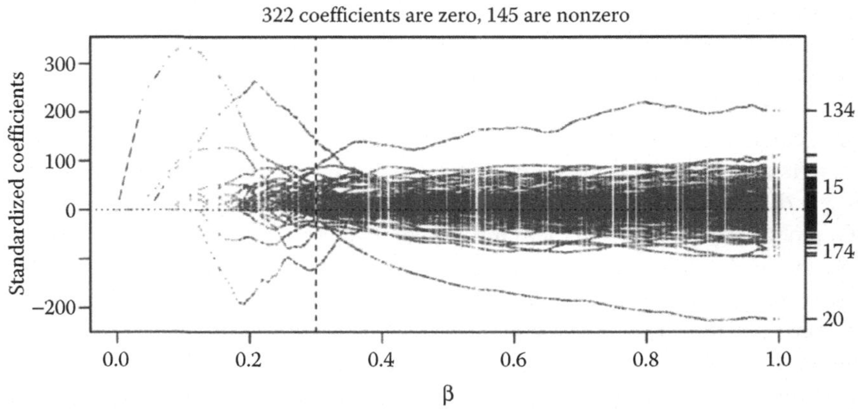

FIGURE 4.40 Lasso regression for the PAC data set. The optimal Lasso parameter is at a fraction of 0.3, and the resulting regression coefficients are the intersections of the curves representing the size of the regression coefficients with the vertical line at 0.3.

The denominator in Equation 4.100 is the (absolute) size of all Lasso regression parameters for a particular choice of λ_L (compare Equation 4.89), and the nominator describes the maximal possible (absolute) size of the Lasso regression parameters (in case there is no singularity problem this would correspond to the OLS solution). The optimal choice is at a fraction of 0.3 which corresponds to a MSEP_{CV} of 63.4 and to a SEP_{CV} of 7.7.

```
ℝ: res_CV=lassoCV(y~X,data=PAC,K=10,fraction=seq(0,1,by=0.05))
   # generates the plot in Figure 4.39
```

Similar to Figure 4.37, an illustration of the size of the regression coefficients is given in Figure 4.40. For an increasing fraction on the horizontal axis, more and more regression coefficients become nonzero. The intersections of the vertical line at 0.3 for the optimal fraction with the curves correspond to the resulting regression coefficients that can be used for prediction.

```
ℝ: res_coef=lassocoef(y~X,data=PAC,sopt=res_CV$sopt)
          # generates the plot in Figure 4.40 and
          # returns the optimal regression coefficients
```

4.9.1.6 Stepwise Regression

An exhaustive search for an optimal variable subset is impossible for this data set because the number of variables is too high. Even an algorithm like leaps-and-bound cannot be applied (Section 4.5.4). Instead, variable selection can be based on a stepwise procedure (Section 4.5.3). Since it is impossible to start with the full model, we start with the empty model (regress the *y*-variable on a constant), with the scope

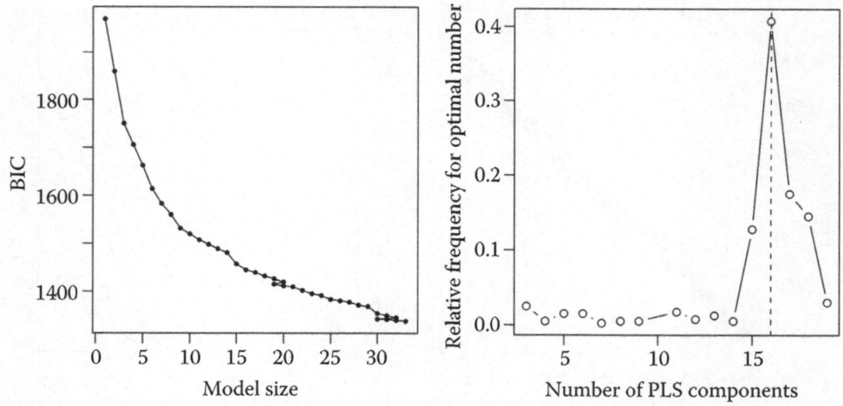

FIGURE 4.41 Stepwise regression for the PAC data set. The BIC measure is reduced within each step of the procedure, resulting in models with a certain number of variables (left). The evaluation of the final model is based on PLS where the number of PLS components is determined by repeated double CV (right).

to the full model. We will allow for both directions, forward and backward selection, and thus variables may enter the model but they can also be removed later on. As a measure of model fit we use the BIC which penalizes larger models more than the AIC criterion (Section 4.2.4).

Figure 4.41 (left) shows the BIC values versus the model size for all steps of the stepwise regression procedure. The model search starts from the point in the upper left corner of the plot, follows the line, and ends in the lower right corner. The BIC measure decreases continuously until it cannot be reduced any more. This final model has 33 variables, and it is visible that during the procedure the model size sometimes gets smaller. Since the absolute value of the BIC measure is not informative, we have to evaluate the final regression model in an appropriate way. There are different possibilities for this purpose: the 33 variables can be directly used for OLS within a CV scheme (second approach), or PLS models can be derived from the 33 variables within a double CV scheme (first approach). In the following we will compare the two strategies.

The first evaluation method is very similar to the repeated double CV used for PLS above in this section, except that we do not use the complete data but only the 33 x-variables that resulted from stepwise regression. The outer CV is done with four segments, and the inner CV for determining the optimal number of PLS components is done with 10 segments. The procedure is repeated 100 times. The resulting distribution of the optimal number of PLS components is shown in Figure 4.41 (right), and 16 components would be appropriate for the final model. The resulting average SEP_{TEST} value from this repeated double CV is 9.60. Figure 4.42 (left) visualizes—in analogy to Figure 4.32 (right)—the predicted values from the 100 repetitions of the double CV using PLS models (gray), as well as their averages (black).

A different evaluation method would be to apply OLS in a simple CV scheme that additionally can be repeated. In each step of the CV, a model with the

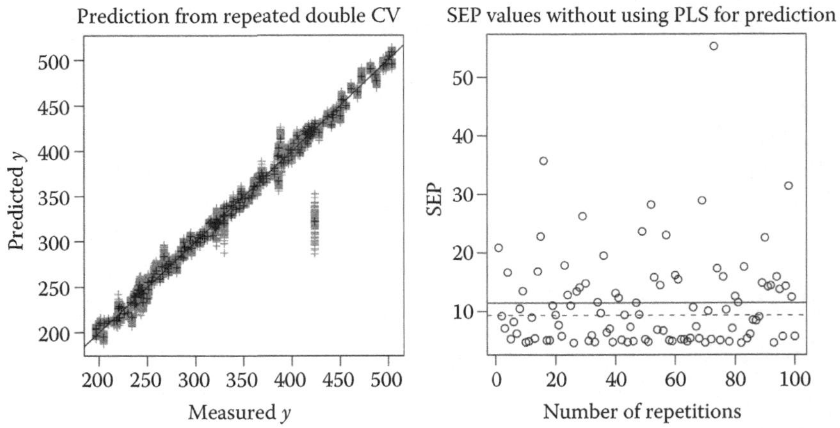

FIGURE 4.42 Evaluation of the final model from stepwise regression. A comparison of measured and predicted y-values (left) using repeated double CV with PLS models for prediction, and resulting SEP values (right) from repeated CV using linear models directly with the 33 selected variables from stepwise regression.

33 regressor variables can be developed from s-1 segments, and the prediction is done for the remaining segment. Thus the selection of an optimal number of PLS components for prediction is avoided. We used this approach with four segments and repeated the CV 100 times. The resulting SEP_{CV} values for each replication are shown in Figure 4.42 (right), together with their mean 11.45 as solid and their median 9.34 as dashed horizontal lines. This evaluation is quite unstable, and the previous evaluation scheme using PLS models for prediction should thus be preferred.

```
R:  resstep=stepwise(y~X,data=PAC)
         # stepwise regression in both directions using the BIC
    plot(apply(resstep$mod,1,sum),resstep$bic)
         # generates the left plot of Figure 4.41
    selvar <- resstep$mod[nrow(resstep$mod),]==TRUE
    # vector TRUE/FALSE for the x-variables in the final model
```

4.9.1.7 Summary

Table 4.3 summarizes the results obtained with the six calibration methods. Only Lasso regression and stepwise variable selection reduce the number of regressor variables, the other methods use all 467 variables in the final model. A direct comparison of the resulting measures SEP_{TEST} and SEP_{CV} is not useful, because the SEP values are sensitive with respect to outliers, and thus can be biased upwards (like for Ridge regression). Since robust PLS was used with 20% trimming of the largest residuals, the SEP values can be recomputed for all methods by eliminating the 20% of the largest residuals. This will result in a fair comparison of all methods.

TABLE 4.3
Results for the PAC Data Set with Six Calibration Methods

Method	m^*	a	SEP_{TEST}	SEP_{CV}	$SEP^{0.2}$
PCR	467	21	14.2	—	7.9
PLS	467	11	12.2	—	5.7
Robust PLS	467	21	—	14.7	6.2
Ridge regression	467	—	—	28.4	4.0
Lasso regression	145	—	—	7.7	5.0
Stepwise variable selection + PLS	33	16	9.6	—	4.4

Note: m^*, number of variables in the final model; a, number of PCR/PLS components. SEP_{TEST}, from repeated CV with outer loop (test sets) using four segments, and inner loop (determination of optimum number of components) using 10 segments; 100 repetitions. SEP_{CV}, single (or repeated—for ridge regression) CV, 10 segments; $SEP^{0.2}$, resulting SEP measure with 20% trimming of the largest absolute residuals.

The results are shown under $SEP^{0.2}$ in the last column of Table 4.3. Ridge regression leads to the best model.

4.9.2 Cereal Data

From 15 cereal flour samples data have been measured as follows: NIR spectra, heating values (HHV, see Section 4.5.8), mass % of the elements carbon, hydrogen, nitrogen, and mass % of starch and ash. The cereal samples are from five groups with three samples from each group, and the groups are denoted by B, barley; M, maize; R, rye; T, triticale; W, wheat. The NIR reflexion data are in the range 1126–2278 nm with intervals of 8 nm; the $m = 145$ values are from the first derivative (Savitzky–Golay method, quadratic polynomial, seven points; Section 7.2, Naes et al. 2004), and they are used as the x-variables. The other six measurements are used as dependent variables forming the $\boldsymbol{Y}$-matrix. The HHVs (kJ/kg) have been determined by a bomb calorimetric method (Friedl et al. 2005). The contents of C, H, and N have been measured by standard methods of elemental analysis. The contents of starch have been measured via the glucose contents after hydrolysis of the starch. The data used in this example are a subset from a broader investigation (Varmuza et al. 2008).

In this example we model all six y-properties together by using PLS2 (Section 4.7), instead of deriving a separate model for each y-variable. The joint treatment of all y-variables is recommended for correlating y-variables and gives a better stability of the models in this case. The ranges of the variables are given in Table 4.4 (upper part) and the Pearson correlation coefficients between the variables in Table 4.5. In order to avoid that single y-variables will dominate the results of PLS2, we use the autoscaled $\boldsymbol{Y}$ data matrix. This will also allow an easier comparison of the prediction errors. The $\mathbb{R}$ code for applying PLS2 with method SIMPLS is as follows.

```
ℝ: library(pls)
   data(cereal,package=="chemometrics") # load data set
   respls2 <- mvr(Ysc~X,data=cereal,method="simpls",
        validation="LOO")
```

For model validation we use leave-one-out CV. Other validation schemes could be used, but in this example we have a severe limitation due to the low number of objects. A plot of the prediction errors from CV versus number of PLS components is shown in Figure 4.43. The dashed lines correspond to the MSE values for the

TABLE 4.4
Basic Statistical Data of the Six *y*-Variables in the Cereal Data Set

	Heating Value	*C*	*H*	*N*	Starch	Ash
Min	18143	40.4	6.51	0.92	59.9	1.18
Max	18594	42.2	6.91	2.15	76.5	2.44
Mean	18389	41.3	6.73	1.58	67.0	1.69
Standard deviation	159	0.480	0.136	0.343	5.61	0.385
PLS2						
R^2_{CV}	0.38	0.63	0.53	0.83	0.64	0.83
SEP_{CV}	137	0.301	0.096	0.145	3.87	0.161
PLS1						
R^2_{CV}	0.34	0.62	0.49	0.84	0.72	0.84
SEP_{CV}	144	0.304	0.101	0.140	3.08	0.155
a_{OPT}	5	5	7	5	3	5

Note: Heating value in kJ/kg, others in mass %. The squared Pearson correlation coefficients, R^2_{CV}, between experimental values and predicted values from leave-one-out CV and the standard error of prediction from leave-one-out CV (SEP_{CV}, see Section 4.2.3) are given for a joint PLS2 model, and for separate PLS models developed for each variable seperately using the optimal number of components a_{OPT} for each model.

TABLE 4.5
Pearson Correlation Coefficients between the Variables in the Cereal Data Set

	Heating Value	*C*	*H*	*N*	Starch
C	0.713				
H	−0.411	−0.427			
N	−0.090	−0.316	0.104		
Starch	0.235	0.161	0.270	−0.728	
Ash	−0.037	−0.013	−0.474	0.667	−0.767

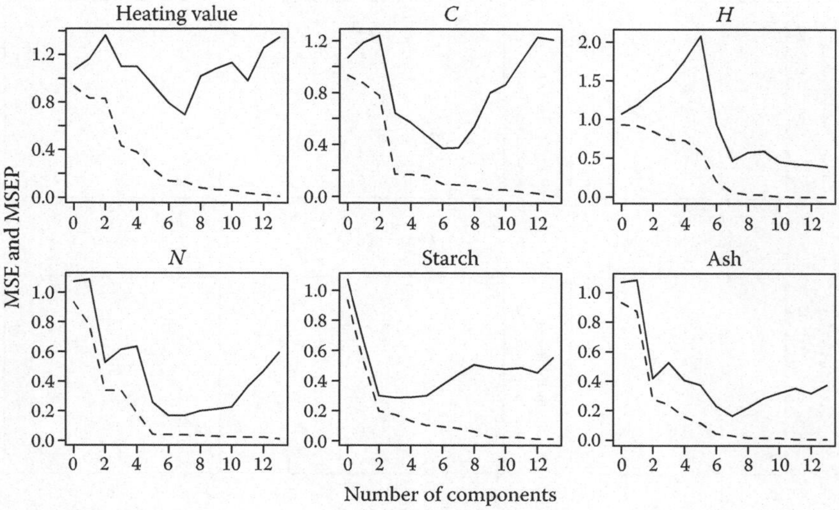

FIGURE 4.43 Prediction errors for the training data (dashed lines) and for the test data (solid lines, leave-one-out CV) of the cereal data set in relation to the number of PLS components.

training data, and the solid lines to the MSEP values of the test data obtained via CV, and the ℝ code is

```
ℝ:  validationplot(respls2,val.type="MSEP",
                         estimate=c("CV","train")
                         # generates the plot in Figure 4.43
```

A decision on the number of PLS components can be made by plotting the averages of the MSEP (and MSE) values over all *y*-variables. This plot in Figure 4.44

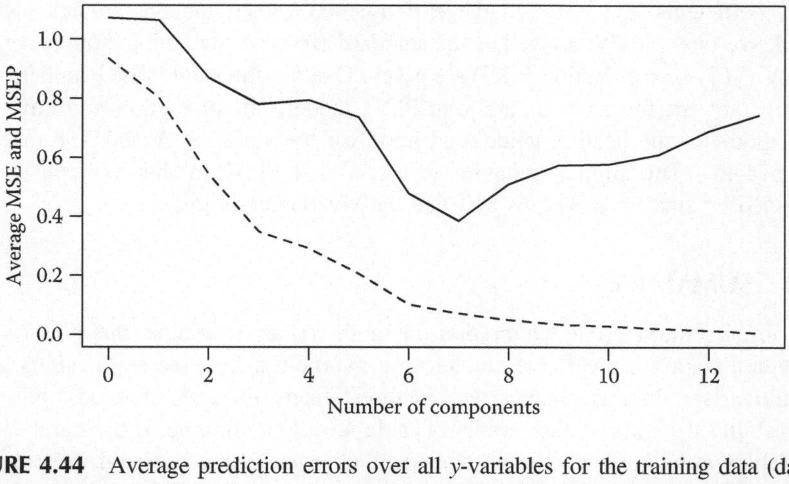

FIGURE 4.44 Average prediction errors over all *y*-variables for the training data (dashed lines) and for the test data (solid lines, leave-one-out CV) of the cereal data set in relation to the number of PLS components.

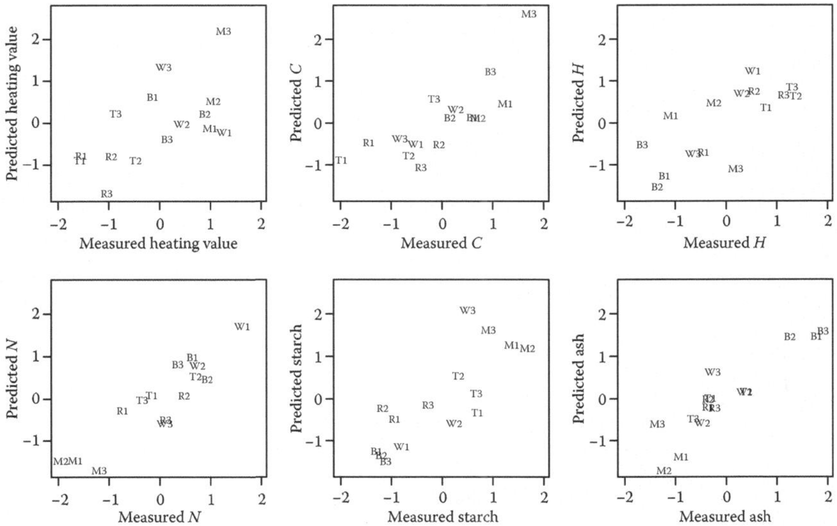

FIGURE 4.45 Measured versus predicted values (in leave-one-out CV) of the six y-variables with predictions based on PLS2 models with seven components. (B, barley; M, maize; R, rye; T, triticale; W, wheat).

suggests that seven components are appropriate. The PLS2 models with the first seven components give—for prediction with leave-one-out CV—the plots shown in Figure 4.45.

Table 4.4 (lower part) compares the performance of the PLS2 models (leave-one-out CV) with single PLS1 models for each variable. For each of the PLS1 models the optimal number of components a_{OPT} is selected using leave-one-out CV. In general, the PLS1 models require a smaller number of components than the joint PLS2 model (compare also Figure 4.43). As performance measures the squared Pearson correlation coefficients R^2_{CV} between the experimental values and the predicted values from leave-one-out CV, as well as the standard errors of prediction from leave-one-out CV (SEP_{CV}, see Section 4.2.3) are given. Overall, the single PLS1 models show a comparable performance as the joint PLS2 model. The prediction performance is rather poor for the heating value, and good for the variables N and Ash (see also Figure 4.45). The similar behavior of PLS1 and PLS2 in this example can be explained by small correlations between the y-variables (Table 4.5).

4.10 SUMMARY

The aim of multivariate calibration methods is to determine the relationships between a response y-variable and several x-variables. In some applications also y is multivariate. In this chapter we discussed many different methods, and their applicability depends on the problem (Table 4.6). For example, if the number m of x-variables is higher than the number n of objects, OLS regression (Section 4.3) or robust regression (Section 4.4) cannot be applied directly, but only to a selection

TABLE 4.6
Comparison of Regression Methods for Calibration

Method	$m > n$	Coll. x	Y	Comp.	All Var.	Opt.	Linear
OLS	N	N	Y	N	Y	N	Y
Robust reg.	N	N	Y	N	Y	N	Y
OLS v. s.	Y	Y	N	N	N	Y	Y
PCR	Y	Y	Y	Y	Y	Y	Y
PLS	Y	Y	N	Y	Y	Y	Y
PLS2	Y	Y	Y	Y	Y	Y	Y
CCA	N	N	Y	Y	Y	N	Y
Ridge	Y	Y	Y	N	Y	Y	Y
Lasso	Y	Y	N	N	N	Y	Y
Tree	Y	Y	Y	N	N	Y	N
ANN	Y	Y	Y	Y	Y	Y	N

Note: Rows: OLS, ordinary least-squares regression (Section 4.3); Robust reg., robust regression (Section 4.4); OLS v. s., ordinary least-squares regression with variable selection (Section 4.5); PCR, principal component regression (Section 4.6); PLS, partial least-squares regression (Section 4.7); CCA, canonical correlation analysis (Section 4.8.1); Ridge, Ridge regression (Section 4.8.2); Lasso, Lasso regression (Section 4.8.2); Tree, regression trees (Section 4.8.3); ANN, artificial neural networks (Section 4.8.3);
Columns: Y, yes; N, no; m, number of variables; n, number of objects; Coll. x, collinear x-variables allowed; Y, more than one y-variable possible; comp., components (latent variables) used for regression; All Var., all variables in final regression model; Opt., optimization of prediction performance; Linear, model is linear in the predictor variables.

of the variables (Section 4.5), or to PCs (Section 4.6) or other mathematical combinations of the x-variables (Section 4.7). Highly correlated x-variables, or multivariate y-information are other facts that automatically exclude some of the discussed methods (see Table 4.6).

A further point for the choice of an appropriate method is, whether the final model needs to be interpreted. In this case, and especially in case of many original regressor variables, the interpretation is in general easier if only a few regressor variables are used for the final model. This can be achieved by variable selection (Section 4.5), Lasso regression (Section 4.8.2), or regression trees (Section 4.8.3.3). Also PCR (Section 4.6) or PLS (Section 4.7) can lead to components that can be interpreted if they summarize the information of thematically related x-variables.

Outliers or inhomogeneous data can affect traditional regression methods, hereby leading to models with poor prediction quality. Robust methods, like robust regression (Section 4.4) or robust PLS (Section 4.7.7), internally downweight outliers but give full weight to objects that support the (linear) model. Note that to all methods discussed in this chapter robust versions have been proposed in the literature.

An evaluation of the resulting regression model is of primary importance. This will finally give preference to a certain selection of variables and/or to a specific method and model for the problem at hand. Measures being used for model comparison are based on the residuals, which should preferably be taken from an independent test set. Since in many cases such a test set is not available, resampling procedures like CV (Section 4.2.5) or bootstrap (Section 4.2.6) need to be used. Various variants of these validation schemes exist, and more sophisticated evaluation methods like REPEATED DOUBLE CV (Section 4.2.5) also result in a higher computational effort. The advantage is that the evaluation can be based on the distribution of the performance measure, and not just on a single value.

Not just by accident PLS regression is the most used method for multivariate calibration in chemometrics. So, we recommend to start with PLS for single y-variables, using all x-variables, applying CV (leave-one-out for a small number of objects, say for $n < 30$, 3–7 segments otherwise). The SEP_{CV} (standard deviation of prediction errors obtained from CV) gives a first idea about the relationship between the used x-variables and the modeled y, and hints how to proceed. Great effort should be applied for a reasonable estimation of the prediction performance of calibration models.

It is advisable not to forget the PARSIMONY principle, also called OCCAM' RAZOR, often paraphrased into

- "All other things being equal, the simplest solution is the best."
- "When you have two competing theories (models) which make exactly the same predictions, the one that is simpler is the better."
- "When you hear hoof beats, think horses, not zebras, unless, you are in Africa."

The reader is invited to transform these phrases into chemometrics.

REFERENCES

Anderssen, E., Dyrstad, K., Westad, F., Martens, H.: *Chemom. Intell. Lab. Syst.* 84, 2006, 69–74. Reducing over-optimism in variable selection by cross-model validation.

Bates, D. M., Watts, D. G.: *Nonlinear Regression Analysis and Its Applications*. Wiley, New York, 1998.

Baumann, K.: *Trends Anal. Chem.* 22, 2003, 395–406. Cross-validation as the objective function for variable-selection techniques.

Broadhurst, D., Goodacre, R., Jones, A., Rowland, J. J., Kell, D. B.: *Anal. Chim. Acta* 348, 1997, 71–86. Genetic algorithms as a method for variable selection in multiple linear regression and partial least squares regression, with applications to pyrolysis mass spectrometry.

Chau, F. T., Liang, Y. Z., Gao, J., Shao, X. G.: *Chemometrics—From Basics to Wavelet Transform*. Wiley, Hoboken, NJ, 2004.

Cheng, B., Titterington, D. M.: *Stat. Sci.* 9, 1994, 2–54. Neural networks: A review from a statistical perspective.

Corina: Software for the generation of high-quality three-dimensional molecular models, by Sadowski, J., Schwab C. H., Gasteiger J. Molecular Networks GmbH Computerchemie, www.mol-net.de, Erlangen, Germany, 2004.

Cummins, D. J., Andrews, C. W.: *J. Chemom.* 9, 1995, 489–507. Iteratively reweighted partial least squares: A performance analysis by Monte Carlo simulations.

Davis, L. (Ed.): *Handbook of Genetic Algorithms*. Van Nostrand-Reinhold, London, United Kingdom, 1991.

de Jong, S.: *Chemom. Intell. Lab. Syst.* 18, 1993, 251–263. SIMPLS: An alternative approach to partial least squares regression.

Dragon: Software for calculation of molecular descriptors, by Todeschini R., Consonni V., Mauri A., Pavan M. Talete srl, www.talete.mi.it, Milan, Italy, 2004.

Du, Y., Liang, Y., Yun, D.: *J. Chem. Inf. Comput. Sci.* 42, 2002, 1283–1292. Data mining for seeking an accurate quantitative relationship between molecular structure and GC retention indices of alkenes by projection pursuit.

Efron, B.: *J. Am. Stat. Assoc.* 78, 1983, 316–331. Estimating the error rate of a prediction rule: Improvement on cross-validation.

Efron, B., Tibshirani, R. J.: *An Introduction to the Bootstrap*. Chapman & Hall, London, United Kingdom, 1993.

Efron, B., Tibshirani, R. J.: *J. Am. Stat. Assoc.* 92, 1997, 548–560. Improvements on cross-validation: The 632 + bootstrap method.

Forina, M., Lanteri, S., Cerrato Oliveros, M.C., Pizarro Millan, C.: *Anal. Bioanal. Chem.* 380, 2004, 397–418. Selection of useful predictors in multivariate calibration.

Frank, I. E., Friedman, J.: *Technometrics* 35, 1993, 109. A statistical view of some chemometrics regression tools.

Frank, I. E., Todeschini, R.: *The Data Analysis Handbook*. Elsevier, Amsterdam, the Netherlands, 1994.

Friedl, A., Padouvas, E., Rotter, H., Varmuza, K.: *Anal. Chim. Acta* 544, 2005, 191–198. Prediction of heating values of biomass fuel from elemental composition.

Furnival, G. M., Wilson, R. W.: *Technometrics* 16, 1974, 499–511. Regressions by leaps and bounds.

Garkani-Nejad, Z., Karlovits, M., Demuth, W., Stimpfl, T., Vycudilik, W., Jalali-Heravi, M., Varmuza, K.: *J. Chromatogr. B*. 1028, 2004, 287–295. Prediction of gas chromatographic retention indices of a diverse set of toxicologically relevant compounds.

Geladi, P., Esbensen, K.: *J. Chemom.* 4, 1990, 337–354. The start and early history of chemometrics: Selected interviews. Part 1.

Gil, J. A., Romera, R.: *J. Chemom.* 12, 1998, 365–378. On robust partial least squares (PLS) methods.

Hastie, T., Tibshirani, R. J., Friedman, J.: *The Elements of Statistical Learning*. Springer, New York, 2001.

Hibbert, D. B.: *Chemom. Intell. Lab. Syst.* 19, 1993, 277–293. Genetic algorithms in chemistry.

Hoerl, A. E., Kennard, R. W.: *Technometrics* 12, 1970, 55–67. Ridge regression: Biased estimation for nonorthogonal problems.

Hoeskuldsson, A.: *J. Chemom.* 2, 1988, 211–228. PLS regression methods.

Hofmann, M., Gatu, C., Kontoghiorghes, E. J.: *Computat. Stat. Data Anal.* 52, 2007, 16–29. Efficient algorithms for computing the best subset regression models for large-scale problems.

Hubert, M., Vanden Branden, K.: *J. Chemom.* 17, 2003, 537–549. Robust methods for partial least squares regression.

Huet, S., Bouvier, A., Poursat, M. A., Jolivet, E.: *Statistical Tools for Nonlinear Regression*. Springer, New York, 2003.

Jalali-Heravi, M., Fatemi, M. H.: *J. Chromatogr. A* 915, 2001, 177–183. Artificial neural network modeling of Kováts retention indices for noncyclic and monocyclic terpenes.

Jansson, P. A.: *Anal. Chem.* 63, 1991, 357–362. Neural networks: An overview.

Johnson, R. A., Wichern, D. W.: *Applied Multivariate Statistical Analysis*. Prentice Hall, Upper Saddle River, NJ, 2002.

Jouan-Rimbaud, D., Massart, D. L., Leardi, R., DeNoord, O. E.: *Anal. Chem.* 67, 1995, 4295–4301. Genetic algorithms as a tool for wavelength selection in multivariate calibration.

Junkes, B. S., Amboni, R. D. M. C., Yunes, R. A., Heinzen, V. E. F.: *Anal. Chim. Acta* 477, 2003, 29–39. Prediction of the chromatographic retention of saturated alcohols on stationary phases of different polarity applying the novel semi-empirical topological index.

Kennard, R. W., Stone, L. A.: *Technometrics* 11, 1969, 137–148. Computer-aided design of experiments.

Körtvelyesi, T., Görgenyi, M., Heberger, K.: *Anal. Chim. Acta* 428, 2001, 73–82. Correlation between retention indices and quantum-chemical descriptors of ketones and aldehydes on stationary phases of different polarity.

Kramer, R.: *Chemometric Techniques for Quantitative Analysis*. Marcel Dekker, New York, 1998.

Leardi, R.: *J. Chemom.* 8, 1994, 65–79. Application of a genetic algorithm for feature selection under full validation conditions and to outlier detection.

Leardi, R.: *J. Chemom.* 15, 2001, 559–569. Genetic algorithms in chemometrics and chemistry: A review.

Leardi, R. (Ed.): *Nature-Inspired Methods in Chemometrics: Genetic Algorithms and Artificial Neural Networks*. Elsevier, Amsterdam, the Netherlands, 2003.

Leardi, R.: *J. Chromatogr. A.* 1158, 2007, 226–233. Genetic algorithms in chemistry.

Lee, M. L., Vassilaros, D. L., White, C. M., Novotny, M.: *Anal. Chem.* 51, 1979, 768–773. Retention indices for programmed-temperature capillary-column gas chromatography of polycyclic aromatic hydrocarbons.

Lindgren, F., Geladi, P., Wold, H.: *J. Chemom.* 7, 1993, 45–59. The kernel algorithm for PLS.

Liu, S., Yin, C., Cai, S., Li, Z.: *Chemom. Intell. Lab. Syst.* 61, 2002, 3–15. Molecular structural vector description and retention index of polycyclic aromatic hydrocarbons.

Looney, C. G.: *Pattern Recognition Using Neural Networks*. Oxford University Press, New York, 1997.

Lucic, B., Trinajstic, N., Sild, S., Karelson, M., Katritzky, A. R.: *J. Chem. Inf. Comput. Sci.* 39, 1999, 610–621. A new efficient approach for variable selection based on multiregression: Prediction of gas chromatographic retention times and response factors.

Mallows, C. L.: *Technometrics* 15, 1973, 661–675. Some comments on Cp.

Manne, R.: *Chemom. Intell. Lab. Syst.* 2 1987, 187–197. Analysis of two partial-least-squares algorithms for multivariate calibration.

Maronna, R., Martin, D., Yohai, V.: *Robust Statistics: Theory and Methods*. Wiley, Toronto, ON, Canada, 2006.

Massart, D. L., Vandeginste, B. G. M., Buydens, L. C. M., De Jong, S., Smeyers-Verbeke, J.: *Handbook of Chemometrics and Qualimetrics: Part A*. Elsevier, Amsterdam, the Netherlands, 1997.

Mevik, B. H., Wehrens, R.: *J. Stat. Software* 18, 2007, 1–24. The pls package: Principal component and partial least squares regression in R.

Miller, A. J.: *Subset Selection in Regression*. CRC Press, Boca Raton, FL, 2002.

Nadler, B., Coifman, R. R.: *J. Chemom.* 19, 2005, 107–118. The prediction error in CLS and PLS: The importance of feature selection prior to multivariate calibration.

Naes, T., Isaksson, T., Fearn, T., Davies, T.: *A User-Friendly Guide to Multivariate Calibration and Classification*. NIR Publications, Chichester, United Kingdom, 2004.

Otto, M.: *Chemometrics—Statistics and Computer Application in Analytical Chemistry*. Wiley-VCH, Weinheim, Germany, 2007.

Pompe, M., Novic, M.: *J. Chem. Inf. Comput. Sci.* 39, 1999, 59–67. Prediction of gas-chromatographic retention indices using topological descriptors.

Rännar, S., Lindgren, F., Geladi, P., Wold, H.: *J. Chemom.* 8, 1994, 111–125. A PLS kernel algorithm for data sets with many variables and fewer objects. Part 1: Theory and algorithm.

Reisinger, K., Haslinger, C., Herger, M., Hofbauer, H.: BIOBIB-Database for Biofuels Institute of Chemical Engineering, Vienna University of Technology, Vienna, Austria, 1996.

Ripley, B. D.: *Pattern Recognition and Neural Networks*. Cambridge University Press, Cambridge, NY, 1996.

Rosipal, R., Krämer, N.: in Saunders, C., Grobelnik, M., Gunn, S. R., Shawe-Taylor, J. (Ed.), *Subspace, Latent Structure and Feature Selection Techniques*. Lecture Notes in Computer Science, Vol. 3940, Springer, Berlin, Germany, 2006, pp. 34–51. Overview and recent advances in partial least squares.

Rosipal, R., Trejo, L. J.: *J. Machine Learn. Res.* 2, 2001, 97–123. Kernel partial least squares regression in reproducing kernel Hilbert space.

Rousseeuw, P. J.: *J. Amer. Stat. Assoc.* 79, 1984, 871–880. Least median of squares regression.

Rousseeuw, P. J., Leroy, A. M.: *Robust Regression and Outlier Detection*. Wiley, New York, 1987.

Schalkoff, R. J.: *Artificial Neural Networks*. McGraw-Hill, New York, 1997.

Seber, G. A. F., Wild, C. J.: *Nonlinear Regression*. Wiley, New York, 2003.

Serneels, S., Croux, C., Filzmoser, P., Van Espen, P. J.: *Chemom. Intell. Lab. Syst.* 79, 2005, 55–64. Partial robust M-regression.

Snee, R. D.: *Technometrics*. 19, 1977, 415–428. Validation of regression models: Methods and examples.

Stone, M., Brooks, R. J.: *J. R. Statist. Soc. B*. 52, 1990, 237–269. Continuum regression: Cross-validated sequentially constructed prediction embracing ordinary least squares, partial least squares and principal component regression.

Tibshirani, R. J.: *J. Royal Stat. Soc. B*. 58, 1996, 267–288. Regression shrinkage and selection via the lasso.

Todeschini, R., Consonni, V.: *Handbook of Molecular Descriptors*. Wiley-VCH, Weinheim, Germany, 2000.

Trygg, J., Wold, S.: Orthogonal Signal Projection, US Patent 6853923, www.freepatentsonline.com/6853923.html (2000).

Trygg, J., Wold, H.: *J. Chemom.* 16, 2002, 119–128. Orthogonal projections to latent structures (O-PLS).

van den Wollenberg, A. L.: *Psychometrika* 42, 1977, 207–218. Redundancy analysis: An alternative for canonical correlation analysis.

Vandeginste, B. G. M., Massart, D. L., Buydens, L. C. M., De Jong, S., Smeyers-Verbeke, J.: *Handbook of Chemometrics and Qualimetrics: Part B*. Elsevier, Amsterdam, the Netherlands, 1998.

Varmuza, K.: in Pomerantsev, A. L. (Ed.), *Progress in Chemometrics Research*, Vol., Nova Science Publishers, New York, 2005, pp. 67–87. Global and local chemometric models of spectra–structure relationships.

Varmuza, K., Liebmann, B., Friedl, A.: *University of Plovdiv "Paisii Hilendarski"—Bulgaria, Scientific Papers—Chemistry* 35[5], 2007, 5–16. Evaluation of the heating value of biomass fuel from elemental composition and infrared data.

Wakeling, I. N., Macfie, H. J. H.: *J. Chemom.* 6, 1992, 189–198. A robust PLS procedure.

Wold, H.: *Chemom. Intell. Lab. Syst.* 14, 1992, 71–84. Nonlinear partial least squares modelling II. Spline inner relation.

Wold, H., Kettaneh-Wold, N., Skagerberg, B.: *Chemom. Intell. Lab. Syst.* 7, 1989, 53–65. Nonlinear PLS modling.

Wold, H., Sjöström, M., Eriksson, L.: *Chemom. Intell. Lab. Syst.* 58, 2001, 109–130. PLS-regression: A basic tool of chemometrics.

Wold, S., Sjöström, M., Eriksson, L.: in Schleyer, P. V. R., Allinger, N. L., Clark, T., Gasteiger, J., Kollman, P.A., Schaefer III, H. F., Schreiner, P. R. (Eds.), *The Encyclo-*

pedia of Computational Chemistry, Vol. 3, Wiley, Chichester, 1998, pp. 2006–2021. Partial least squares projections to latent structures (PLS) in chemistry.

Woloszyn, T. F., Jurs, P. C.: *Anal. Chem.* 64, 1992, 3059–3063. Quantitative structure–retention relationship studies of sulfur vesicants.

Xu, Q. S., Massart, D. L., Liang, Y. Z., Fang, K. T.: *J. Chromatogr. A* 998, 2003, 155–167. Two-step multivariate adaptive regression splines for modeling a quantitative relationship between gas chromatographic retention indices and molecular descriptors.

Yan, A., Jiao, G., Hu, Z., Fan, B. T.: *Comp. Chem.* 24, 2000, 171–179. Use of artificial neural networks to predict the gas chromatographic retention index data of alkylbenzenes on carbowax-20M.

Yoshida, H., Leardi, R., Funatsu, K., Varmuza, K.: *Anal. Chim. Acta* 446, 2001, 485–494. Feature selection by genetic algorithms for mass spectral classifiers.

Zupan, J., Gasteiger, J.: *Neural Networks in Chemistry and Drug Design*. Wiley-VCH, Weinheim, Germany, 1999.

5 Classification

5.1 CONCEPTS

Statistical data often arise from two or more different types or groups (classes, categories), where the grouping effect is known in advance. As mentioned in Section 1.3, a main root of chemometrics has been trying to solve chemical classification problems, especially the automatic recognition of substance classes from molecular spectral data (Crawford and Morrison 1968; Jurs et al. 1969) and the assignment of the origin of samples (Kowalski and Bender 1972). Such applications have been called PATTERN RECOGNITION IN CHEMISTRY (Brereton 1992; Varmuza 1980) before the term chemometrics has been introduced. Recently, classification problems are gaining increasing interest, for instance for the classification of technological materials using near infrared data (Naes et al. 2004) or in biochemistry, in medical applications, and in multivariate image analysis (Xu et al. 2007). The IDENTIFICATION of objects can be considered as a special case of classification, so to say with only one object in each group. An important task of this type is the identification of chemical compounds from spectral data; however, this topic will only be marginally touched in Section 5.3.3 (*k*-nearest neighbor [*k*-NN] classification).

In classification problems, it is thus known to which group the objects belong, and the working hypothesis is that the characteristics of the groups are described by the multivariate data structure of the groups' objects. The task for the statistical analysis is to summarize this multivariate structure appropriately in order to establish rules for correctly assigning new observations for which the group membership is not known. The rules used for classification should be as reliable as possible such that the number of misclassified objects is as small as possible.

Since the group membership of new objects in unknown, it is impossible to evaluate an existing classification rule in terms of number of misclassifications. Therefore, such an evaluation can only be undertaken for data where the grouping information is available. However, as already noted in calibration (Chapter 4), it is not recommended to establish classification rules at the same data where the evaluation is made, because the result of the evaluation will be too optimistic in general. A strict separation into training and test data is necessary, and the concepts of optimizing the complexity of models, using cross validation (CV) and bootstrap, and of validation schemes discussed in Section 4.2 apply here analogously (Figure 5.1).

Two concepts of probabilities are important in classification. The groups to be distinguished have so-called PRIOR PROBABILITIES (PRIORS), the theoretical or natural probabilities (before classification) of objects belonging to one of the groups. After classification, we have POSTERIOR PROBABILITIES of the objects to belong to a group, which are in general different from the prior probabilities, and hopefully allow a clear

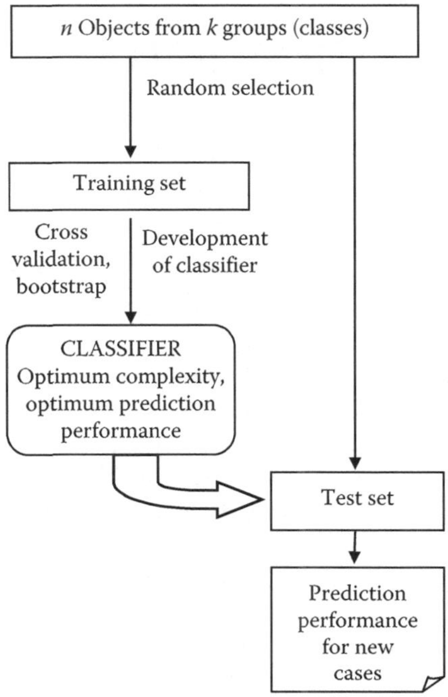

FIGURE 5.1 Scheme of classifier development and test.

assignment to one of the groups. Note that for a given classifier the posterior probabilities depend on the (assumed) prior probabilities (Section 5.7.2).

For mutually exclusive groups, the sum of all prior probabilities must be equal to one. For instance, the probability of snowfall on any day in month May in Vienna may be 0.05 based on long-term weather data. The prior probability p_1 of group 1 (snowfall) can be set to 0.05, and the complementary prior probability p_2 of group 2 (no snowfall) to 0.95. A classifier that tries to predict "snowfall yes or no"—based for instance on recent weather data—may consider the prior probabilities for the two groups and may give more weight to group 2. In this case, the resulting discriminant rule and any estimated misclassification rate depend on the assumed priors; however, for not too small data sets, the dependence should be small. Considering prior probabilities in the development of classifiers is often called the BAYESIAN APPROACH.

Estimation of prior probabilities from existing data is often difficult and weak; furthermore in many classification problems it is only reasonable to assume EQUAL PRIOR PROBABILITIES—so to say not to consider priors during classifier development and estimation of the misclassification rate. Consider the classification problem to decide whether the chemical structure of a compound contains a benzene ring or not, based on spectroscopic data. What priors should be used for the two groups "benzene ring" and "no benzene ring:" the probabilities in a spectroscopic data base (from which the classifier has been developed), or in the Beilstein database, or in any distribution of compounds in a laboratory or on Earth? The appropriate strategy in this case is to develop a classifier without considering (different) priors, and in an application to unknown compounds to assume equal probabilities for belonging to the two possible groups (corresponding to maximum uncertainty).

To summarize, the goal of classification is to establish rules—also called CLASSIFIERS—on the basis of objects with known group memberships, which can be reliably used for predicting the group membership of new observations, as well as evaluating the performance of the rules. Note that classification is a supervised technique, while the automatic identification of a grouping structure in the data is an unsupervised technique belonging to CLUSTER ANALYSIS (see Chapter 6). In cluster analysis, the group information is not required, and usually not even available.

There are various different ways for finding classification rules. The main approaches are based on

- Appropriate latent variables (discriminant variables defined by regression models, Section 5.2)
- Modeling the density functions of the classes and the use of distances (Section 5.3)
- Classification trees (Section 5.4)
- Artificial neural networks (ANNs) (Section 5.5)
- Support vector machines (SVMs) (Section 5.6)

5.2 LINEAR CLASSIFICATION METHODS

Given a data set $\boldsymbol{X}$, the joint idea of linear classification methods is to find one or several linear functions of the x-variables that can be used for classification; we called them linear latent variables, see Figure 2.18 in Section 2.6.3. The traditional statistical methods for this purpose are LINEAR DISCRIMINANT ANALYSIS (LDA) and various other linear regression methods, as well as LOGISTIC REGRESSION (LR). For high-dimensional problems (many x-variables), singularity problems arise if the variables are highly correlating or if less objects in a group than variables are available. In this case, the information contained in the x-variables can be summarized by (intermediate) latent variables that allow for dimension reduction. Most commonly, principal component analysis (PCA) (Chapter 3) or partial least-squares (PLS) regression (Section 4.7) are used for this purpose. LDA, regression, or LR are then carried out with the PCA scores or PLS scores instead of the original variables.

5.2.1 Linear Discriminant Analysis

We assume that n objects have been measured for m characteristics (variables), and that the objects originate from k different groups. Suppose that the groups consist of $n_1, \ldots, n_k$ objects, where $n_1 + \cdots + n_k = n$. There are two different approaches to derive a rule for discrimination between the groups, the Bayesian and the Fisher approach (Huberty 1994; Johnson and Wichern 2002).

5.2.1.1 Bayes Discriminant Analysis

For the Bayesian discriminant rule, an underlying data distribution f_j for each group $j = 1, \ldots, k$ is required, which is usually assumed to be a multivariate normal distribution with mean $\boldsymbol{\mu}_j$ and covariance matrix $\boldsymbol{\Sigma}_j$. Moreover, we assume that there exists a certain PRIOR PROBABILITY p_j for each group, and $p_1 + \cdots + p_k = 1$. The Bayesian rule uses the POSTERIOR PROBABILITY $P(l|\boldsymbol{x})$ that an object $\boldsymbol{x}$ belongs to group l, which is given by

$$P(l|\boldsymbol{x}) = \frac{f_l(\boldsymbol{x})p_l}{\sum_{j=1}^{k} f_j(\boldsymbol{x})p_j} \tag{5.1}$$

Since the denominator in Equation 5.1 is the same for each group, we can directly compare the posterior probabilities $P(j|\boldsymbol{x})$ for all groups. Observation $\boldsymbol{x}$ will be assigned to that group for which the posterior probability is the largest. Thus the decision boundary between two classes h and l is given by objects $\boldsymbol{x}$ for which the posterior probabilities are equal, i.e., $P(h|\boldsymbol{x}) = P(l|\boldsymbol{x})$.

This rule is visualized in Figure 5.2 for three groups in the univariate case. For simplicity, we assume that the prior probabilities are equal, i.e., $p_1 = p_2 = p_3 = 1/3$. Thus the decision boundaries are at the values of x for which the density functions $f_j(x)$ are equal. In this case (equal prior probabilities), an object x will be assigned to that group where the density function $f_j(x)$ is the largest. Accordingly, we obtain a discriminant rule which can be used to assign any object x to one of the groups. If the prior probabilities would not be equal, the decision boundaries (dashed lines) would be moved towards the group with smaller prior probability.

Maximizing the posterior probabilities in case of multivariate normal densities will result in quadratic or linear discriminant rules. However, the rules are LINEAR if we use the additional assumption that the covariance matrices of all groups are equal, i.e., $\boldsymbol{\Sigma}_1 = \cdots = \boldsymbol{\Sigma}_k = \boldsymbol{\Sigma}$. In this case, the classification rule is based on LINEAR DISCRIMINANT SCORES d_j for groups j

$$d_j(\boldsymbol{x}) = \boldsymbol{\mu}_j^{\mathrm{T}} \boldsymbol{\Sigma}^{-1} \boldsymbol{x} - \boldsymbol{\mu}_j^{\mathrm{T}} \boldsymbol{\Sigma}^{-1} \boldsymbol{\mu}_j / 2 + \log(p_j) \quad \text{for } j = 1, \ldots, k \tag{5.2}$$

which are directly derived from plugging in the density of the multivariate normal distribution into the equation for posterior probabilities (Johnson and Wichern 2002). An object $\boldsymbol{x}$ is assigned to that group for which the discriminant score is the largest.

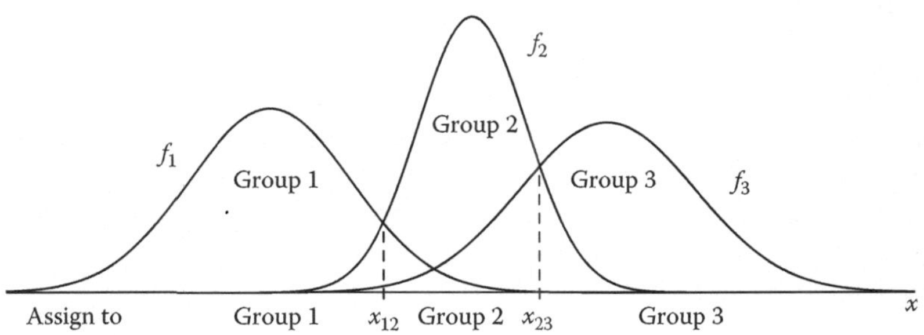

FIGURE 5.2 Visualization of the Bayesian decision rule in the univariate case, where the prior probabilities of the three groups are equal. The dashed lines are at the decision boundaries between groups 1 and 2 (x_{12}) and between groups 2 and 3 (x_{23}).

The discriminant scores (Equation 5.2) are linear functions in $\boldsymbol{x}$, and therefore the resulting rule leads to an LDA with linear separation boundaries between the groups. The first term in Equation 5.2 can be considered as a scalar product between a loading vector (discriminant vector) $\boldsymbol{b}_{\text{BAYES}} = \boldsymbol{\mu}_j^{\text{T}}\boldsymbol{\Sigma}^{-1}$ and an object vector $\boldsymbol{x}$ (Figure 5.4, left). This scalar product is adjusted by the second term which is a constant for each class j that adjusts for the mean. The third term log (p_j) adjusts for the group prior probabilities; the lower the p_j, the more is the final discriminant score reduced—groups with a small prior probability (minorities) are diminished in order to minimize the TOTAL probability of misclassification. Note, if the prior probabilities are equal and the covariance matrices are considered to be equal, the Bayes classifier is identical to Fisher discriminant analysis (see below in this section).

Figure 5.3 (left) shows the adjustment of the discriminant line for varying prior probabilities. The artificial data set contains two variables: group means are denoted by ×, and ellipses indicate the shape of the distribution. If the prior probabilities of the two groups are equal, the group separation is symmetric. However, if the prior probabilities become unequal, the separation line moves to the group with the smaller prior probability resulting in a rule which minimizes the total probability of misclassification. In other words, if for instance group 2 becomes less probable ($p_2 < 0.5$), assignment to class 1 becomes more frequent—independent from the discriminating properties of the data.

The right plot in Figure 5.3 shows a linear discrimination of three groups. Here all three groups have the same prior probability, but their covariance matrices are not equal (different shape and orientation). The resulting rule is no longer optimal in the sense defined above. An optimal rule, however, could be obtained by quadratic discriminant analysis which does not require equality of the group covariances.

For given data, the population parameters $\boldsymbol{\mu}_j$, $\boldsymbol{\Sigma}$, and p_j in Equation 5.2 have to be estimated. If the group sizes n_j reflect the population groups sizes, the prior

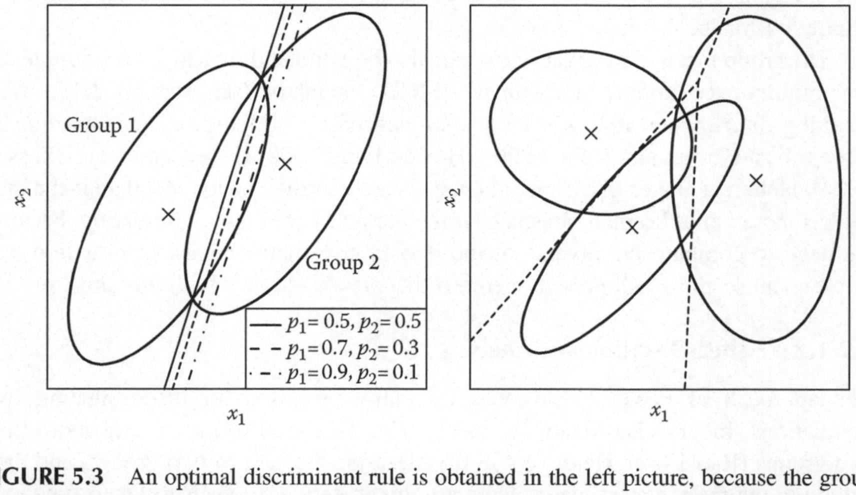

FIGURE 5.3 An optimal discriminant rule is obtained in the left picture, because the group covariances are equal and an adjustment is made for different prior probabilities. The linear rule shown in the right picture is not optimal—in terms of a minimum probability of misclassification—because of the different covariance matrices.

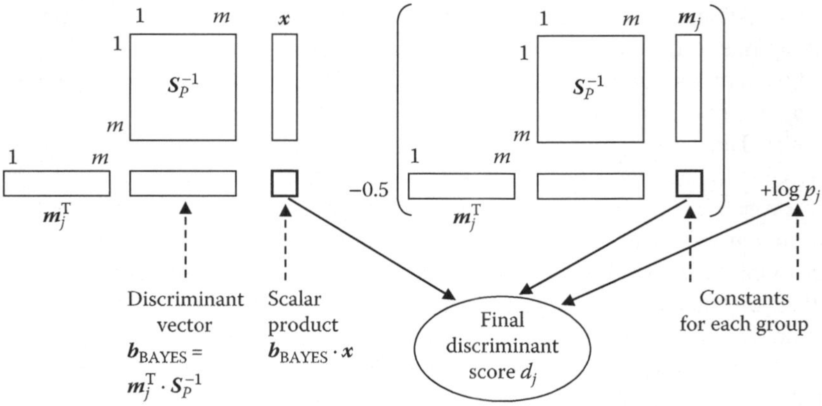

FIGURE 5.4 Linear discriminant scores d_j for group j by the Bayesian classification rule based on (Equation 5.2). $\boldsymbol{m}_j^{\mathrm{T}}$, mean vector of all objects in group j; $\boldsymbol{S}_P^{-1}$, inverse of the pooled covariance matrix (Equation 5.3); $\boldsymbol{x}$, object vector (to be classified) defined by m variables; p_j, prior probability of group j.

probabilities p_j can be estimated by n_j/n. The population means $\boldsymbol{\mu}_j$ can be estimated by the arithmetic means of the data in the groups, and the population covariances $\boldsymbol{\Sigma}_j$ by the sample covariance matrices $\boldsymbol{S}_j$. Since we assumed that the group covariances are equal, we have to provide a joint estimation of the covariance for $\boldsymbol{\Sigma}$, and this can be done by the POOLED COVARIANCE MATRIX (a weighted sum of the group covariance matrices).

$$\boldsymbol{S}_P = \frac{(n_1 - 1)\boldsymbol{S}_1 + \cdots + (n_k - 1)\boldsymbol{S}_k}{n_1 + \cdots + n_k - k} \tag{5.3}$$

The calculation of the discriminant scores in Equation 5.2 is schematically shown in Figure 5.4 for the estimated quantities.

The group means and covariances can also be estimated robustly, for example, by the minimum covariance determinant (MCD) estimator (see Section 2.3.2). The resulting discriminant rule will be less influenced by outlying objects and thus be more robust (Croux and Dehon 2001; He and Fung 2000; Hubert and Van Driessen 2004). Note that Bayes discriminant analysis as described is not adequate if the data set has more variables than objects or if the variables are highly correlating, because we need to compute the inverse of the pooled covariance matrix in Equation 5.2. Subsequent sections will present methods that are able to deal with this situation.

5.2.1.2 Fisher Discriminant Analysis

The approach of Fisher (1938) was originally proposed for discriminating two populations (binary classification), and later on extended to the case of more than two groups (Rao 1948). Here we will first describe the case of two groups, and then extend to the more general case. Although this method also leads to linear functions for classification, it does not explicitly require multivariate normal distributions of the groups with equal covariance matrices. However, if these assumptions are not fulfilled, the Fisher rule will no longer be optimal in terms of minimizing the total probability of misclassification.

5.2.1.2.1 Binary Classification

In case of two groups, the Fisher method transforms the multivariate data to a univariate discriminant variable such that the transformed groups are separated as much as possible. For this transformation, a linear combination of the original x-variables is used, in other words a latent variable.

$$y = b_1x_1 + b_2x_2 + \cdots + b_mx_m \tag{5.4}$$

which provides a maximum group separation. The coefficients $b_1, \ldots, b_m$ form a decision vector $\boldsymbol{b}$ (a loading vector as described in Section 2.6). Projecting the objects on an axis defined by $\boldsymbol{b}$ gives discriminant scores y_{1h} with $h = 1, \ldots, n_1$ for the first group, and values y_{2l} with $l = 1, \ldots, n_2$ for the second group. Denote $\bar{y}_1$ and $\bar{y}_2$ as the arithmetic means of the discriminant scores of the first and second group, respectively. Then the criterion for group separation is formulated as

$$\frac{|\bar{y}_1 - \bar{y}_2|}{s_y} \rightarrow \max \tag{5.5}$$

where s_y is the square root of the POOLED VARIANCE

$$s_y^2 = \frac{(n_1 - 1)s_1^2 + (n_2 - 1)s_2^2}{n_1 + n_2 - 2} \tag{5.6}$$

The pooled variance s_y^2 is a weighted sum of the variances s_1^2 and s_2^2 of y for groups 1 and 2, respectively. This criterion is equivalent to the test criterion used in the two-sample t-test (Section 1.6.5), which examines the difference of group means considering the standard deviations of the groups. In other words, a latent variable is defined which has maximum t-value in a two-sample t-test of the discriminant scores for both groups. Fisher has shown that the decision vector $\boldsymbol{b}$ maximizing the criterion (Equation 5.5) is given by

$$\boldsymbol{b}_{\text{FISHER}} = \boldsymbol{S}_P^{-1}(\bar{\boldsymbol{x}}_1 - \bar{\boldsymbol{x}}_2) \tag{5.7}$$

where

$\bar{\boldsymbol{x}}_1$ and $\bar{\boldsymbol{x}}_2$ are the arithmetic mean vectors of the data from groups 1 and 2, respectively
$\boldsymbol{S}_P$ is the pooled covariance matrix defined in Equation 5.3

Thus, a new object $\boldsymbol{x}_i$ is classified by calculating the discriminant score y_i (projection on the direction defined by the decision vector)

$$y_i = \boldsymbol{b}_{\text{FISHER}}^{\text{T}}\boldsymbol{x}_i \tag{5.8}$$

The discriminant score y_i is compared with a classification threshold

$$y_0 = \frac{\boldsymbol{b}_{\text{FISHER}}^{\text{T}}\bar{\boldsymbol{x}}_1 + \boldsymbol{b}_{\text{FISHER}}^{\text{T}}\bar{\boldsymbol{x}}_2}{2} \tag{5.9}$$

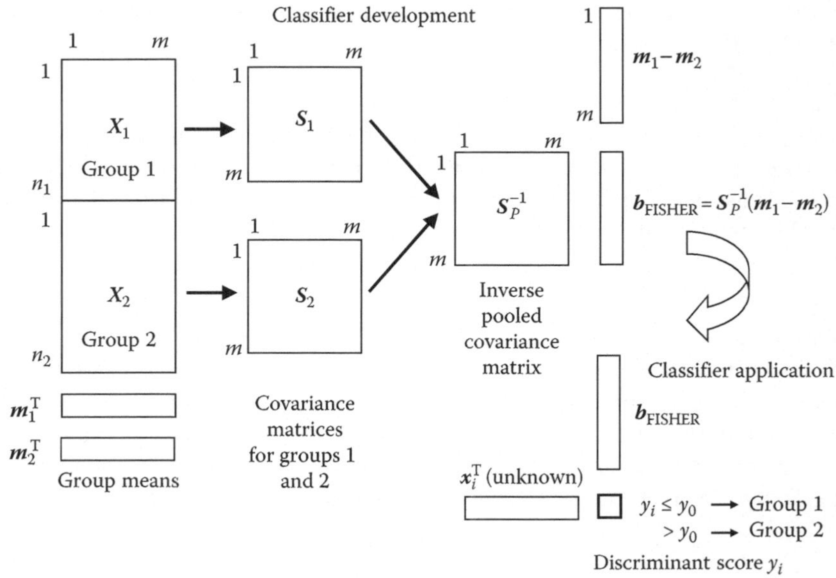

FIGURE 5.5 Scheme of Fisher discriminant analysis.

which is the mean of the scores obtained by projecting the group means on the discriminant direction. If $y_i \leq y_0$, then object i is assigned to group 1, else to group 2. The scheme of Fisher discriminant analysis for two groups is summarized in Figure 5.5.

Figure 5.6 visualizes the idea of Fisher discriminant analysis for two groups in the two-dimensional case. The group centers (filled symbols) are projected on the discriminant variable, giving $\bar{y}_1$ and $\bar{y}_2$; the mean of both is the classification threshold y_0.

By rearranging Equation 5.2 for the Bayesian rule, it can be seen that in case of two groups the solution is exactly the same as for the Fisher rule if the prior probabilities are equal. However, since the prior probabilities p_1 and p_2 are not considered in the Fisher rule, the results will be different for the Bayesian rule if $p_1 \neq p_2$.

5.2.1.2.2 *Multicategory Classification*

An extension of the Fisher rule to more than two groups can be described as follows. Using the same notation as above, we define the VARIATION BETWEEN THE GROUPS, $\boldsymbol{B}$, by

$$\boldsymbol{B} = \sum_{j=1}^{k} p_j(\boldsymbol{\mu}_j - \boldsymbol{\mu})(\boldsymbol{\mu}_j - \boldsymbol{\mu})^{\mathrm{T}} \tag{5.10}$$

where $\boldsymbol{\mu} = \sum_{j=1}^{k} p_j \boldsymbol{\mu}_j$ is the overall weighted mean for all populations. Moreover, we define the WITHIN GROUPS COVARIANCE MATRIX $\boldsymbol{W}$ by

$$\boldsymbol{W} = \sum_{j=1}^{k} p_j \boldsymbol{\Sigma}_j \tag{5.11}$$

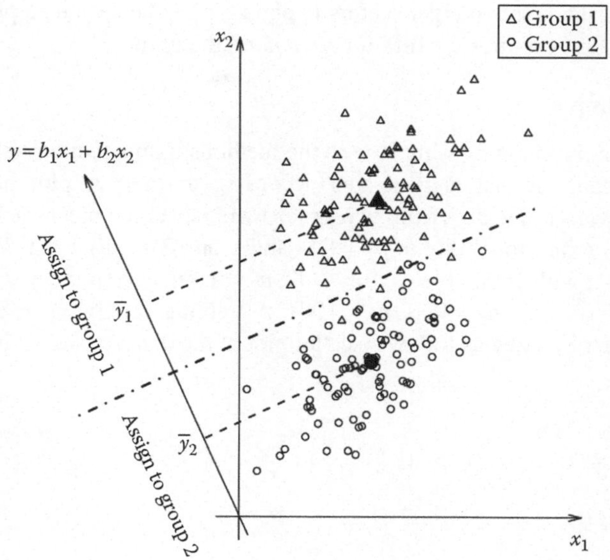

FIGURE 5.6 Visualization of the Fisher discriminant rule in two dimensions. A latent variable (a discriminant variable) y is computed, and the group assignment is according to the average of the projected group means $\bar{y}_1$ and $\bar{y}_2$ (dashed–dotted line). The filled symbols represent the group means; they are projected (dashed lines) on the discriminant variable y.

which can be seen as a pooled version of the group covariance matrices. Using the assumption of equal group covariance matrices, it can be shown that the group centers can be best separated by maximizing

$$\frac{\boldsymbol{b}^{\mathrm{T}}\boldsymbol{B}\boldsymbol{b}}{\boldsymbol{b}^{\mathrm{T}}\boldsymbol{W}\boldsymbol{b}} \tag{5.12}$$

where $\boldsymbol{b} \neq \boldsymbol{0}$ is an m-dimensional vector. The solution of this maximization problem is given by the eigenvectors $\boldsymbol{v}_1, \ldots, \boldsymbol{v}_l$ of the matrix $\boldsymbol{W}^{-1}\boldsymbol{B}$, scaled so that $\boldsymbol{v}_h^{\mathrm{T}}\boldsymbol{W}\boldsymbol{v}_h = 1$ for $h = 1, \ldots, l$. Here, the number l of positive eigenvalues turns out to be $l \leq \min(k-1, m)$. By combining the eigenvectors $\boldsymbol{v}_1, \ldots, \boldsymbol{v}_l$ in the matrix $\boldsymbol{V}$, we can define the FISHER DISCRIMINANT SCORES d_j^{F} for an object $\boldsymbol{x}$ as

$$d_j^{\mathrm{F}}(\boldsymbol{x}) = \left[\left(\boldsymbol{x} - \boldsymbol{\mu}_j\right)^{\mathrm{T}} \boldsymbol{V}\boldsymbol{V}^{\mathrm{T}} \left(\boldsymbol{x} - \boldsymbol{\mu}_j\right) - 2\log\left(p_j\right)\right]^{1/2} \tag{5.13}$$

for $j = 1, \ldots, k$. Here we included the penalty term $-2\log(p_j)$, which adjusts the decision boundary appropriately in case of unequal prior probabilities. Thus, a new object $\boldsymbol{x}$ is assigned to that group for which the Fisher discriminant score is the smallest. For given data, the population quantities can be estimated in the same way as above for the Bayesian rule.

If the assumptions (multivariate normal distributions with equal group covariance matrices) are fulfilled, the Fisher rule gives the same result as the Bayesian rule. However, there is an interesting aspect for the Fisher rule in the context of visualization, because this formulation allows for dimension reduction. By projecting the data in the space of the first two eigenvectors $\boldsymbol{v}_1$ and $\boldsymbol{v}_2$, one obtains a data presentation in the plane that best captures the differences among the groups.

5.2.1.3 Example

A demonstration of how to using $\mathbb{R}$ with the methods in this section will be done for an artificial data set that is generated according to the left plot in Figure 5.3. Accordingly, we sample data from two groups with specified means and covariance. Group 1 has a prior probability of $p_1 = 0.9$, and group 2 has $p_2 = 0.1$. We generate a training data set with 1000 objects ($n_1 = 900$, $n_2 = 100$), and an independent test data set with 1000 objects ($n_1 = 900$, $n_2 = 100$). A training set `dtrain` and a test set `dtest` can be generated as follows (scatter plot of the data are shown in Figure 5.7).

```
R: mu1 <- c(0,0)                                      # mean group 1
   mu2 <- c(3.5,1)                                    # mean group 2
   sig <- matrix(c(1.5,1,1,1.5),ncol=2)               # cov. matrix
   library(mvtnorm)
   n1 <- 900
   n2 <- 100
   group <- c(rep(1,n1),rep(2,n2))
   set.seed(130)                            # set any random seed
   X1train <- rmvnorm(n1,mu1,sig)           # multiv. normal distr.
   X2train <- rmvnorm(n2,mu2,sig)
   Xtrain <- rbind(X1train,X2train)
   dtrain <- data.frame (X=Xtrain,group=group)    # training set
   set.seed(131)
   X1test <- rmvnorm(n1,mu1,sig)
   X2test <- rmvnorm(n2,mu2,sig)
   Xtest <- rbind(X1test,X2test)
   dtest <- data.frame(X=Xtest,group=group)       # test set
```

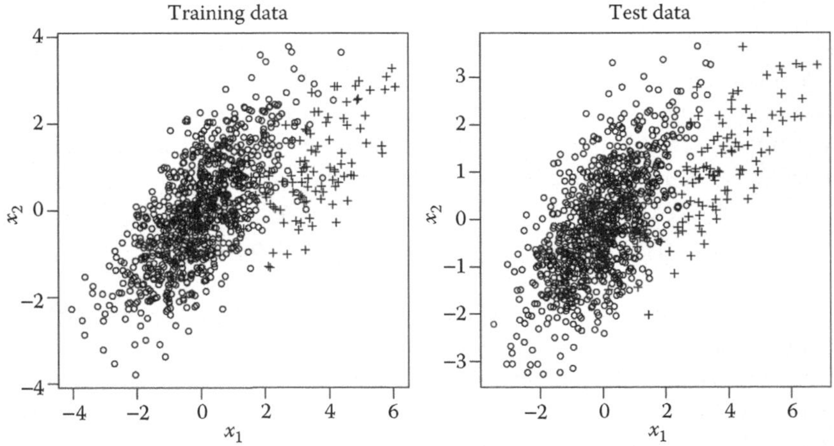

FIGURE 5.7 Training and test data generated according to Figure 5.3 (left). This example is used to demonstrate the use of $\mathbb{R}$ for methods described in this section. Group 1 is denoted by circles ($n_1 = 900$) and group 2 by plus signs ($n_2 = 100$).

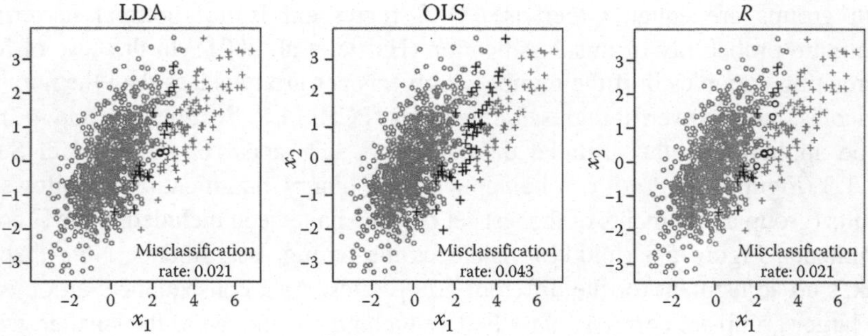

FIGURE 5.8 LDA with Bayesian discriminant analysis (left), OLS (middle), and logistic regression (right) are applied to the training data shown in Figure 5.7 (left). The plots show the results from applying the models to the test data shown in Figure 5.7 (right). Assignment to group 1 is denoted by circles, to group 2 by plus signs. The dark points are the misclassified objects, leading to the given total misclassification rates.

The model is built for the training data, and evaluated for the test data as follows with the results (group assignments of the test set objects) included in the ℝ code as comments. Figure 5.8 (left) visualizes the wrong assignments for LDA.

```
ℝ: library(MASS)  # includes functions for LDA
   resLDA <- lda(group~.,data=dtrain)
                                      # LDA for training data
   predLDA <- predict(resLDA,newdata=dtest)$class
                      # predicted class memberships for test data
   table(group,predLDA)          # summarizes the assignments

   # results for test set
   # group   0    1
   #     0  84   16
   #     1   5  895
   # these are 5+16 wrong assignments,
   # misclassification rate 0.021
```

5.2.2 Linear Regression for Discriminant Analysis

5.2.2.1 Binary Classification

In case of classifying objects into two groups, one could use any regression method (see Chapter 4), where the x-variables describe the features of the objects, and the y-variable contains the group information. It is not relevant how the y-variable is coded; using a coding -1 and $+1$ for the different groups will result in a decision boundary at 0. Thus, if the predicted value for a new object is positive, it will be assigned to the group labeled by $+1$, and if it is negative the object is assigned to the other group. For ordinary least-squares (OLS), the resulting discrimination rule is identical with the Bayesian rule or Fisher rule as long as the prior probabilities of both groups are equal. Otherwise, OLS regression is not optimal in terms of minimum probability of misclassification (Hastie et al. 2001). In the case of highly correlating variables or if the number of objects per group is less than the number of variables, more powerful regression methods (PCR, PLS, Sections 4.6 and 4.7) have to be applied. For the artificial data sets `dtrain` and `dtest`, used in Section 5.2.1.3 (Figure 5.7), OLS can be applied for a binary classification as follows. The results (group assignments of the test set objects) have been included in the ℝ code as comments; Figure 5.8 (middle) visualizes the wrong assignments. Note that OLS makes no adjustment for the different group sizes. As a consequence, all objects of the larger group are correctly classified, but a high percentage of the smaller group is classified incorrectly.

ℝ:
```
resOLS <- lm(group~.,data=dtrain)
                              # OLS regression for training data
predOLS <- predict(resOLS,newdata=dtest)>0.5
                     # predicted class memberships for test data
   # decision boundary 0.5 because 0/1 coding for groups
table(group,predOLS)        # summarizes the assignments

# results for test set
# group   0    1
#     0  57   43
#     1   0  900
# these are 43 wrong assignments,
# misclassification rate 0.043
```

5.2.2.2 Multicategory Classification with OLS

If there are $k > 2$ groups for classification, one can use a binary coding for each group separately. Thus, for each object, we have to define one y-variable for each group by $y_{ij} = 1$ if object i belongs to group j, and $y_{ij} = -1$ otherwise. The resulting matrix $\boldsymbol{Y}$ with n rows and k columns is then used in a multivariate regression model (see Section 4.3.3)

$$\boldsymbol{Y} = \boldsymbol{X}\boldsymbol{B} + \boldsymbol{E} \tag{5.14}$$

where

- $\boldsymbol{X}$ contains the variables of the training data and additionally values of 1 in the first column for the intercept.
- $\boldsymbol{B}((m+1) \times k)$ is the matrix with the regression coefficients for m variables and k groups
- $\boldsymbol{E}$ represents the errors

The estimated regression coefficients using OLS are $\boldsymbol{B} = (\boldsymbol{X}^{\mathrm{T}}\boldsymbol{X})^{-1}\boldsymbol{X}^{\mathrm{T}}\boldsymbol{Y}$. For a new object $\boldsymbol{x}$, the prediction is

$$\hat{\boldsymbol{y}} = (\hat{y}_1, \ldots, \hat{y}_k)^{\mathrm{T}} = [(1, \boldsymbol{x}^{\mathrm{T}})\boldsymbol{B}]^{\mathrm{T}} \tag{5.15}$$

which is a vector of length k, and thus a prediction for each group is obtained. The object is finally assigned to that group with the largest component $\hat{y}_j$, for $j=1, \ldots, k$. Instead of multivariate regression (one classification model for all groups) for each group (each column in $\boldsymbol{Y}$), a separate classification model can be calculated.

```
R: resOLS <- lm(Y~X)               # OLS regression with binary matrix Y
   predict(resOLS, newdata)  # prediction for new data
```

Besides the nonoptimality of the resulting classification rule, there is the usual problem with OLS regression in the context of high-dimensional data. In case of many x-variables, the resulting rule can have poor performance, and thus all alternative strategies discussed in Chapter 4 can be used. A reduction of the regressor variables by variable selection can improve the performance, but also Ridge or Lasso regression are alternatives; for the use of PLS, see below.

5.2.2.3 Multicategory Classification with PLS

An approach that has become popular in chemometrics is the use of PLS regression for classification, particularly the use of PLS2 for multicategory classification. The method is commonly referred to as PLS DISCRIMINANT ANALYSIS (PLS-DA or D-PLS) (Brereton 2007; Naes et al. 2004). It works in the same way as described above with binary coding of the y-variables. Also the classification of new objects can be done in the same way as mentioned above, although difficulties arise if for instance more than one group is indicated by a positive value of $\hat{y}_j$. For such cases heuristic decision schemes can be applied which may either assign the object to more than one group (if possible) or select the group with the maximum $\hat{y}_j$. An advantage of D-PLS is the availability of scores and loadings which can be used for a graphical presentation of the results and may be helpful in the interpretation of the group characteristics.

5.2.3 Logistic Regression

Similar to linear classification methods (Section 5.2.2), logistic regression (LR) uses the x-variables for building a regression model. In this method, the posterior probabilities of the groups are modeled by linear functions of the x-variables, with the constraints that the probabilities remain in the interval [0, 1] and sum up to 1. In the following, we introduce the LR model for the case of $k=2$ groups, but it can be easily extended to more groups (Kleinbaum and Klein 2002).

We consider for an object $\boldsymbol{x} = (x_1, \ldots, x_m)^{\mathrm{T}}$ the posterior probability $P_1 = P(1\,|\,\boldsymbol{x})$ that an object $\boldsymbol{x}$ belongs to group 1 and $P_2 = P(2\,|\,\boldsymbol{x})$ for belonging to group 2. The log-ratio of the posterior probabilities is then modeled by a linear combination of the x-variables

$$\log\,[P_1/P_2] = b_0 + b_1x_1 + b_2x_2 + \cdots + b_mx_m = z \tag{5.16}$$

Since the sum of the posterior probabilities should be 1, we can reformulate Equation 5.16 by

$$P_1(z) = \frac{e^z}{1 + e^z} \tag{5.17}$$

$$P_2(z) = \frac{1}{1 + e^z} \tag{5.18}$$

This shows that the posterior probabilities are indeed in the interval [0, 1] and that they sum up to 1. The function for $P_1(z)$, Equation 5.17, is known as the LOGISTIC FUNCTION. Both functions are graphically shown in Figure 5.9. The input z of the functions can take any value between minus and plus infinity, and the output is always limited to the interval [0, 1]. For z one could think of an appropriately summarized contribution of all x-variables, and a high value of z yields a high probability for one of the groups. In this example, a classification threshold of 0 is reasonable for assigning new objects to one of the two groups. The definition of a REJECTION INTERVAL (dead zone) around zero may be useful if the two groups overlap considerably; thereby a smaller misclassification rate can be achieved—of course at the cost of "no answer" for some objects (no answer may be better than a wrong answer).

Note that there is a strong similarity to LDA (Section 5.2.1), because it can be shown that also for LDA the log-ratio of the posterior probabilities is modeled by a linear function of the x-variables. However, for LR, we make no assumption for the data distribution, and the parameters are estimated differently. The estimation of the coefficients $b_0, b_1, \ldots, b_m$ is done by the MAXIMUM LIKELIHOOD METHOD which leads to an ITERATIVELY REWEIGHTED LEAST SQUARES (IRLS) algorithm (Hastie et al. 2001).

An advantage of LR in comparison to LDA is the fact that statistical inference in the form of tests and confidence intervals for the regression parameters can be derived (compare Section 4.3). It is thus possible to test whether the jth regression coefficient $b_j = 0$. If the hypothesis can be rejected, the jth regressor variable x_j

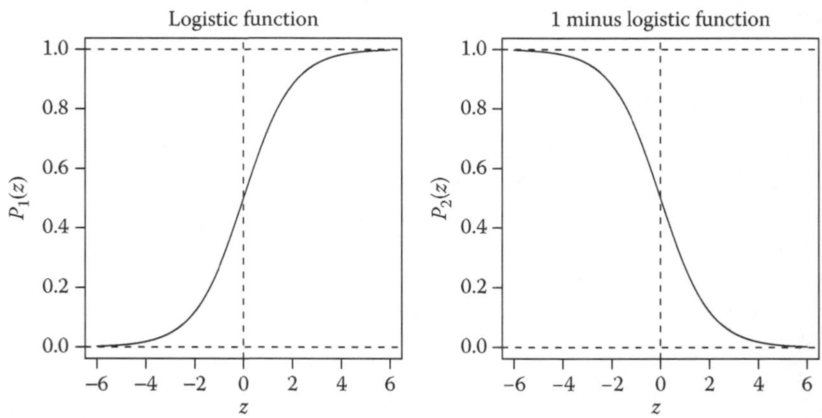

FIGURE 5.9 Visualization of Equations 5.17 and 5.18 for modeling the posterior probabilities with LR. The left-hand plot pictures the logistic function $P_1(z) = e^z/(1 + e^z)$; the right-hand plot shows $P_2(z) = 1 - P_1(z)$.

contributes to the explanation of the discriminant variable and can thus be regarded as important for the group separation. However, as already noted in Section 4.3 for OLS regression, the interpretation of the importance of a variable for discrimination is only valid within the context of the other variables, since it can happen that a variable is considered as unimportant only because the information of this variable is contained in one or several other regressor variables.

The same limitation as in OLS concerning highly correlated x-variables holds for LR, because the IRLS algorithm performs OLS on weighted x-objects. Possible solutions are for instance (1) use of PCA scores instead of the original variables (Section 4.6), or (2) PLS logistic regression (Bastien et al. 2005; Esposito Vinci and Tenenhaus 2001) where the regression coefficients in Equation 5.16 are computed by PLS.

For the artificial data sets `dtrain` and `dtest` used in Section 5.2.1 (Figure 5.7), LR can be applied for a binary classification as follows. The results (group assignments of the test set objects) have been included in the ℝ code as comments; Figure 5.8 (right) visualizes the wrong assignments.

```
ℝ:  resLR <- glm(group~.,data=dtrain,family=binomial)
                              # logistic regression for training data
    predLR <- predict(resLR,newdata=dtest)>0.5
                       # predicted class memberships for test data
         # decision boundary 0.5 because 0/1 coding for groups
    table(group,predLR)          # summarizes the assignments

    # results for test set
    # group   0    1
    #     0  87   23
    #     1   8  892
    # these are 23+8 wrong assignments,
    # misclassification rate 0.021
```

5.3 KERNEL AND PROTOTYPE METHODS

5.3.1 SIMCA

The acronym SIMCA stands for *Soft Independent Modeling of Class Analogy*, and it denotes a method based on disjoint principal component models proposed by Svante Wold (Wold 1976). The idea is to describe the multivariate data structure of each group separately in a reduced space using PCA (see Chapter 3). The special feature of SIMCA is that PCA is applied to each group separately and also the number of PCs is selected individually and not jointly for all groups (Figure 5.10). A PCA model is an envelope, in the form of a sphere, ellipsoid, cylinder, or rectangular box optimally enclosing a group of objects. This allows for an optimal dimension reduction in each group in order to reliably classify new objects. Due to the use of PCA, this approach works even for high-dimensional data with rather a small number of samples. In addition to the group assignment for new objects, SIMCA provides also information about the relevance of different variables to the classification, or measures of separation. Historically, SIMCA is a milestone in

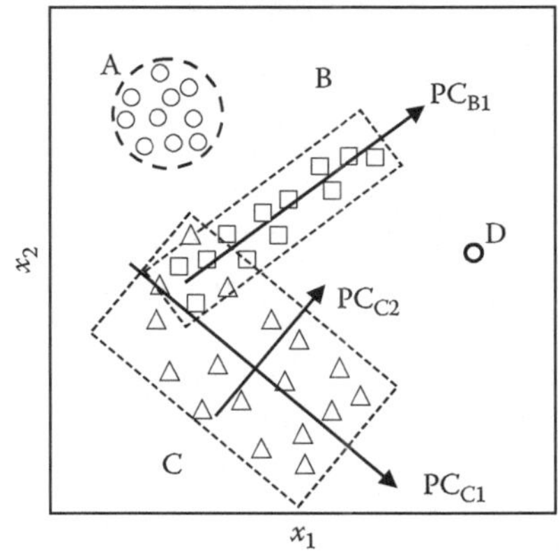

FIGURE 5.10 Principle of SIMCA modeling and classification. Group A can be modeled by a single prototype point (usually the center of the group) and a sphere with an appropriate radius. Group B is distributed along a straight line, and one principal component, PC_{B1}, together with an appropriate radius defines a model with a cylindrical shape. Group C requires two principal components, PC_{C1} and PC_{C2}, and the geometric model is a rectangular box. The single object *D* would be recognized not to belong to any of the groups A–C. Note that the distances OD and SD (Equations 5.22 and 5.23) assume ellipsoidal shapes of the groups.

chemometrics by introducing the concept of SOFT MODELING in contrary to hard modeling as used in most approaches of discriminant analysis. In most classification methods, a new object is always assigned to one of the defined groups; SIMCA is capable to find that an object belongs to more than one group or does not belong to any of the defined groups (Brereton 2006; Eriksson et al. 2006; Vandeginste et al. 1998).

Suppose that objects from k different groups are given, and that the number of objects in each group j is n_j, with $n_1 + \cdots + n_k = n$. Denote the data matrix containing the objects of class j by $\boldsymbol{X}_j$, thus having n_j rows and m columns. We assume that $\boldsymbol{X}_j$ is mean-centered, which is done by subtracting the mean vector $\bar{\boldsymbol{x}}_j$ computed from the objects of group j, from each original object in group j. Depending on the application, the data matrices $\boldsymbol{X}_j$ need to be autoscaled. Applying PCA to the matrices $\boldsymbol{X}_j$ results in

$$\boldsymbol{T}_j = \boldsymbol{X}_j \boldsymbol{P}_j \quad \text{for } j = 1, \ldots, k \tag{5.19}$$

(see Chapter 3) where $\boldsymbol{T}_j$ is the score matrix and $\boldsymbol{P}_j$ is the loadings matrix for group j. Since each group is modeled by an individual number of PCs, say a_j, both score and loadings matrix have a_j columns. The choice of the dimensions a_j will be explained below.

Since the goal of SIMCA is to classify a new object $\boldsymbol{x}$, a measure for the closeness of the object to the groups needs to be defined. For this purpose, several proposals have been made in the literature. They are based on the ORTHOGONAL DISTANCE, which represents the Euclidean distance of an object to the PCA space (see Section 3.7.3). First we need to compute the score vector $\boldsymbol{t}_j$ of $\boldsymbol{x}$ in the jth PCA space, and using Equation 5.19 and the group center $\bar{\boldsymbol{x}}_j$ we obtain

$$\boldsymbol{t}_j = \boldsymbol{P}_j^{\mathrm{T}} (\boldsymbol{x} - \bar{\boldsymbol{x}}_j) \tag{5.20}$$

Then the scores have to be back-transformed to the original space, yielding an estimation $\hat{\boldsymbol{x}}_j$ of $\boldsymbol{x}$ that is obtained using the jth PCA scores:

$$\hat{\boldsymbol{x}}_j = \boldsymbol{P}_j \boldsymbol{t}_j + \bar{\boldsymbol{x}}_j \tag{5.21}$$

Finally, the orthogonal distance OD_j of the new object $\boldsymbol{x}$ to the jth PCA space is given by

$$\mathrm{OD}_j = \|\boldsymbol{x} - \hat{\boldsymbol{x}}_j\| \quad \text{for } j = 1, \ldots, k \tag{5.22}$$

The task now is to find an appropriate classification rule for a new object $\boldsymbol{x}$. The orthogonal distances cannot directly be used for this purpose because also the spread of the data groups has to be considered. Thus, one can use an F-test (see Section 1.6.5) based on the orthogonal distances. However, since this approach does not optimally use the information of the PCA spaces of the groups, a geometrically based classification rule that incorporates the distance to the boundary of the disjoint PCA spaces was defined (Albano et al. 1978; Wold 1976). Recently, another classification rule was introduced that turned out to be more effective (Vanden Branden and Hubert 2005). While the previous rule uses distances to rectangular PCA spaces (see Figure 5.10), the new approach considers elliptical spaces where the shape of the ellipses corresponds to the covariance structure of the groups. This distance measure, the SCORE DISTANCE, was already used in Section 3.7.3, and it measures the Mahalanobis distance of an object $\boldsymbol{x}$ to the center of the PCA space (see Figure 3.15). The score distance SD_j of a new object $\boldsymbol{x}$ to the jth group is given by

$$\mathrm{SD}_j = \left[\sum_{l=1}^{a_j} \frac{\boldsymbol{t}_{jl}^2}{\boldsymbol{v}_{jl}} \right]^{1/2} \tag{5.23}$$

where

t_{jl} are the components of the scores $\boldsymbol{t}_j = \left(t_{j1}, \ldots, t_{ja_j}\right)^{\mathrm{T}}$ from Equation 5.20
v_{jl} are the largest eigenvalues for $l = 1, \ldots, a_j$ in the jth group

Using the cutoff values for score distance $c_{\mathrm{SD}_j} = \sqrt{\chi^2_{a_j,0.975}}$ and orthogonal distance $c_{\mathrm{OD}_j} = \left[\mathrm{median}\left(\mathrm{OD}_j^{2/3}\right) + \mathrm{MAD}\left(\mathrm{OD}_j^{2/3}\right) z_{0.975}\right]^{3/2}$ (see Section 3.7.3), the distance measures can be standardized and combined to result in a score value d_j^{D}

$$d_j^{\mathrm{D}}(\boldsymbol{x}) = \gamma \left(\frac{\mathrm{OD}_j}{c_{\mathrm{OD}_j}} \right) + (1 - \gamma) \left(\frac{\mathrm{SD}_j}{c_{\mathrm{SD}_j}} \right) \quad \text{for } j = 1, \ldots, k \tag{5.24}$$

Here, γ is a tuning parameter taking values in the interval [0, 1] that adjusts the importance of the score and orthogonal distance for the classification. One can use CV to find the optimum value of γ. Equation 5.24 results in a score value of an object $\boldsymbol{x}$ for each group. A "soft" classification rule defines that an object $\boldsymbol{x}$ is assigned to

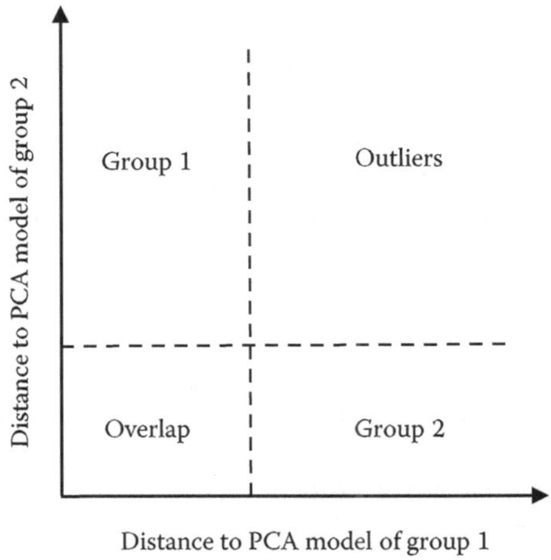

FIGURE 5.11 The Coomans plot uses the distances of the objects to the PCA models of two groups. It visualizes whether objects belong to one of the groups, to both, or to none.

all groups for which the score value is smaller than 1, thereby covering the case of overlapping group models. Moreover, objects that do not fit to any of the groups (the standardized score and/or orthogonal distances are larger than 1 for all groups) are not assigned and treated as outliers. Reliable outlier identification, however, is only possible in the context of robust estimation (see Section 3.7.3). Therefore, a robust SIMCA method, based on robust estimation of the PCA spaces, has been suggested (Vanden Branden and Hubert 2005; Verboven and Hubert 2005). The COOMANS PLOT (Vandeginste et al. 1998) is a scatter plot with a point for each object using as coordinates the (orthogonal) distances to the PCA models of two groups (Figure 5.11).

We come back to the problem of selecting the optimum dimensions $a_1, \ldots, a_k$ of the PCA models. This can be done with an appropriate evaluation technique like CV, and the goal is to minimize the total probability of misclassification. The latter can be obtained from the evaluation set, by computing the percentage of misclassified objects in each group, multiplied by the relative group size, and summarized over all groups.

5.3.2 GAUSSIAN MIXTURE MODELS

The idea of mixture models is to consider the overall density function as a sum of the density functions of the single groups. Usually, the group density functions are modeled by Gaussian densities $\varphi(\boldsymbol{x}; \boldsymbol{\mu}_j, \boldsymbol{\Sigma}_j)$, with mean $\boldsymbol{\mu}_j$ and covariance matrix $\boldsymbol{\Sigma}_j$, leading to a model

$$f(\boldsymbol{x}) = p_1\varphi(\boldsymbol{x}; \boldsymbol{\mu}_1, \boldsymbol{\Sigma}_1) + \cdots + p_k\varphi(\boldsymbol{x}; \boldsymbol{\mu}_k, \boldsymbol{\Sigma}_k) \qquad (5.25)$$

for the overall density function f at positions $\boldsymbol{x}$ in the multivariate variable space. The mixing proportions p_j sum up to 1. The task is to estimate all parameters $\boldsymbol{\mu}_j$, $\boldsymbol{\Sigma}_j$, and p_j

for $j=1,\ldots, k$ on the basis of the available data. This will then allow a group assignment of new objects.

There is an important special case, namely the restriction of the covariance matrices $\boldsymbol{\Sigma}_j$ to be spherical, i.e., $\boldsymbol{\Sigma}_j=\sigma_j \boldsymbol{I}$. Thus, the shape of the classes is spherically symmetric (in three dimensions, these are balls) with the individual sizes controlled by the parameters σ_j. The method should be applied only if the shapes of the groups follow these restrictions. The resulting joint density then has the form

$$f(\boldsymbol{x}) = p_1\varphi(\boldsymbol{x}; \boldsymbol{\mu}_1, \sigma_1\boldsymbol{I}) + \cdots + p_k\varphi(\boldsymbol{x}; \boldsymbol{\mu}_k, \sigma_k\boldsymbol{I}) \tag{5.26}$$

which is the form of the RADIAL BASIS FUNCTIONS (RBF) discussed in Section 4.8.3.2, see Equation 4.95. This again shows again the close relation of regression and classification problems. The main difference here is that the number of basis functions is given by the number of classes, k. The parameter estimation for the model (Equation 5.26) can be done as described in Section 4.8.3 in the context of RBF, and the group assignment is based on a distance measure of a new object to the groups which incorporates the group densities.

The parameter estimation for the mixture model (Equation 5.25) is based on maximum likelihood estimation. The likelihood function L is defined as the product of the densities for the objects, i.e.,

$$L = f(\boldsymbol{x}_1)\, f(\boldsymbol{x}_2)\cdots f(\boldsymbol{x}_n) \tag{5.27}$$

with the density function defined in Equation 5.25. The parameters are estimated by maximizing the logarithm of the likelihood function,

$$\log L = \sum_{i=1}^{n}\left[p_1\varphi(\boldsymbol{x}_i; \boldsymbol{\mu}_1, \boldsymbol{\Sigma}_1) + \cdots + p_k\varphi(\boldsymbol{x}_i; \boldsymbol{\mu}_k, \boldsymbol{\Sigma}_k)\right] \tag{5.28}$$

A direct maximization, however, is difficult, and therefore the problem is split into two parts that are alternated until convergence:

- E-step: In this EXPECTATION STEP, each object is assigned a weight for each cluster. The weights are based on the likelihood $p_j\varphi(\boldsymbol{x}_i;\ \boldsymbol{\mu}_j,\ \boldsymbol{\Sigma}_j)$ of each object i for group j, resulting in a number in the interval [0, 1]. An object that is located close to the center of a group will receive a weight close to 1 for this group and weights close to 0 for all other groups. If the object is further away from the center, the weights are adjusted accordingly.
- M-step: In this MAXIMIZATION STEP, the parameters $\boldsymbol{\mu}_j$, $\boldsymbol{\Sigma}_j$, and p_j are estimated. For $\boldsymbol{\mu}_j$ and $\boldsymbol{\Sigma}_j$ weighted means and covariances are computed, with weights derived in the E-step. p_j can be estimated by summing up all weights for group j and dividing by the sample size n.

This algorithm is well-known under the name EXPECTATION MAXIMIZATION ALGORITHM (EM) (McLachlan and Pee 2000). Since the parameters $\boldsymbol{\mu}_j$, $\boldsymbol{\Sigma}_j$, and p_j are already needed in the E-step for computing the likelihoods, these parameters have to be

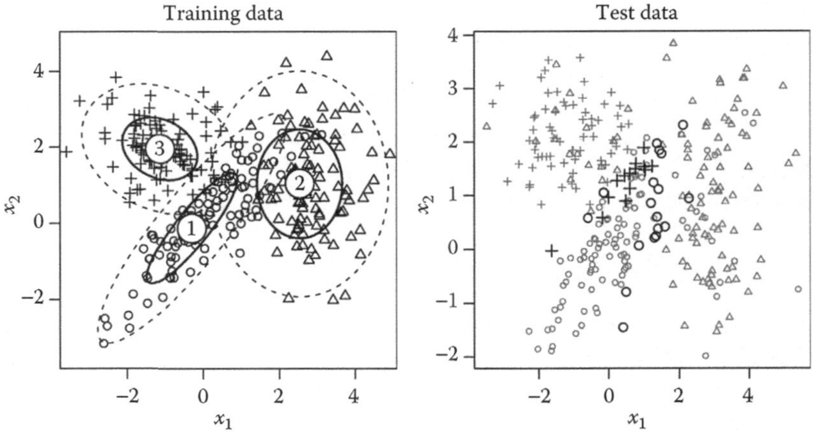

FIGURE 5.12 Training and test data with three groups, generated according to Figure 5.3 (right). The symbols refer to the true group membership of the generated data. A Gaussian mixture model is fit to the training data (left), where the information of group membership is provided. The model is applied to an independent test data set (right). The dark symbols are misclassified objects in the test data; they are located in the overlapping region of the groups.

initialized. This can be done in the classification setup because the group memberships are known for the training data, and thus weights of 0 or 1 can be used.

In the following example, Gaussian mixture models will be fit for a data set generated according to the right plot in Figure 5.3 (three slightly overlapping groups; compare ℝ code given in Section 5.2.1.3). We sample data from the three groups with specified means and covariances. Each group of the training data consists of 100 objects. An independent test data set is generated with 100 objects for each group. The model is built for the training data, and evaluated for the test data. Training and test data are shown in Figure 5.12, together with the classification results from Gaussian mixture model fitting.

```
ℝ:  library(flexmix)          # includes functions for mixture
                              # modeling
    resMix <- flexmix(xtrain~1,k=3,model=
              FLXMCmvnorm(diag=FALSE),cluster=classMatrix)
        # "model" is defined as Gaussian mixture model
        # "cluster" is assigned a binary membership matrix
    plotEll(resMix,xtrain)  # generates Figure 5.12 (left)
    resclass <- cluster(resMix,newdata=xtest)
        # computes class memberships for the test data
```

5.3.3 *k*-NN Classification

In contrast to the previous methods discussed in this section, *k*-nearest neighbor (*k*-NN) classification methods require no model to be fit because they can be considered as memory based. These methods work in a local neighborhood around a considered test data point to be classified. The neighborhood is usually determined by the Euclidean distance, and the closest k objects are used for estimation of the group membership of a new object. For binary variables (for instance molecular descriptors for chemical structures), the Tanimoto index (Section 6.2) is often used as distance/similarity measure. If the different variables are measured in different units, it is advisable that the data are first autoscaled such that each variable has mean 0 and variance 1. Of course, the choice of the neighborhood size k determines the quality of the results, and an optimal choice of k is usually made via CV by testing k-values from 1 to k_{max}.

For k-NN classification, the task is to predict the class membership of a new object $\boldsymbol{x}$. Using for instance the Euclidean distance measure, the k-NNs (of the training data) to $\boldsymbol{x}$ are determined. The neighbors are found by calculating the distances between the new object and all objects in the training set. The closest k objects are the k nearest neighbors to $\boldsymbol{x}$, and they will be denoted by $\boldsymbol{x}_{(1)}, \ldots, \boldsymbol{x}_{(k)}$. The predicted class membership $\hat{y}(\boldsymbol{x})$ of the new object $\boldsymbol{x}$ is obtained from the known class memberships $y(\boldsymbol{x}_{(1)}), \ldots, y(\boldsymbol{x}_{(k)})$ of the k nearest neighbors, and can be taken as the class that occurs most frequently among the k neighbors. Thus, the prediction corresponds to a "majority vote" among the neighbors, but also decision schemes considering the distances between the neighbors and the unknown have been suggested. The decision boundary between different groups can thus be very rough, and it strongly depends on the parameter k.

For $k=1$ (1-NN), a new object would always get the same class membership as its next neighbor. Thus, for small values of k, it is easily possible that classes do no longer form connected regions in the data space, but they can consist of isolated clouds. The classification of new objects can thus be poor if k is chosen too small or too large. In the former case, we are concerned with overfitting, and in the latter case with underfitting.

k-NN classification has some advantages: it neither requires linearly separable groups nor compact clusters for the groups; it can be easily applied to multiclass problems and is conceptually very simple; thus k-NN is used as a reference method ("the best description of the data is the data themselves"). No training of classifiers is required; however, the whole training set (or a representative selection of it) is required for the classification of new objects; and calculation of the many distances may be time consuming for large training sets and many variables.

Figure 5.13 shows a two-dimensional example with two groups. The data points shown are the training data and the symbols correspond to their group memberships. The lines represent the decision boundaries for $k=1$ (left picture) and $k=15$ (right picture). Any new point in the plane would be classified according to these boundaries. The boundaries are rough, and in the case $k=1$ several isolated regions for the two groups occur.

The above example demonstrates that the choice of k is crucial. As mentioned above, k should be selected such that the smallest misclassification rate for the test data occurs. If no test data are available, an appropriate resampling procedure (CV or bootstrap) has to be used. Figure 5.14 shows for the example with three overlapping groups, used above in Figure 5.12, how k can be selected. Since we have independent training and test data available, and since their group membership is known, we

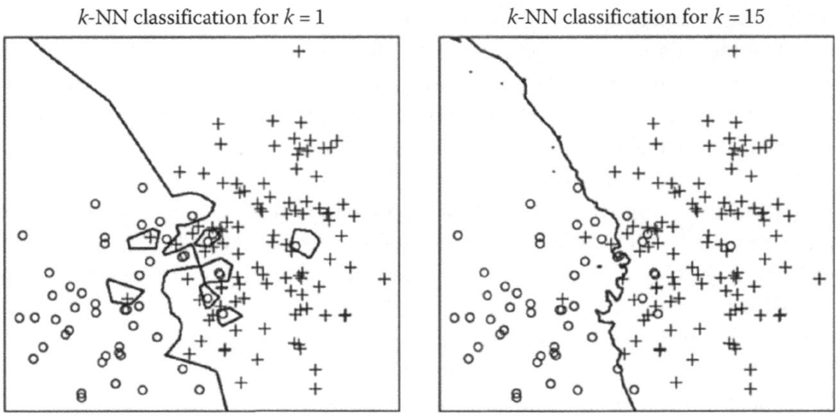

FIGURE 5.13 k-NN classification for two groups of two-dimensional data. The training data are shown with the symbol corresponding to the group membership. Any new data point would be classified according to the presented decision boundaries, where $k = 1$ in the left plot, and $k = 15$ in the right plot has been used.

can compute for different choices of k the misclassification rate for the test data. The result is shown in Figure 5.14 (left), and an approximate value of $k = 25$ is obtained to be optimal for the test data set. The plot on the right hand side shows the resulting misclassifications for this number of neighbors. In the used $\mathbb{R}$ function, ties during classification are broken at random. If there are ties for the kth neighbor, all equidistant objects are included in the vote.

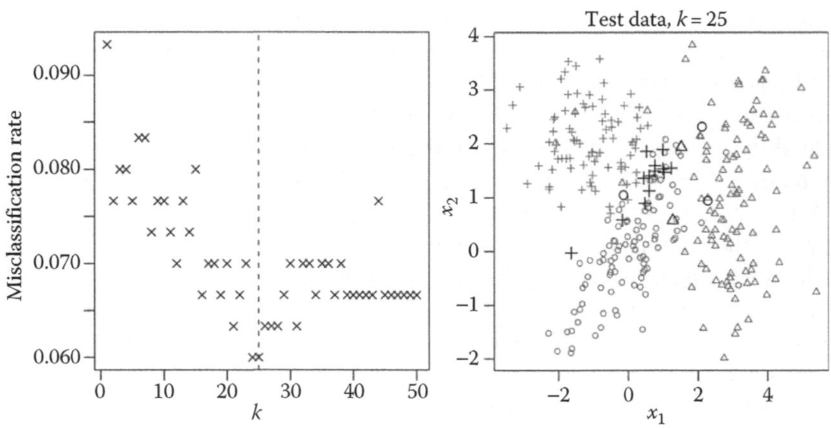

FIGURE 5.14 k-NN classification for the training and test data used in Figure 5.12. The left plot shows the misclassification rate for the test data with varying value of k for k-NN classification, and the right plot presents the result for $k = 25$. The misclassified objects are shown by dark symbols.

```
R: library(class)  # includes k-NN classification
   resknn <- knn(xtrain,xtest,grp,k=25)
          # k-NN classification for training data "xtrain",
          # test data "xtest", true classification "grp" for
          # training data, and desired value for k.
          # The results are the predicted group memberships
          # for the test data.
```

An interesting version of *k*-NN classification is the use of POTENTIAL FUNCTIONS. For a two group classification problem, one can assume at each data point of group 1 a positive electrical charge, and at each data point of group 2 a negative electrical charge. The potential field z of an electrical charge decreases with increasing distance d, for instance by the potential function

$$z(d) = \frac{1}{1 + qd^2} \tag{5.29}$$

with q defining the width of the potential function (Forina et al. 1991; Meisel 1972; Vandeginste et al. 1998). The width parameter has to be optimized; other—not physics based—equations for potential functions have been suggested. Superposition of the potentials from all objects belonging to the same group gives the cumulative potential of that group. Usually the group potential is normalized by the number of objects in the group, and it can be considered as an estimated probability density for that group. For the two-group problem, the decision boundary is given by all points with zero potential. For more than two classes, a formal extension to more than two "types of charges" is easily possible. Because the potential of a charge becomes almost zero at a certain distance, only neighbors up to this distance need to be considered for classification of an unknown.

SPECTRAL SIMILARITY SEARCH is a routine method for identification of compounds, and is similar to *k*-NN classification. For molecular spectra (IR, MS, NMR), more complicated, problem-specific similarity measures are used than criteria based on the Euclidean distance (Davies 2003; Robien 2003; Thiele and Salzer 2003). If the unknown is contained in the used data base (spectral library), identification is often possible; for compounds not present in the data base, *k*-NN classification may give hints to which compound classes the unknown belongs.

The *k*-NN approach can also be applied for the prediction of a continuous property y as an alternative to regression methods as described in Chapter 4. The property of an unknown can be simply computed as the average of the properties of k neighbors, or a LOCAL REGRESSION MODEL can be computed from them.

5.4 CLASSIFICATION TREES

In Section 4.8.3.3, we already mentioned REGRESSION TREES which are very similar to classification trees. The main difference is that the response y-variable now represents the class membership of the training data. The task is again to partition the space of the x-variables along the coordinates into regions $R_1, \ldots, R_r$, such that a measure for misclassification is as small as possible. The resulting trees are called CLASSIFICATION TREES. The abbreviation CART is frequently used as a synonym for CLASSIFICATION AND REGRESSION TREES (Breiman et al. 1984).

Suppose we have given training data $\boldsymbol{x}_1, \ldots, \boldsymbol{x}_n$ and their group memberships $y_1, \ldots, y_n$, where y_i takes a value $1, 2, \ldots, k$ for k groups. Suppose that n_l objects fall into region R_l, where $n_1 + \cdots + n_r = n$. Furthermore, let $I(y_i = j)$ be the index function with the result 1 if $y_i = j$ and 0 otherwise. In other words, the index function gives a positive count if the group membership of object $\boldsymbol{x}_i$ is j, and zero otherwise. We consider all objects $\boldsymbol{x}_i$ falling into region R_l, and use their group memberships y_i. Then we can compute the relative frequency p_{lj} of the jth group in the lth region by

$$p_{lj} = \frac{1}{n_l} \sum_{\boldsymbol{x}_i \in R_l} I(y_i = j) \tag{5.30}$$

The sum in Equation 5.30 counts how many objects from region R_l are from group j. Dividing by the number of objects in region R_l results in the proportion of objects from the jth group in the lth region. By varying j, we can compute the relative frequencies of each group in region R_l. The objects in region R_l are then classified to the majority class, i.e., to the group $j(l)$ with the largest relative frequency. The goal is to keep a measure for misclassification as small as possible. In practice, different measures $Q_l(T)$ for quantifying the misclassification in region R_l of a tree T are used:

- MISCLASSIFICATION ERROR: $\frac{1}{n_l} \sum_{\boldsymbol{x}_i \in R_l} I(y_i \neq j(l)) = 1 - p_{l,j(l)}$

  This corresponds to the fraction of objects in the lth region that does not belong to the majority class.

- GINI INDEX: $\sum_{j=1}^{k} p_{lj}(1 - p_{lj})$

  The Gini index is the sum of products of the relative frequencies of one class with the relative frequencies of all other classes.

- CROSS ENTROPY OR DEVIANCE: $-\sum_{j=1}^{k} p_{lj} \log(p_{lj})$

  The idea is similar to the Gini index.

The Gini index and the cross entropy measure are differentiable which is of advantage for optimization. Moreover, the Gini index and the deviance are more sensitive to changes in the relative frequencies than the misclassification error. Which criterion is better depends on the data set, some authors prefer Gini which favors a split into a small pure region and a large impure one.

Figure 5.15 illustrates the use of the different measures for a simple example in two dimensions with three groups. By dividing the plane along the horizontal axis at a split point s, we obtain the two regions R_1 and R_2. For each region, the measures are

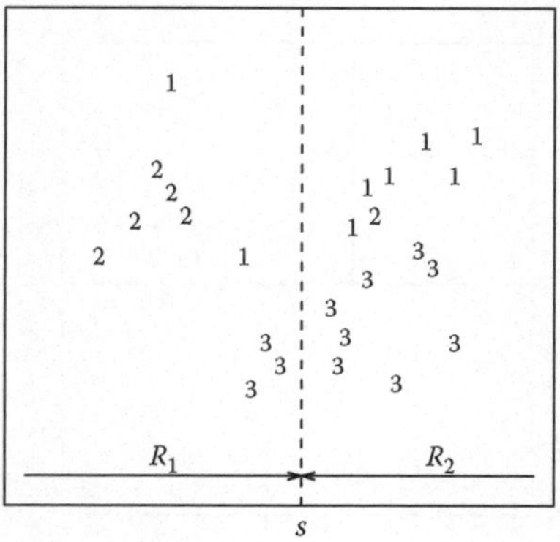

FIGURE 5.15 Different misclassification measures are used for two-dimensional data with three groups. They all depend on the choice of the split variable (here the horizontal axis) and the split point (here the point s). The task is to find the variable and the split point which minimize a chosen error measure.

computed. The goal is then to choose the split point s such that the measure for misclassification is as small as possible. However, not only the split point has to be selected, but also the split variable for which the split is actually performed.

Both tasks, finding the best split variable and finding the optimal split point in order to minimize the error measure can be done very quickly by scanning through all x-variables. The variable with the best split point gives the first branch in the decision tree. Once the optimal split has been performed, one has to repeat the procedure in each of the resulting two regions. Thus, the classification grows tree-like which motivates the name of the method. As a result, the space of the x-variables is split step by step into smaller regions, and the error measure in each region also becomes smaller. However, it would not make sense to continue until each object falls into its own region. This would just result in an error measure of zero for the training data, but in an overfit for the test data. Thus, there is an optimal size of the tree where the classification error for test data becomes a minimum. This tree size can be controlled by a complexity parameter (CP), and the optimal complexity is derived for instance by a CV scheme.

In more detail, let $|T|$ denote the size of a tree T, i.e., $|T|$ is the number of regions defined by the tree. Using a criterion $Q_l(T)$ from above (misclassification error, Gini index, or deviance) for quantifying the misclassification in region R_l of the tree T, a complexity criterion can be defined by

$$CP_\alpha(T) = \sum_{l=1}^{|T|} n_l Q_l(T) + \alpha |T| \tag{5.31}$$

which has to be minimized. The parameter $\alpha \geq 0$ controls the size of the tree. A choice of $\alpha = 0$ will result in the full tree where the measure Q_l of misclassification will be a minimum. Increasing the value α penalizes larger trees, and thus this parameter regulates the compromise between misclassification and tree size.

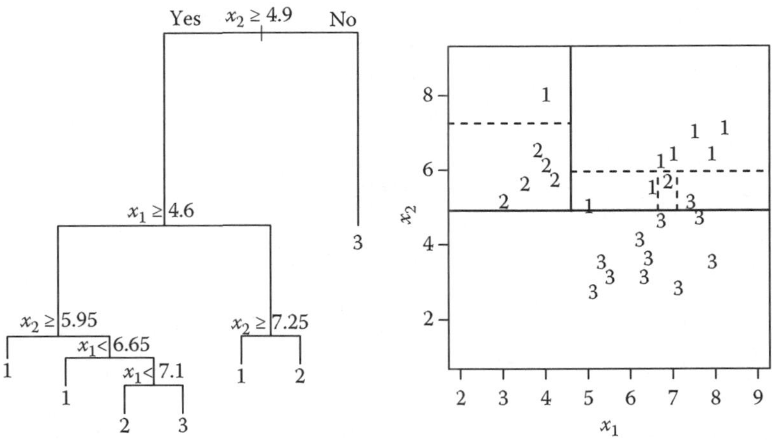

FIGURE 5.16 Full classification tree for the data example in Figure 5.15 in the left panel, and the resulting classification lines in the right panel. The dashed lines will not be used when the tree is pruned to its optimal complexity.

As a simple example, the data from Figure 5.15 are used for building a complete classification tree which gives a perfect separation of all groups. The full tree is shown in Figure 5.16 (left), and the resulting separation lines are shown in Figure 5.16 (right). Some of the separation lines are dashed and they will be eliminated later on when the tree is pruned to its optimal complexity.

Since the full tree is not suitable for prediction, 10-fold CV was performed to obtain the optimal size of the tree. Figure 5.17 (left) shows the resulting errors (MSE_{CV}, mean squared error from CV) using Equation 5.31 with the Gini index, together with lines that represent the standard errors. The horizontal axis shows the values of CP (see Equation 5.31), which serves as a CP. With the one-standard-error

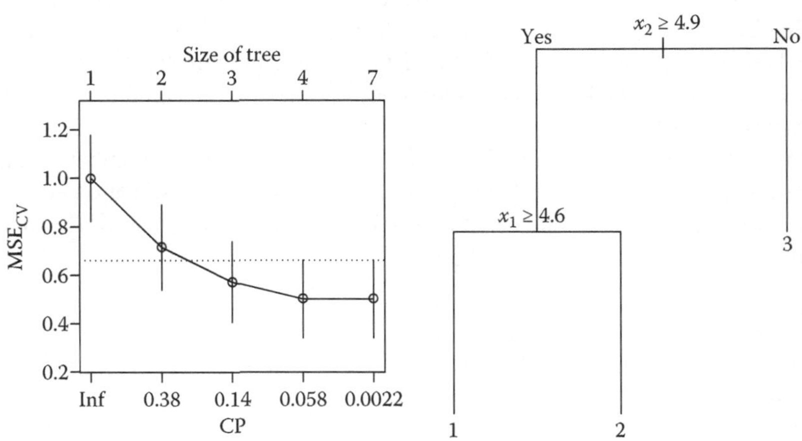

FIGURE 5.17 Optimal complexity of the tree shown in Figure 5.16 is obtained by CV, and a tree of size 3 (CP 0.14) will be used (left). The resulting tree is shown in the right panel.

rule (see Section 4.2.2), the optimal size of the tree is 3, with a CP of 0.14. Thus, the full tree is pruned to this complexity, resulting in Figure 5.17 (right). The ℝ code to this example is below.

```
ℝ:  library(rpart)  # classification trees
    tree1 <- rpart(grp~.,data=dat,method="class")
          # use all remaining variables in "dat" for the tree
    plot(tree1)          # plots the tree, see Figure 5.16
    text(tree1)          # adds text labels to the plot
    plotcp(tree1)  # plot results of cross validation for
                 # finding the tree complexity, see Fig. 5.17
    tree2 <- prune(tree1,cp=0.14)       # pruning of the above
                       # tree using the optimal cp
                       # plot, text can be applied to tree2
```

As a summary, classification trees are a simple but powerful technique for group separation.

- They are not based on distributional assumptions or other strict data requirements.
- They work also for categorical data, and they can be used in the two- and multiple-group case.
- They use binary partitions along the original data coordinates in a sequential manner, leading to an easy interpretation.
- They use CV to find the optimal tree size.

The main limitation of classification trees is their instability. Small changes in the data can result in a completely different tree. This is due to the hierarchical structure of the binary decisions, where a slightly different split on top can cause completely different splits subsequently. A procedure called BAGGING can reduce this instability by averaging many trees (Hastie et al. 2001).

5.5 ARTIFICIAL NEURAL NETWORKS

The basic concepts of ANN were already described in Section 4.8.3.4 and used there for regression problems. They can be used also for the classification case, and the structure and concepts remain the same. While in the regression case the y-variable was the regressor response variable, it will be a binary target variable representing two groups in the classification case. For a more general situation with $k > 2$ classes, we need to use k target variables $y_1, y_2, \ldots, y_k$ which are again binary variables that describe the group memberships. More specifically, suppose we have given a data matrix $\boldsymbol{X}$ with n rows containing the training data and m columns for the variables. Then the target variables form a binary matrix $\boldsymbol{Y}$ with n rows and k columns. If an object $\boldsymbol{x}_i$ belongs to group j, then the ith row of the target matrix has an entry of 1 at position j and zeros on all other positions. The setup of the data is the same as used in multicategory classification with OLS or with PLS discriminant analysis (see Section 5.2.2).

The outcome from the neural network is a prediction of the class membership for each object (either training objects or test objects). It is a matrix $\hat{\boldsymbol{Y}}$ with the same dimensions as $\boldsymbol{Y}$, its elements $\hat{y}_{ij}$ are in the interval [0, 1], and they can be seen as somewhat like a probability for the assignment of the ith object $\boldsymbol{x}_i$ to the jth group. The final assignment of $\boldsymbol{x}_i$ is made to the group with the largest value $\hat{y}_{i1}, \hat{y}_{i2}, \ldots, \hat{y}_{ik}$, but also other decision schemes can be applied that for instance include the possibility of "no assignment" if the $\hat{y}$-values do not clearly indicate one group.

While in the regression case the optimization criterion was based on residual sum of squares, this would not be meaningful in the classification case. A usual error function in the context of neural networks is the CROSS ENTROPY or DEVIANCE, defined as

$$-\sum_{i=1}^{n}\sum_{j=1}^{k}\hat{y}_{ij}\log\hat{y}_{ij} \rightarrow \min \tag{5.32}$$

(see also Section 5.4). Neural networks are sensitive to overfitting, and therefore often a regularization is introduced which is called WEIGHT DECAY. The idea is similar to Ridge or Lasso regression (Section 4.8.2) by adding a penalty term to the criterion. Thus the modified criterion has the form

$$-\sum_{i=1}^{n}\sum_{j=1}^{k}\hat{y}_{ij}\log\hat{y}_{ij} + \lambda\sum(\text{parameters})^2 \rightarrow \min \tag{5.33}$$

where "parameters" stands for the values of all parameters that are used within the neural network. Thus the additional term brings the size of all parameters into consideration. The size of the term $\lambda \geq 0$ determines how much emphasis is on the constraint of shrinking the parameters. Using no weight decay ($\lambda = 0$) results in criterion (Equation 5.32) for optimization.

For fitting a neural network, it is often recommended to optimize the values of λ via CV. An important issue for the number of parameters is the choice of the number of hidden units, i.e., the number of variables that are used in the hidden layer (see Section 4.8.3). Typically, 5–100 hidden units are used, with the number increasing with the number of training data variables. We will demonstrate in a simple example how the results change for different numbers of hidden units and different values of λ.

In Section 5.3.3, we used a two-dimensional example with two overlapping groups for k-NN classification (see Figure 5.13). The same example is used here in Figure 5.18 for a neural network classification where the number of hidden units and the weight decay are varied. The resulting classification boundaries for the two groups are shown, and any new data point in the plane would be classified according to these boundaries. In the top row of the figure, only five hidden units were used, leading to relatively smooth classification boundaries. The bottom row shows the results for 30 hidden units, and probably too complex (overfitted) boundaries are obtained. A weight decay of zero (left column) has the effect of edges and non-smooth boundaries, while increasing the weight decay results in smoother boundaries. As mentioned above, the optimal parameters can be obtained via CV.

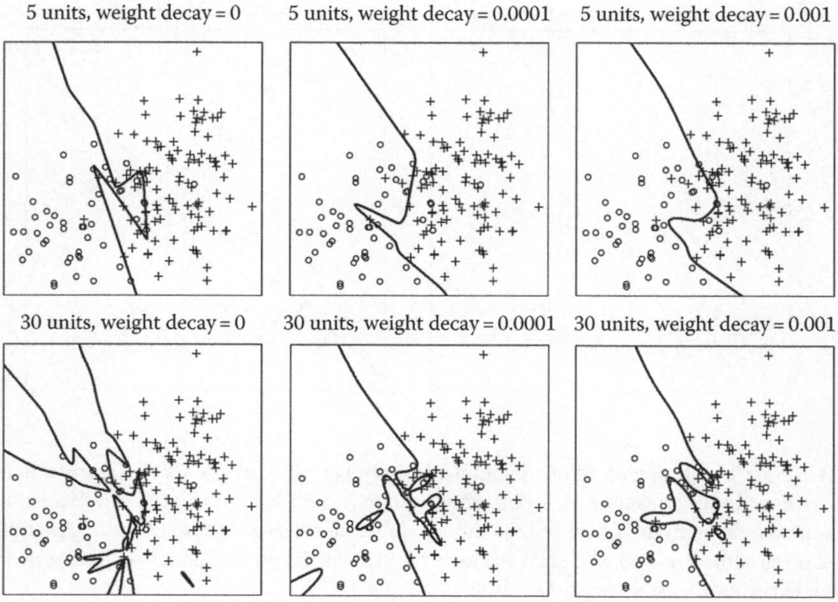

FIGURE 5.18 Classification with neural network for two groups of two-dimensional data. The training data are shown with the symbol corresponding to the group membership. Any new data point would be classified according to the presented decision boundaries. The number of hidden units and the weight decay were varied.

```
ℝ: library(nnet)  # neural networks
   resNN <- nnet(Xtrain,grp,size=5,decay=0)
   # fits neural network with size=5 hidden units and
   # weight decay 0
   predict(resNN,Xtest)  # prediction of the class
```

5.6 SUPPORT VECTOR MACHINE

The term support vector machine (SVM) stands for a statistical technology that can be used for both classification and regression (Christianini and Shawe-Taylor 2000; Vapnik 1995). In the context of classification they are producing linear boundaries between object groups in a transformed space of the x-variables which is usually of much higher dimension than the original x-space. The idea of the transformed higher-dimensional space is to make groups linearly separable. Furthermore, in the transformed space, the class boundaries are constructed in order to maximize the margin between the groups (MAXIMUM MARGIN CLASSIFIER). The back-transformed boundaries are nonlinear. SVMs find increasing interest in chemistry and biology (Brereton 2007; Ivanciuc 2007; Thissen et al. 2004; Xu et al. 2006). A comparison of SVMs with other classification and regression methods found out that they show mostly good performances, although other methods proved to be very competitive (Meyer et al. 2003).

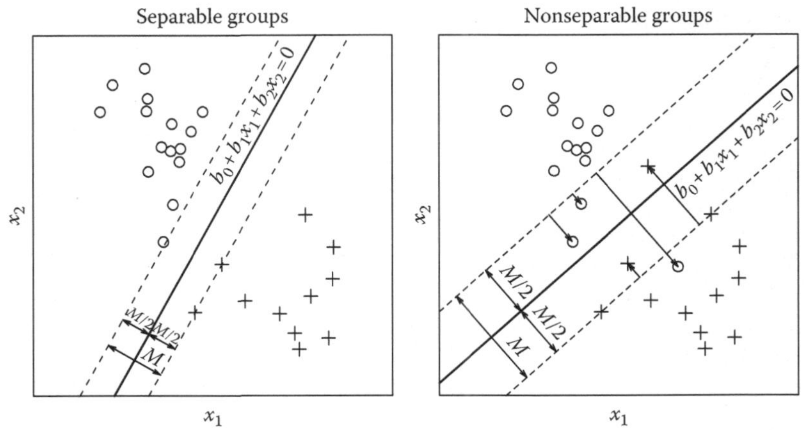

FIGURE 5.19 Illustration of obtaining the decision boundary by SVMs for classification of two groups. In the separable case (left) the separating solid line is between the dashed lines maximizing the margin M between the groups. In the nonseparable case (right) the margin is maximized subject to a constant that accounts for the total distance of points on the wrong side of the hyperplanes with margin M.

In the following, we will illustrate the concepts of SVMs for the two-group case. First, we explain the situation where the groups do not overlap (in the transformed space), and then we outline the procedure for overlapping groups (not linearly separable groups). In our example, the transformed space is only two-dimensional, allowing for a visualization of the concepts. Thus Figure 5.19 shows two data groups that are separable (left) and two groups with a certain overlap (right). Since the group separation is done by a linear function in the transformed space, we are looking for lines in Figure 5.19 that are separating the groups. In the three-dimensional case, the lines have to be generalized to planes, and in higher dimension one is looking for hyperplanes that separate the groups. A line in two dimensions is defined by all pairs of points (x_1, x_2) for which $b_1x_1 + b_2x_2$ is a constant, with fixed coefficients b_1 and b_2. This can be written as $b_0 + b_1x_1 + b_2x_2 = 0$, where the coefficient b_0 is the negative constant. Depending on the values of the coefficients, there are infinitely many possible lines, and two special (solid) lines are shown in Figure 5.19 left and right. The line in the left plot separates both point clouds completely, whereas in the right plot it is not possible to find a line that allows a complete separation.

Extended to a higher-dimensional case, say, to the r-dimensional case, a HYPERPLANE (DECISION PLANE) is defined by the condition

$$b_0 + \boldsymbol{b}^{\mathrm{T}}\boldsymbol{x} = 0 \qquad (5.34)$$

with the coefficients b_0 and $\boldsymbol{b}^{\mathrm{T}} = (b_1, b_2, \ldots, b_r)$ and the vector $\boldsymbol{x} = (x_1, x_2, \ldots, x_r)$ for the r variables. For a unique definition, we assume that $\boldsymbol{b}$ is a unit vector, i.e., $\boldsymbol{b}^{\mathrm{T}}\boldsymbol{b} = 1$.

Let us consider now a training data set of n objects in the r-dimensional space, i.e., the vectors (points) $\boldsymbol{x}_1, \ldots, \boldsymbol{x}_n$. In the two-group case, we have the information of the group membership for each object, i.e., given by the values y_i which are for example either −1 (first group) or +1 (second group), for $i = 1, \ldots, n$. Using Equation 5.34, the hyperplane can be applied for classification as follows:

$$\text{If } b_0 + \boldsymbol{b}^{\mathrm{T}}\boldsymbol{x}_i < 0, \text{ assign } \boldsymbol{x}_i \text{ to the first group, otherwise to the second group.} \quad (5.35)$$

In fact, $b_0 + \boldsymbol{b}^{\mathrm{T}}\boldsymbol{x}_i$ gives the signed distance of an object $\boldsymbol{x}_i$ to the decision plane, and for classification only the sign is primarily important (although the distance from the decision plane may be used to measure the certainty of classification). If the two groups are linearly separable, one can find a hyperplane which gives a perfect group separation as follows:

$$y_i(b_0 + \boldsymbol{b}^{\mathrm{T}}\boldsymbol{x}_i) > 0 \quad \text{for } i = 1, \ldots, n \tag{5.36}$$

Note that for correct classifications the sign of the true group membership y_i is always the same as the "predicted" sign. An optimum position of the decision plane is assumed if the margin between the groups is maximal. The margin is given by two parallel hyperplanes, and it is assumed that a maximum margin for the training set is also a good choice for test set objects. In the two-dimensional case, this optimum hyperplane is uniquely defined by three data points, two belonging to one group and the third to the other group, see Figure 5.19 (left). Analogously, in higher dimension, one has to consider more data points for a unique definition of the parallel hyperplanes with maximum margin.

The data points that define the position of the separating hyperplanes can be seen as vectors in the transformed space, and they are called SUPPORT VECTORS. The resulting largest margin is denoted by M in Figure 5.19 (left), and it is visualized by the two dashed lines. The separating line is at a distance of $M/2$, exactly between the dashed lines, and it is shown by the solid line. This is the line, or in the more general case, the hyperplane defined in Equation 5.34 we were looking for, that creates the biggest margin between both groups. The optimization problem can now be formulated as

$$\begin{aligned} &M \rightarrow \max \text{ for coefficients } b_0 \text{ and } \boldsymbol{b} \text{ with } \boldsymbol{b}^{\mathrm{T}}\boldsymbol{b} = 1 \\ &\qquad \text{subject to } y_i(b_0 + \boldsymbol{b}^{\mathrm{T}}\boldsymbol{x}_i) \geq M/2 \quad \text{for } i = 1, \ldots, n \end{aligned} \tag{5.37}$$

In the linearly nonseparable case (Figure 5.19, right) where the groups are overlapping in the transformed space, Equation 5.37 for finding the optimal hyperplane has to be modified. We can again maximize the margin but have to allow for points to be on the wrong side of the resulting hyperplane. This concept is realized by introducing so-called SLACK VARIABLES ξ_i, for $i = 1, \ldots, n$. They correspond to distances from the hyperplanes with margin M, and are 0 for objects that are on the correct side of the hyperplanes, and positive otherwise. Figure 5.19 (right) shows the values of the slack variables with arrows from the hyperplanes with margin M. In a criterion, we constrain the sum of these distances, and thus the criterion for linearly nonseparable groups is

$$M \rightarrow \max \text{ for coefficients } b_0 \text{ and } \boldsymbol{b} \text{ with } \boldsymbol{b}^{\mathrm{T}}\boldsymbol{b} = 1$$

$$\text{subject to } y_i(b_0 + \boldsymbol{b}^{\mathrm{T}}\boldsymbol{x}_i) \geq (M/2)(1 - \xi_i)$$
$$\text{for } i = 1, \ldots, n, \quad \xi_i \geq 0, \quad \sum \xi_i \leq \text{constant} \tag{5.38}$$

In the criterion given by Equation 5.38, ξ_i is expressed in units of $M/2$, and is thus the proportional amount of the points on the wrong side of the hyperplanes with margin M. The total proportion of these points is bounded by the condition $\sum \xi_i \leq$ constant. By fixing the constant, the maximization problem (Equation 5.38) can be solved, and the result for the overlapping example data set is shown in Figure 5.19 (right). The decision boundary is the hyperplane (here the solid line) in the middle of the hyperplanes with margin M. This illustration also shows that the points that do not create the overlap (i.e., points that are well inside their class boundary) have not much effect on the position of the decision boundary. This seems to be an advantage, and it is quite in contrast to methods like LDA or PLS where information from all data points is incorporated for the decision boundary, by using the pooled covariance matrix and the group centers. Thus the decision boundary from an SVM is mainly oriented at objects that are difficult to classify, and not at objects that are clearly distinct. However, this can also lead to instabilities, especially if data outliers are used as support vectors (Steinwart and Christmann 2008).

The concepts described so far are usually not applied in the original data space but in a transformed and enlarged space using basis expansions (see Section 4.8.3 about nonlinear regression). Thus each observation $\boldsymbol{x}_i$ is expressed by a set of basis functions (object vectors $\boldsymbol{x}_i$ with m dimensions are replaced by vectors $\boldsymbol{h}(\boldsymbol{x}_i)$ with r dimensions)

$$\boldsymbol{h}(\boldsymbol{x}_i) = (h_1(\boldsymbol{x}_i), h_2(\boldsymbol{x}_i), \ldots, h_r(\boldsymbol{x}_i)) \quad \text{for } i = 1, \ldots, n \tag{5.39}$$

that are used in Equation 5.35 for classification. This leads to nonlinear classification boundaries in the original space. The dimension r can get very large, and in some cases it is even infinite. Since the optimization problem formulated above is already relatively complicated (the problem formulated in Equation 5.38 is quadratic with linear inequality constraints and therefore it requires quadratic programming), one would assume that the basis expansion leads to infeasible computations. However, it can be shown that for particular choices of the basis functions the computations even simplify, because the so-called KERNEL TRICK can be applied (Boser et al. 1992). The main ideas are as follows—for more details see Hastie et al. 2001. The optimization problem (Equation 5.38) can be formulated in a more compact way, where the objects $\boldsymbol{x}_i$ and $\boldsymbol{x}_j$ (for $i, j = 1, \ldots, n$) enter the function for the optimization only via the product $\boldsymbol{x}_i^{\mathrm{T}}\boldsymbol{x}_j$. If the objects are transformed by the basis functions (Equation 5.39), this product becomes

$$K(\boldsymbol{x}_i, \boldsymbol{x}_j) = \boldsymbol{h}(\boldsymbol{x}_i)^{\mathrm{T}}\boldsymbol{h}(\boldsymbol{x}_j) \tag{5.40}$$

where K denotes the kernel function that computes products in the transformed space (compare KERNEL METHODS in Section 4.8.3.2). Thus it is not necessary to explicitly specify the basis functions (Equation 5.39) but only to specify the kernel functions (Equation 5.40). Standard choices are

- Polynomial of degree d: $K(\boldsymbol{x}_i, \boldsymbol{x}_j) = (1 + \boldsymbol{x}_i^{\mathrm{T}}\boldsymbol{x}_j)^d$
- Radial basis: $K(\boldsymbol{x}_i, \boldsymbol{x}_j) = \exp(-c||\boldsymbol{x}_i - \boldsymbol{x}_j||^2)$ with $c > 0$
- Neural network: $K(\boldsymbol{x}_i, \boldsymbol{x}_j) = \tanh(c_1\boldsymbol{x}_i^{\mathrm{T}}\boldsymbol{x}_j + c_2)$ with $c_1 > 0$ and $c_2 < 0$

An algorithm for computing the decision boundary thus requires the choice of the kernel function; frequently chosen are radial basis functions (RBFs). A further input parameter is the priority of the size constraint for $\sum\xi_i$ used in the optimization problem (Equation 5.38). This constraint is controlled by a parameter that is often denoted by γ. A large value of γ forces the size of $\sum\xi_i$ to be small, which can lead to an overfit and to a wiggly boundary in the original data space. The choice $\gamma = \infty$ leads to the separable case. On the other hand, a small value of γ allows for a larger size of $\sum\xi_i$ and leads to smoother separation boundaries. As a default value for γ, one can use $1/m$, i.e., 1 divided by the number of x-variables. However, the choice of γ can also be based on CV.

In the following example, the effect of these parameter choices will be demonstrated. We use the same example as in Section 5.5 with two overlapping groups. Figure 5.20 shows the resulting decision boundaries for different kernel functions

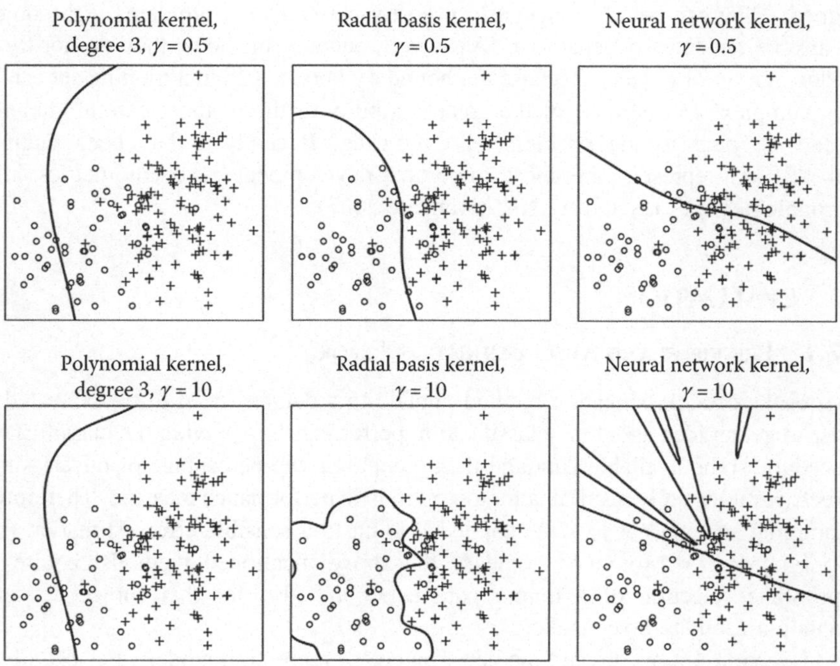

FIGURE 5.20 Classification with SVMs for two groups of two-dimensional data. The training data are shown with the symbol corresponding to the group membership. Any new data point would be classified according to the presented decision boundaries. The results are obtained by using different kernel functions and by changing the parameter γ.

and different choices of the tuning parameter γ. The choice of $\gamma = 0.5$ (top row) results in very smooth decision boundaries, whereas $\gamma = 10$ (bottom row) gives rough and complex boundaries.

```
R:  library(e1071)  # includes SVMs
    resSVM <- svm(Xtrain,grp,kernel = "radial",gamma = 0.5)
      # fits SVM with radial basis kernel and γ = 0.5
    predict(resSVM,Xtest)  # prediction of the class
```

The idea of a decision plane with maximum thickness (margin) is not new. In early time of pattern recognition in chemistry, so-called LINEAR LEARNING MACHINES (LLM) have been popular (Isenhour and Jurs 1973; Jurs and Isenhour 1975; Nilsson 1965; Varmuza 1980). The ideas of LLMs and ANNs date back to the start of artificial intelligence and the concept of perceptrons (Minsky and Papert 1969; Rosenblatt 1960). An LLM is an iterative procedure that tries to find a decision plane for a complete separation of two groups of objects. To find a position of the decision plane in the middle between the groups, a finite thickness (dead zone) can be given to the plane. To find out the maximum thickness, it is increased stepwise until no complete separation is possible (Preuss and Jurs 1974). For not linearly separable groups, a "negative thickness" can be used, and for new objects situated in the dead zone the classification is rejected.

SVMs are an alternative to ANNs and have the advantage to be more strictly defined. They can be advantageously used for classification problems if the object groups are not linearly separable. SVMs are boundary methods; they do not try to model a group of objects. Because the boundary obtained from a training set can be very complex, and because of the great flexibility of the method, careful attention should be paid to the problem of overfitting. Recently it has been claimed that "SVMs represent the most important development in chemometrics after (chronologically) PLS and ANNs" (Ivanciuc 2007).

5.7 EVALUATION

5.7.1 PRINCIPLES AND MISCLASSIFICATION ERROR

The results of a classification method applied to a data set need to be evaluated in order to get an idea about the classification performance. Applying the classification procedure to the available data and then computing the percentage of misclassified objects would usually result in a far too optimistic performance measure. Therefore a resampling scheme needs to be applied leading to a sensible error estimation (see also Figure 5.1). Moreover, some of the above mentioned methods require an appropriate selection of a tuning parameter, and also for this choice a useful evaluation needs to be consulted.

In Section 4.2, we treated the performance of regression models. For classification methods, these concepts remain valid and can be directly used. However, there is an important difference concerning the performance measures to be used. While for regression the basic information for the evaluation measures are the residuals, i.e., the difference between observed and predicted y-values, this would not make much sense in the context of classification. Here the y-values are categorical values representing the group memberships, and they need to be treated differently. Typical error measures were already mentioned above, and the most widely used measure is the MISCLASSIFICATION ERROR which is the fraction of objects that were assigned to a wrong group. A formal definition is

$$\frac{1}{n}\sum_{i=1}^{n} I(\hat{y}_i \neq y_i) \tag{5.41}$$

where

y_i denotes the group number of the ith object
$\hat{y}_i$ is the estimated group number
the index function I gives 1 if the group numbers are not the same and 0 otherwise

As already mentioned above, this error measure should be computed for test data because otherwise it will be too optimistic.

Instead of the misclassification error, a LOSS FUNCTION can be used that allows to consider different risks for the different types of wrong classifications; for instance assigning healthy people to be sick can be given another risk than assigning sick people to be healthy.

One has to be careful with the use of the misclassification error as a performance measure. For example, assume a classification problem with two groups with prior probabilities $p_1 = 0.9$ and $p_2 = 0.1$, where the available data also reflect the prior probabilities, i.e., $n_1 \approx np_1$ and $n_2 \approx np_2$. A stupid classification rule that assigns all the objects to the first (more frequent) group would have a misclassification error of about only 10%. Thus it can be more advisable to additionally report the misclassification rates per group, which in this case are 0% for the first group but 100% for the second group which clearly indicates that such a classifier is useless.

5.7.2 PREDICTIVE ABILITY

In the following, we assume that all objects come from two groups (denoted as 1 and 2), and that they are classified to one of the two groups. We thus have a BINARY CLASSIFIER without rejections. The goal is to derive a measure for the prediction performance of the classifier. This concept can be easily extended to the multiple group case.

Since the group memberships of the objects from the test set are known, these can be used to evaluate the classifier. Suppose we have n objects in the test set, where n_1 belong to group 1 and n_2 are from group 2, with $n_1 + n_2 = n$. Then we can count how many objects are correctly classified and how many are misclassified. For objects from group 1, the classifier assigns $n_{1\to1}$ objects to the correct and $n_{1\to2}$ objects to the wrong group. Similarly, $n_{2\to2}$ objects from group 2 are correctly, and $n_{2\to1}$ are wrongly classified. The total numbers of group assignments to groups 1 and 2 are denoted by $n_{\to1}$ and $n_{\to2}$, respectively. Table 5.1 shows the resulting frequencies in a GROUP ASSIGNMENT TABLE.

The PREDICTIVE ABILITIES P_1 and P_2 for groups 1 and 2, respectively, are defined as proportions of objects in each group that were correctly assigned:

TABLE 5.1
Number of Objects n_1 and n_2 in the Groups 1 and 2 of a Test Set and Counts of Group Assignments Resulting from the Classifier, Represented in the Group Assignment Table

		Group Assignment		
		1	2	Sum
Group membership	1	$n_{1\to 1}$	$n_{1\to 2}$	n_1
	2	$n_{2\to 1}$	$n_{2\to 2}$	n_2
Sum		$n_{\to 1}$	$n_{\to 2}$	n

$$P_1 = \frac{n_{1\to 1}}{n_1} \tag{5.42}$$

$$P_2 = \frac{n_{2\to 2}}{n_2} \tag{5.43}$$

The predictive abilities are values in the interval [0, 1] and are in fact estimated probabilities for a correct classification per group; often they are given in percent. They are independent from the group sizes in the test set, and therefore good measures of the prediction performance. The predictive abilities of the groups can be combined to a single value for instance by taking the arithmetic mean, resulting in the AVERAGE PREDICTIVE ABILITY $\bar{P}$

$$\bar{P} = \frac{P_1 + P_2}{2} \tag{5.44}$$

A classifier is formally informative if $\bar{P} > 0.5$; otherwise something went wrong in the development of the classifier and a simple change of the classification answers would improve it. For the OVERALL PREDICTIVE ABILITY, P, the proportion of all correct classifications is considered

$$P = \frac{n_{1\to 1} + n_{2\to 2}}{n} = \frac{n_{1\to 1}}{n} P_1 + \frac{n_{2\to 2}}{n} P_2 \tag{5.45}$$

This measure depends on the relative group sizes (estimated prior probabilities), and thus it is in general not suited to characterize the prediction performance of a classifier.

As mentioned in Section 5.7.1, it is recommended to use an appropriate resampling procedure for the evaluation. A separate classifier is developed from each training set, and its performance is evaluated from the corresponding test set. This will result in several values for the predictive abilities, and statistical measures can be used to describe their distributions.

5.7.3 Confidence in Classification Answers

While the predictive ability—as defined in Equations 5.42 and 5.43—relates the number of correct answers of the classifier to the number of objects per group, we could also build the relation to the number of decisions for each group. Using the notation of Table 5.1, the number of objects that were assigned to the first group is $n_{\rightarrow 1}$, and for the second group we have $n_{\rightarrow 2}$ assignments. The fraction of correct assignments to the first group on all assignments to this group is the CONFIDENCE IN CLASSIFICATION ANSWERS for group 1,

$$C_1 = \frac{n_{1\rightarrow 1}}{n_{\rightarrow 1}} \tag{5.46}$$

and for group 2

$$C_2 = \frac{n_{2\rightarrow 2}}{n_{\rightarrow 2}} \tag{5.47}$$

Since the resulting measures are in the interval [0, 1], they can be interpreted as probabilities that a given answer from the classifier is correct; often they are given in percent. Taking the formula for the first group, the relation to the predictive ability (Section 5.7.2) becomes visible by the representation

$$C_1 = \frac{n_{1\rightarrow 1}}{n_{1\rightarrow 1} + n_{2\rightarrow 1}} = \frac{n_{1\rightarrow 1}/n_1}{n_{1\rightarrow 1}/n_1 + n_{2\rightarrow 1}/n_1} = \frac{P_1}{P_1 + n_{2\rightarrow 1}/n_1} \tag{5.48}$$

The confidence in classification answers and the predictive ability for group 1 are equal if and only if $n_{2\rightarrow 1} = n_{1\rightarrow 2}$, i.e., if the numbers of misclassified objects in both groups are the same, otherwise $C_1 \neq P_1$. Note that the confidence in classification answers depends on the relative group sizes in the test set, and therefore this measure should only be used if n_1 is comparable to n_2. In this case, an average of both values, $\bar{C}$, can be used as a measure of classification performance.

5.8 EXAMPLES

5.8.1 ORIGIN OF GLASS SAMPLES

We will compare most of the methods mentioned in this chapter for a data set that is well known in the machine-learning literature. This data set consists of 214 samples of glass from six different types (Table 5.2). The available nine variables are the refractive index, and the mass percentages of the elements Al, Ba, Ca, Fe, K, Mg, Na, and Si. The data set was created by B. German (Central Research Establishment, England) and is available for instance in the ℝ package MASS as data set *fgl*. The motivation behind classifying the glass types comes from forensic investigations; a correct classification of the glass left at the scene of a crime can be helpful for clearing up the crime. Since some of the classification methods are not invariant with respect to the data scale (e.g., *k*-NN), the data are first autoscaled.

5.8.1.1 Linear Discriminant Analysis

For LDA (Section 5.2.1) we select a training data set randomly (2/3 of the objects) and use the derived classification rule to predict the group membership of the

TABLE 5.2
Glass Identification Data: Information on the Glass Type, the Abbreviation Used in This Section, and the Number of Samples per Glass Type

Glass Type	Abbreviation	Number of Objects
Building windows, float processed	*WinF*	70
Building windows, nonfloat processed	*WinNF*	76
Vehicle windows, float processed	*Veh*	17
Containers	*Con*	13
Tableware	*Tabl*	9
Headlamps	*Head*	29
	Sum	214

remaining data (test data). As LDA method we used the Bayesian rule, where the prior probabilities are estimated by the relative frequencies of the groups for the training data. The misclassification rate is then computed for each group separately, and also for all groups jointly (total number of misclassified objects of the test data divided by the number of test data points, Equation 5.41). Since the result will depend on the choice of the test data, the procedure is repeated 100 times. This gives a distribution for the misclassification rate of each group, as well as for the overall misclassification error. The distributions are shown in Figure 5.21 by boxplots. The median of the total misclassification error is 0.39, which is a poor result. For some groups, like *Head* the misclassification error is considerably lower, however, for other groups like *Veh* it is higher, even 100%. The reason is that small groups are

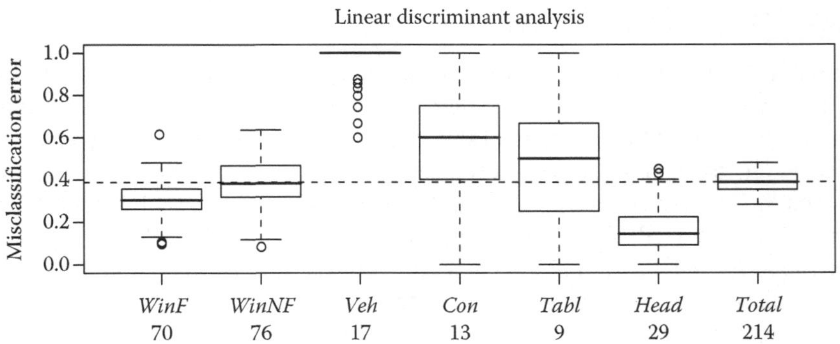

FIGURE 5.21 LDA applied to the glass data with six glass types. The boxplots represent the misclassification errors (for test sets) for the separate groups and the overall misclassification error (right boxplot) obtained by 100 random selections of training and test data. The dashed horizontal line is the median of the total misclassification error (0.39). The numbers below the group levels are the numbers of objects in the groups for the complete data set.

underrepresented in the randomly selected training sample. A better strategy would thus be to select the training sample in such a way that a fixed fraction (e.g., 1/2 or 2/3) of the objects of each group has to be taken. When using this strategy for LDA, the misclassification error improves slightly for the small groups, and also a marginal improvement of the total misclassification error can be achieved (median 0.37 for a fraction 2/3).

The ℝ code for the evaluation follows the outline of the ℝ code as shown in Section 5.2.1.3. We describe how the data are prepared (also for the following analyses), how training data are selected, and how the prediction of the group membership for test data is obtained. The LDA procedure needs to be repeated 100 times in a loop for obtaining the misclassification errors.

```
ℝ: library(MASS)      # contains the Glass data
   data(fgl)          # Glass data set
   grp <- fgl$type    # grouping variable (glass type)
   X <- scale(fgl[,1:9])        # autoscaled data
   dat <- data.frame(grp,X)     # create data frame
   train <- sample(1:nrow(dat),143)
                                # 2/3 of objects, randomly
   resLDA <- lda(X[train,],grp[train])  # Bayes-LDA
   predLDA <- predict(resLDA,newdata=X[-train,])$class
     # predicted group membership for test data
   table(grp[-train],predLDA)   # table with assignments
```

5.8.1.2 Logistic Regression

The same scheme as for LDA is used: 2/3 of the objects are used as training data set and the rest forms the test data set. Each object has the same chance to enter the training set, thus the different group sizes in the data are ignored. Compared to LDA, LR (Section 5.2.3) gives better results for the small groups *Veh, Con*, and *Tabl*, but slightly worse results for *WinF* (Figure 5.22). The median of the total misclassification error is the same as for LDA (0.39).

LR for more than two groups is often called multinomial logistic regression. The ℝ functions mentioned in Section 5.2.3 cannot be applied in this case. Below is the outline for the multiple group case.

```
ℝ: library(VGAM)  # contains functions for LR with k>2 groups
   resmix <- vglm(grp~.,data=dat[train,],family=multinomial)
           # multinomial logistic regression for training data
   predmix <- predict(resmix,newdata=dat[-train,],
             type="response")  # matrix with predictions
             # (probabilities) of each object to each group
   predgrp <- apply(predmix,1,which.max)
       # search the index with the maximum prediction
       # for each object (row of predmix)
   table(grp[-train],predgrp)  # table with assignments
```

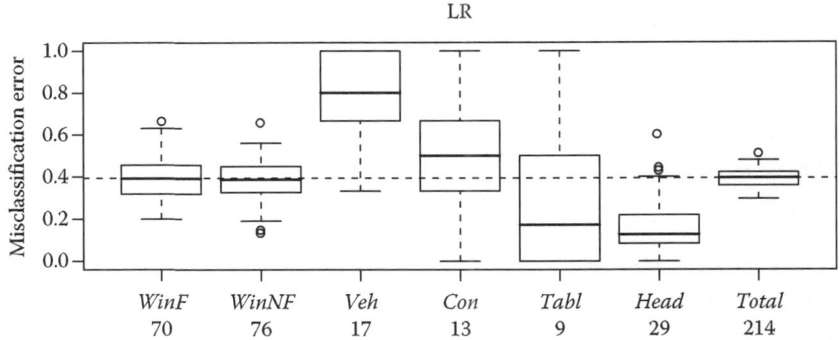

FIGURE 5.22 Logistic regression applied to the glass data with six glass types. The boxplots represent the misclassification errors (for test sets) for the separate groups and the overall misclassification error (right boxplot) obtained by 100 random selections of training and test data. The dashed horizontal line is the median of the total misclassification error (0.39). The numbers below the group levels are the numbers of objects in the groups for the complete data set.

5.8.1.3 Gaussian Mixture Models

Training and test data are randomly generated as mentioned above. Gaussian mixture models (Section 5.3.2) are fitted to the training data, which are then used to predict the group membership of the test data. Following the same style as above, the misclassification errors are presented by boxplots in Figure 5.23. This method also works well for the small groups, and the median of the total misclassification error is now 0.35.

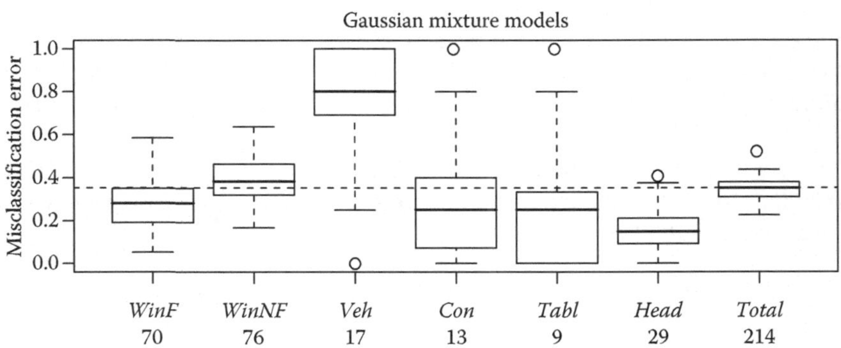

FIGURE 5.23 Gaussian mixture models are fitted to the glass data with six glass types. The boxplots represent the misclassification errors (for test sets) for the separate groups and the overall misclassification error (right boxplot) obtained by 100 random selections of training and test data. The dashed horizontal line is the median of the total misclassification error (0.35). The numbers below the group levels are the numbers of objects in the groups for the complete data set.

For computational reasons, we used a different program as in Section 5.3.2 to fit the mixture models. This program is based on the ideas of Hastie and Tibshirani (1996) to use several prototype objects representing each group in order to obtain more accurate predictions.

```
ℝ: library(mda)  # mixture discriminant analysis
   resmix <- mda(grp~.,data=dat[train,])
       # Gaussian mixture models for training data
   predmix <- predict(resmix,newdata=dat[-train,],
              type="post")  # matrix with posterior
              # probabilities of each object to each group
   predgrp <- apply(predmix,1,which.max)
             # search the index with the maximum prediction
             # for each object (row of predgrp)
   table(grp[-train],predgrp)  # table with assignments
```

5.8.1.4 *k*-NN Methods

k-NN and the following methods require a tuning of one or more parameters (Section 5.3.3). In case of *k*-NN, this parameter is *k*, the number of neighbors to be considered for predicting the group membership of a test data point. The parameter tuning is done by CV, and for reasons of comparability we always follow the same scheme which is described in the following.

The data are randomly split into a calibration and a test set using the proportions 2/3 and 1/3, respectively. Within the calibration set, 10-fold CV is performed by applying the classification method with a certain parameter choice on nine segments (training set) and predicting the 10th segment (evaluation set) with the derived classification rule. This evaluation scheme thus follows method (2) as described in Section 4.2.1. The optimum value of the parameter used in a classification method is derived from the CV results. For simplicity, the misclassification errors are not presented separately for the groups but in the form of total misclassification errors. Thus for each parameter setting the overall misclassification error is computed, and we obtain

- *Training error*: This is the misclassification error for the objects of the training sets that are fitted within the 10-fold CV. For example, for *k*-NN with $k=1$ this error will be zero, because the classification rule is derived from nine segments of the training data, and evaluated at the same nine segments, resulting in a perfect classification.
- *CV error*: This refers to the misclassification error of the evaluation sets within the CV. For 10-fold CV, the misclassification error can be computed for each of the 10 evaluation sets. For each parameter choice, we will represent the mean of the 10 errors as well as their standard error. This allows to apply the one-standard-error rule as outlined in Section 4.2.2.

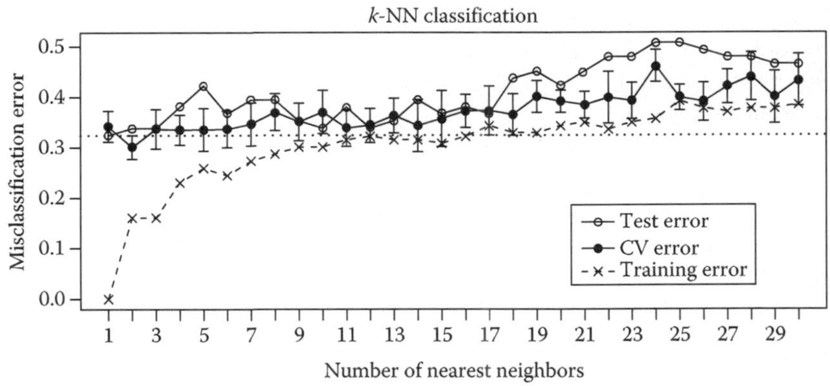

FIGURE 5.24 *k*-NN classification for the glass data with six glass types. The optimal parameter for *k*, the number of nearest neighbors, is 2. The test error for this parameter choice is 0.34.

- *Test error*: The classification rule is derived from the whole calibration set with a certain parameter choice. Then the rule is applied to the test set, and the test error is the resulting misclassification error of the test set. Note that in principle it would be sufficient to compute the test error only for the optimal parameter choice.

Different selections of calibration and test data set may lead to different answers for the errors. In the following, we present results from one random split; however, in the final overall comparison (Section 5.8.1.8) the evaluation scheme is repeated 100 times to get an idea of the distribution of the test error for the optimal parameter choice.

Figure 5.24 shows the results for *k*-NN classification for a range of *k* from 1 to 30. As mentioned above, the training error must be zero for $k=1$, and it increases with *k*. The CV error is visualized by black dots for the means and vertical bars for mean plus/minus one standard error. The dotted horizontal line is drawn at the mean plus 1 standard error for the smallest mean CV error, and it is used for the selection of the optimal parameter (see Section 4.2.2). Accordingly, $k=2$ is the optimal solution, most likely because of the small data groups. The resulting test error for $k=2$ is 0.34.

```
R: library(chemometrics)
   calibr <- sample(1:nrow(dat),143)  # 2/3 of objects
   resknn <- knnEval(X,grp,calibr,knnvec=seq(1,30))
          # generates Figure 5.24 for k=1,2, . . . ,30
```

5.8.1.5 Classification Trees

The same evaluation scheme as described above for *k*-NN is used for classification trees (Section 5.4). The parameter to be optimized is the tree complexity. Figure 5.25

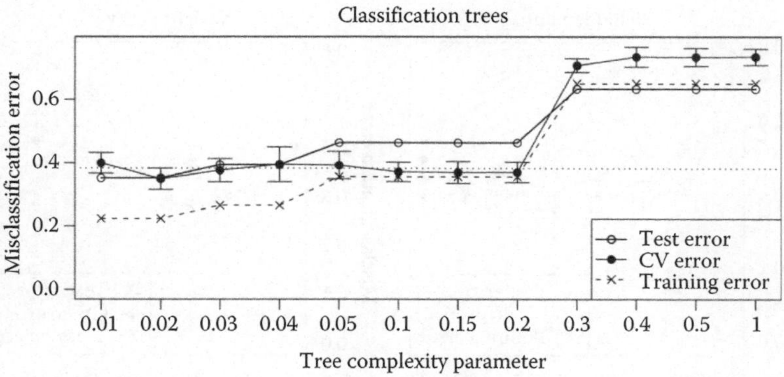

FIGURE 5.25 Classification trees for the glass data with six glass types. The optimal parameter for the tree complexity is 0.02. The test error for this parameter choice is 0.35.

shows the outcome of the evaluation in form of training, CV, and test errors, depending on the selected tree complexity parameter (horizontal axis). According to the one-standard-error rule, the optimal value is 0.02. For this choice, the classification error for the test set is 0.35.

```
ℝ:  library(chemometrics)
    calibr <- sample(1:nrow(dat),143)  # 2/3 of objects
    cpsel <- c(0.01:0.05,0.1,0.15,0.2:0.5,1)
        # selected parameters for the tree complexity
    restree <- treeEval (X,grp,calibr,cp=cpsel)
         # generates Figure 5.25
```

5.8.1.6 Artificial Neural Networks

Here we need to select two parameters, the optimal weight decay and the optimal number of hidden units (Section 5.5). We use the same evaluation scheme as described for *k*-NN. The results are shown in Figure 5.26. The left plot shows the dependency of the error rate from the weight decay for 20 hidden units, and for the right plot the number of hidden units is varied and the weight decay is fixed at 0.2. It seems that these two choices, a weight decay of 0.2 and 20 hidden units, are optimal. However, it needs several trials with different choices to come to a final good selection. This is also because the ANN algorithm does not give unique solutions, thus suggesting slightly different optimal parameter choices each time of a trial. For example, in Figure 5.26, the resulting misclassification errors for a weight decay of 0.2 and 20 hidden units are slightly different: the test error in the left plot is just above 0.4, and in the right plot it is well below 0.4. A more sophisticated investigation would repeat the procedure for different random splits into calibration and test set to overcome this problem; furthermore an optimum pair of weight decay and number of hidden units could be found by a sequential simplex optimization.

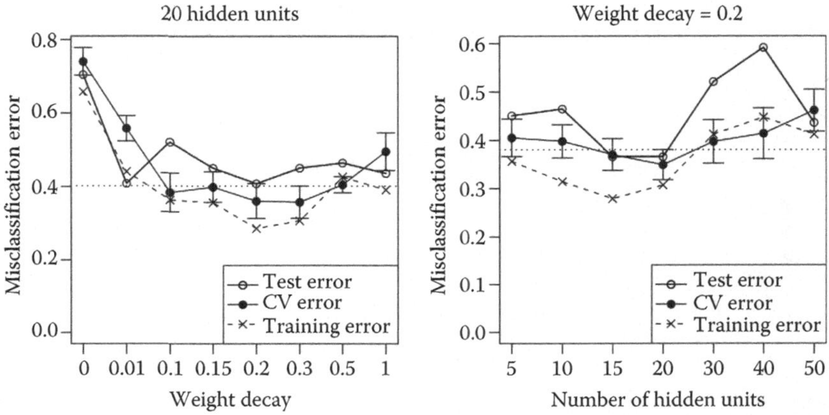

FIGURE 5.26 ANNs applied to the glass data with six glass types. The optimal parameter choices are (probably) 20 hidden units and a weight decay of 0.2. The plots show the misclassification errors by fixing one of these parameters. Since the result is not unique, we obtain two answers for the test error: 0.41 in the left plot and 0.37 in the right plot.

```
R:  library(chemometrics)
    calibr <- sample(1:nrow(dat),143)  # 2/3 of objects
    weightsel <- c(0,0.01,0.1,0.15,0.2,0.3,0.5,1)
           # selected parameters for the weight decay
    resANN <- nnetEval(X,grp,calibr,decay=weightsel,size=20)
            # generates Figure 5.26 (left) with a fixed
            # number of hidden layers (size=20)
```

5.8.1.7 Support Vector Machines

The most important parameter choices for SVMs (Section 5.6) are the specification of the kernel function and the parameter γ controlling the priority of the size constraint of the slack variables (see Section 5.6). We selected RBFs for the kernel because they are fast to compute. Figure 5.27 shows the misclassification errors for varying values of γ by using the evaluation scheme described above for *k*-NN classification. The choice of $\gamma=0.1$ is optimal, and it leads to a test error of 0.34.

```
R:  library(chemometrics)
    calibr <- sample(1:nrow(dat),143)  # 2/3 of objects
    gamsel <- c(0,0.05,0.1,0.2,0.3,0.5,1,2,5)
           # selected parameters for gamma
    resSVM <- svmEval(X,grp,calibr,gamvec=gamsel)
           # generates Figure 5.27
```

5.8.1.8 Overall Comparison

As mentioned above, the test errors depend on the selection of training (calibration) and test data sets. We can get an idea about the distribution of the test errors by

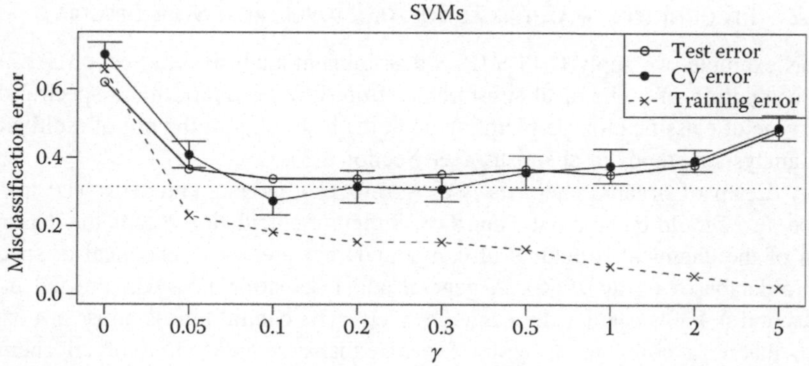

FIGURE 5.27 SVMs applied to the glass data with six glass types. The optimal choice of the parameter γ is 0.1, leading to a test error of 0.34.

repeating the evaluation several times. This has already been done above for LDA, logistic regression and Gaussian mixture models, and the distribution was visualized by boxplots. We repeated the evaluation scheme outlined above (see Section 5.3.3) also for the other methods, however, only by using the optimal parameter choices obtained from a single split. The resulting test errors for 100 replications are presented by notched boxplots in Figure 5.28. The notches around the medians represent the widths of the confidence intervals for the medians. The best methods for this data set are *k*-NN, classification trees (Tree), and SVMs. In summary, the performances of the different classification methods are not very different and are all rather poor with at best about 1/3 of the test glass samples incorrectly assigned—applicability for forensic purposes seems doubtful.

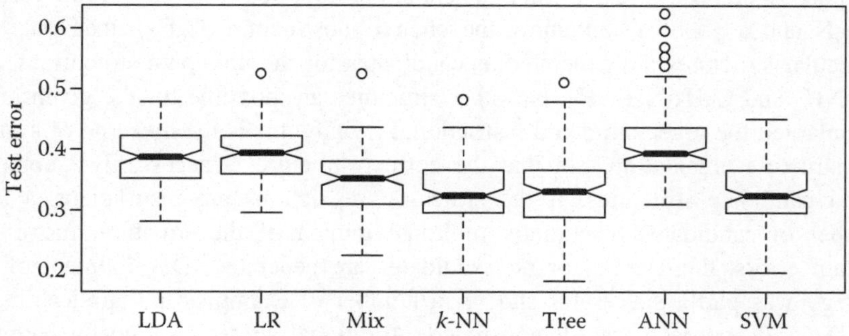

FIGURE 5.28 Comparison of the test errors for the glass data using different classification methods. One hundred replications of the evaluation procedure (described in the text) are performed for the optimal parameter choices (if the method depends on the choice of a parameter). The methods are LDA, LR, Gaussian mixture models (Mix), *k*-NN classification, classification trees (Tree), ANN, and SVMs.

5.8.2 Recognition of Chemical Substructures from Mass Spectra

In this example, we apply D-PLS (PLS discriminant analysis, see Section 5.2.2) for the recognition of a chemical substructure from low-resolution mass spectral data. This type of classification problems stood at the beginning of the use of multivariate data analysis methods in chemistry (see Section 1.3).

A dream of organic chemists is a SYSTEMATIC CHEMICAL STRUCTURE ELUCIDATION which also should be automatic and fast. Structure elucidation means the determination of the chemical structures of compounds not present in chemical or spectroscopic databases (Gray 1986). A general and fascinating idea was created in the 1960s and is known under the name DENDRAL. The central tool is an ISOMER GENERATOR, that is a computer program for an exhaustive generation of all chemical structures from a given brutto formula. For a compound with unknown structure, the brutto formula (or a rather small set of possible brutto formulae) can in principle be determined by high-resolution mass spectrometry. For instance $C_4H_{10}O$ has seven isomers: two primary alcohols, one secondary alcohol, one tertiary alcohol, and three ethers. In this case, all isomers are stable compounds; however, in general the isomer generator only considers the valence rules but not the stability of chemical structures. Larger brutto formulae have a high number of valence isomers; for instance the relatively small brutto formula C_7H_9N has 24,312 isomers. The isomer generator software MOLGEN (Benecke et al. 1997) needs only 0.15 s on a 2.2 GHz personal computer for counting and storing the isomers of C_7H_9N (Molgen 1997).

According to the Dendral approach, one also needs restrictions for the unknown chemical structure, usually given as substructures that are

- Present in the unknown molecular structure (collected in a GOODLIST)
- Absent in the unknown molecular structure (collected in a BADLIST)

Such restrictions are most often derived from molecular spectra that have been measured on the unknown; deduced either by spectroscopic experience or by computer methods. If the isomer generator is, e.g., fed with the brutto formula C_7H_9N and a goodlist containing the phenyl substructure (C_6H_5), then only two molecular structures are generated as candidates for the unknown structure: C_6H_5-CH_2NH_2 and C_6H_5-NH-CH_3. No other structures are possible for the given brutto formula and the given structural restriction. Actually this is the meaning of systematic structure elucidation, not that the correct structure is necessarily among the found candidate structures. If the structural restrictions are insufficient, a large number of candidates (eventually millions) remain; if the structural restrictions contain errors, then wrong or no candidates are generated. Development of this strategy was partly successful and for a number of examples the applicability of the Dendral approach was demonstrated. Additional strategies—mostly based on modern NMR spectroscopic techniques—are necessary for larger molecules, say above about 15 carbon atoms (Elyashberg et al. 2008; Funatsu and Sasaki 1996; Munk 1998; Neudert and Penk 1996; Robien 2003).

Providing structural restrictions for the goodlist and the badlist has been a challenging task in chemometrics. A lot of effort has gone into the development of multivariate classifiers that recognize the presence or absence of substructures from spectroscopic data, especially from infrared spectra (Penchev et al. 1999; Thiele and Salzer 2003) and from mass spectra (Varmuza 2000; Varmuza and Werther 1996); for NMR spectra search, routines are more successful (Will et al. 1996). Development of SPECTRAL SUBSTRUCTURE CLASSIFIERS—based on multivariate techniques—requires a set of spectra from compounds with known chemical structures; about half of them containing a certain substructure and the rest not containing it. The spectral data have to be transformed into a set of vectors which define the $\boldsymbol{X}$-matrix. For low-resolution mass spectra, the peak intensities at integer mass numbers can be directly used as vector components; however, it has been shown that so-called MASS SPECTRAL FEATURES (Section 7.4) give better classification results (Crawford and Morrison 1968; Drablos 1992; Werther et al. 2002). IR spectra are usually represented by vectors with the components given by the average absorbances of wavelength intervals. As an example for a spectral substructure classifier, we investigate the recognition of a phenyl substructure (C_6H_5) from low-resolution mass spectral data (peak lists with integer masses).

The data used are from $n_1 = 300$ compounds containing a phenyl substructure (group 1) and $n_2 = 300$ compounds not containing this substructure (group 2); the compounds have been selected randomly from about 100,000 compounds in the NIST Mass Spectral Database (NIST 1998). Each mass spectrum has been transformed into $m = 658$ variables (mass spectral features) as described in Section 7.4; all variables are in the range 0–100. The data are randomly split into a calibration set with 150 objects from each group, and a test set also including 150 objects from each group. Thus we use scheme (2) from Section 4.2.1 for evaluation: A model is created and optimized from the calibration set by CV. Then the test set is used for evaluating the model performance.

The data set includes a large number of highly correlated x-variables. This fact excludes covariance-based methods like LDA for classification, but D-PLS is appropriate and will be used. The y-variable is a binary variable with the coding -1 and $+1$ for group 1 (phenyl) and group 2 (nonphenyl), respectively. The decision boundary for the group assignments is at zero. This means that if the predicted y-value for an object is smaller than zero, the object will be assigned to the group with code -1, otherwise it will be assigned to the group with code 1. No attempt will be made to optimize the value of the decision boundary. The essential parts of the R-code are provided.

```
R:  library(chemometrics)
    data(Phenyl)  # load Phenyl data
    Phenyl$grp    # group information with values -1 and 1
    # objects 1 to 150 are from group 1 of calibration set
    # objects 151 to 300 are from group 2 of calibration set
    # objects 301 to 450 are from group 1 of test set
    # objects 451 to 600 are from group 2 of test set
```

A crucial point for D-PLS is the determination of the number of PLS components that allow best classification of new objects. Figure 5.29 shows the mean squared

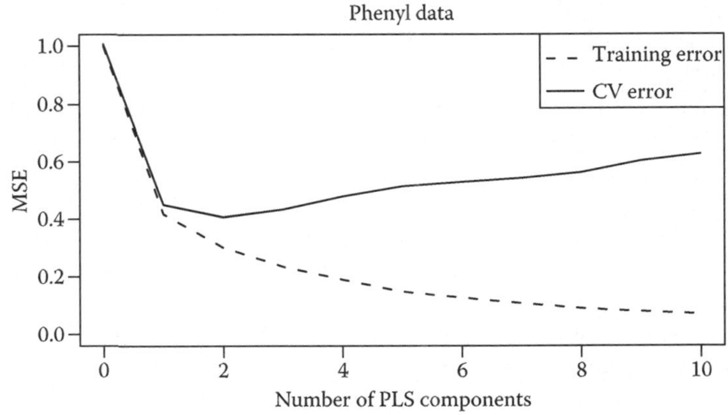

FIGURE 5.29 Mean squared errors for different numbers of PLS components used for D-PLS.

errors for different numbers of PLS components. These results are obtained via 10-fold CV using the SIMPLS algorithm (Section 4.7.5) for the calibration set. The dashed line refers to the mean squared errors for the training set objects, and the solid line to the mean squared errors for CV; the latter has the minimum at two PLS components. Note that a more careful evaluation could be done for deciding on the number of PLS components, like an evaluation based on repeated double CV (see Section 4.2.5). Moreover, this evaluation is only based on the mean squared error (MSE), but not on classification performance measures like misclassification rate or the predictive abilities (see Section 5.7). For this reason we will consider classification models based on one to three PLS components, and computed from the whole calibration set. These models are then applied to the objects of the test set.

```
ℝ: pls.Ph <- mvr(grp~.,data=Phenyl,subset=1:300,ncomp=10,
              method="simpls",validation="CV",segments=10)
   plot(pls.Ph,plottype="validation",val.type="MSEP")
              # generates Figure 5.29
```

Figure 5.30 shows the density functions of the estimated y-values (discriminant variable) of the calibration set (left) and the test set (right). The solid lines refer to a PLS model with two components, the dashed lines to a PLS model with one component, and the dashed-dotted lines to a PLS model with three components. The maxima of the density functions are approximately at the values -1 and 1, corresponding to the group codes. The overlap of the density functions indicates wrong group assignments; for the calibration set, the overlap is smallest for the model with two components, for the test set two or three components are better than one component.

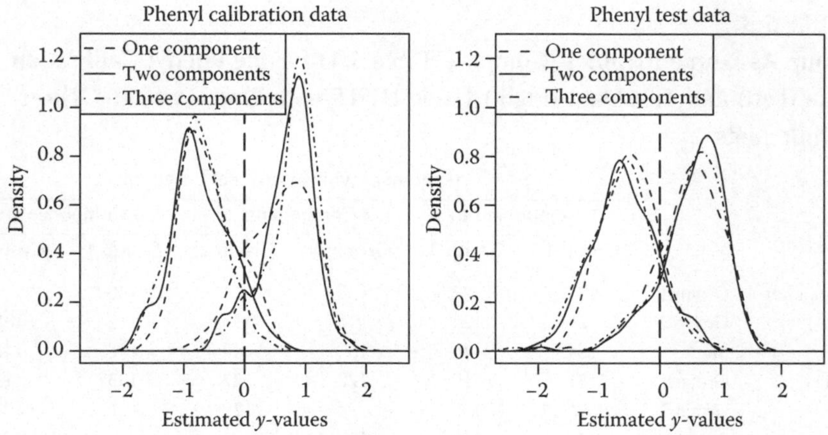

FIGURE 5.30 D-PLS is applied to the phenyl data using a model with one, two, and three components calculated from the calibration set. The density functions of the estimated y-values (discriminant variable) are shown for the calibration set (left) and for the test set (right).

```
R:  yhat.Ph <- drop(predict (pls.Ph,newdata=Phenyl,ncomp=1:3)
    plot(density(yhat.Ph[1:150,2]))        # plot density function
        # for group 1 of calibration set using 2 components
    lines(density(yhat.Ph[151:300,2]))  # plot density function
        # for group 2 of calibration set using 2 components
    yhat2.Ph <- drop(predict(pls.Ph,newdata=Phenyl,ncomp=2)
    assignm <- yhat2.Ph>0
    table(Phenyl$grp[301:600],assignm[301:600])
    # group assignment table for test data (Tab. 5.3, bottom)
```

On the basis of the group assignment tables shown in Table 5.3, misclassification rates, predictive abilities, and confidence in classification answers can be computed (see Section 5.7). The results are shown in Table 5.4. While the misclassification rate measures the error, all other measures evaluate the correctness. The different evaluation measures guide to the same conclusions. As can be expected, the classification performance is better for the calibration set—from which the D-PLS model was built—than for the independent test set. Also the overlap of the density functions is for the test set larger than for the calibration set (Figure 5.30). Since both groups have the same number of objects, the average predictive ability, $\bar{P}$, and the average confidence in the classification answers, $\bar{C}$, are very similar and reach about 86% for a model with two PLS components.

Predictive abilities below 90% are not acceptable for substructure classifiers because any wrong substructure in the goodlist or badlist prohibits that the correct structure is among the candidates. On the other hand, a missing substructure only increases the number of candidates. Therefore the overlapping region of the density functions (or a part of it) can be declared as a REJECTION REGION, defined by two thresholds y_{LOW} and y_{HIGH}. If the value of the discriminant variable is below y_{LOW}, the object is assigned to group 1, if it is above y_{HIGH} it is assigned to group 2, and in between no classification is made—in other words the classification is rejected.

TABLE 5.3
Group Assignment Tables (Compare Table 5.1) for the Phenyl Calibration Data (Left) and Test Data (Right) Using D-PLS with One, Two, and Three Components

		Assignment with PLS Model Based on					
		1 Component to		2 Components to		3 Components to	
		Group 1	Group 2	Group 1	Group 2	Group 1	Group 2
Calibration	Group 1	128	22	137	13	146	4
	Group 2	26	124	15	135	12	138
	Sum	154	146	152	148	158	142
Test	Group 1	131	19	128	22	133	17
	Group 2	25	125	21	129	22	128
	Sum	156	144	149	151	155	145

Notes: The numbers are the frequencies of the assignments. The rows correspond to the true group memberships, the columns to the assignments.

TABLE 5.4
Evaluation for the Phenyl Calibration and Test Data for a Model with One, Two, and Three PLS Components

	No. of PLS Components	Misclass. Rate	P_1	P_2	$\bar{P}$	C_1	C_2	$\bar{C}$
Calibration	1	16.0	85.3	82.7	84.0	83.1	84.9	84.0
	2	9.3	91.3	90.0	90.7	90.1	91.2	90.7
	3	5.3	97.3	92.0	94.7	92.4	97.1	94.8
Test	1	14.7	87.3	83.3	85.3	84.0	86.8	85.4
	2	14.3	85.3	86.0	85.7	85.9	85.4	85.7
	3	13.0	88.7	85.3	87.0	85.8	88.3	87.0

Notes: Misclassification rate, predictive abilities P_1 for group 1 and P_2 for group 2, overall predictive ability $\bar{P}$, confidence in classification answers C_1 for assignment to group 1 and C_2 for assignment to group 2, and their average $\bar{C}$. All values are given in percent.

The thresholds can be set for instance so that 95% of the classifications for groups 1 and group 2 are correct, and of course they have to be determined from the calibration set. The confidence in the classification answers is thereby enhanced, however, at the cost of a certain percentage of no assignments—of course the percent rejections must not be too high (e.g., $<30\%$) for a useful classifier.

When using a model with two PLS components for the phenyl data, the thresholds corresponding to 95% are $y_{\mathrm{LOW}} = -0.22$ and $y_{\mathrm{HIGH}} = 0.15$, respectively. In this way, 36 out of 300 objects from the calibration set are not classified to any of the groups, corresponding to 12.0%. Using the same thresholds for the test set, 37 out of 300 objects (12.3%) are not classified. The evaluation measures for the classified objects are shown in Table 5.5, and they clearly improve compared to Table 5.4.

TABLE 5.5
Evaluation for the Phenyl Calibration and Test Data for a Model with Two PLS Components Applying a Rejection Region

	No. of PLS Components	Misclass. Rate	P_1	P_2	$\bar{P}$	C_1	C_2	$\bar{C}$
Calibration	2	4.9	95.3	94.9	95.1	94.5	95.6	95.1
Test	2	10.3	88.9	90.5	90.0	89.6	89.9	89.7

Notes: Misclassification rate, predictive abilities P_1 for group 1 and P_2 for group 2, overall predictive ability $\bar{P}$, confidence in classification answers C_1 for assignment to group 1 and C_2 for assignment to group 2, and their average $\bar{C}$. All values are given in percent. This rejection region has been defined for 95% correct classifications of group 1 and 2 objects in the calibration set.

The applied scheme for the development of substructure classifiers can be applied to any substructures for which appropriate sets of mass spectra are available. Usually the same spectral features (variables) are used for different substructure classifiers.

From the point of view of a spectroscopist, other strategies—based on spectroscopic knowledge—may be also considered. However, the rules for relating substructures to spectral data are usually weak, especially in mass spectrometry. For instance, presence of a peak at m/z 77 (ion $C_6H_5^+$) with a high intensity may be expected to be characteristic for phenyl compounds. Figure 5.31 shows histograms of the peak heights at m/z 77 for both groups (each 300 spectra). We see a severe overlap, and only if the peak intensity at m/z 77 is above 50 (% base peak intensity) a safe decision for the presence of a phenyl substructure is possible; however, only a

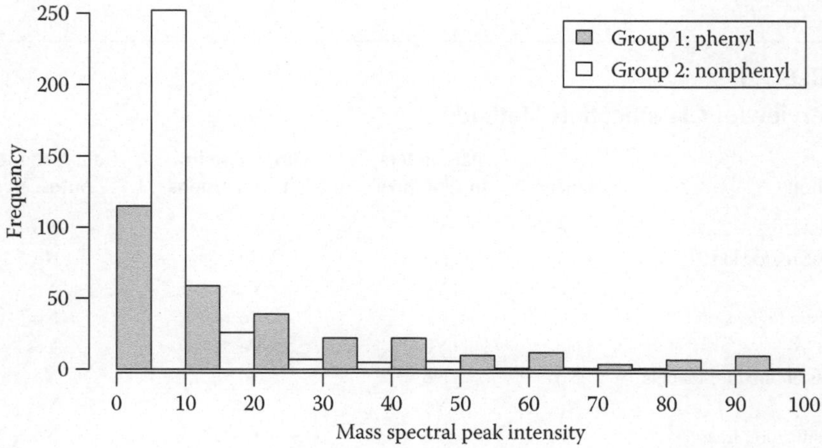

FIGURE 5.31 Histograms of mass spectral peak intensities at m/z 77 for 300 compounds with a phenyl substructure and 300 compounds without a phenyl substructure. The peak intensities are divided into 10 intervals between 0% and 100% base peak intensity. For each interval, the frequency of peaks is given for both groups.

small percentage of the phenyl compounds would be detected. Most phenyl compounds have peak intensities at m/z 77 between 0 and 10, just as nonphenyl compounds. Thus a successful univariate classifier for the phenyl substructure, which is solely based on this "key ion," is not feasible.

5.9 SUMMARY

A great variety of different methods for multivariate classification (pattern recognition) is available (Table 5.6). The conceptually most simply one is k-NN classification (Section 5.3.3), which is solely based on the fundamental hypothesis of multivariate data analysis, that the distance between objects is related to the similarity of the objects. k-NN does not assume any model of the object groups, is nonlinear, applicable to multicategory classification, and mathematically very simple; furthermore, the method is very similar to spectral similarity search. On the other hand, an example for a rather sophisticated classification method is the SVM (Section 5.6).

Most used methods for typical data in chemometrics, containing more variables than objects per group and highly correlating variables, are

- LDA (Section 5.2.1) in various versions, usually with a preceding PCA to overcome the collinearity problem
- D-PLS (discriminant PLS, Section 5.2.2), for binary and for multicategory classifications
- SVMs (Section 5.6), a powerful approach which finds increasing interest

A number of methods model the object groups by PCA models of appropriate complexity (SIMCA, Section 5.3.1) or by Gaussian functions (Section 5.3.2). These methods are successful if the groups form compact clusters, they can handle

TABLE 5.6
Overview of Classification Methods

Method	Linear	Parameters to Optimize	Direct Use in High Dimensions	Data Need to Be Autoscaled
LDA	Yes	No	No	No
Linear regression	Yes	No	No	No
PLS	Yes	Yes	Yes	No
Logistic regression	Yes	No	No	No
SIMCA	No	Yes	Yes	Yes
Gaussian mixture models	No	No	No	No
k-NN	No	Yes	Yes	Yes
Classification trees	No	Yes	Yes	No
ANNs	No	Yes	Yes	Yes
SVMs	No	Yes	Yes	Yes

outliers (objects not belonging to any of the defined groups) and objects belonging to more than one group.

Two groups of objects can be separated by a decision surface (defined by a discriminant variable). Methods using a decision plane and thus a linear discriminant variable (corresponding to a linear latent variable as described in Section 2.6) are LDA, PLS, and LR (Section 5.2.3). Only if linear classification methods have an insufficient prediction performance, nonlinear methods should be applied, such as classification trees (CART, Section 5.4), SVMs (Section 5.6), or ANNs (Section 5.5).

Comparison of the success of different classification methods requires a realistic estimation of performance measures for classification, like misclassification rates (% wrong) or predictive abilities (% correct) for new cases (Section 5.7)—together with an estimation of the spread of these measures. Because the number of objects with known class memberships is usually small, appropriate resampling techniques like repeated double CV or bootstrap (Section 4.2) have to be applied. A difficulty is that performance measures from regression (based on residuals) are often used in the development of classifiers but not misclassification rates.

REFERENCES

Albano, C., Dunn, W. I., Edlund, U., Johansson, E., Nordén, B., Sjöström, M., Wold, S.: *Anal. Chim. Acta*. 103, 1978, 429–443. Four levels of pattern recognition.

Bastien, P., Esposito Vinci, V., Tenenhaus, M.: *Comput. Stat. Data Anal.* 48, 2005, 17–46. PLS generalised linear regression.

Benecke, C., Grüner, T., Grund, R., Hohberger, R., Kerber, A., Laue, R., Wieland, T.: Molgen: Software. University of Bayreuth, Mathematical Institute; www.molgen.de, Bayreuth, Germany, 1997.

Boser, B. E., Guyon, I. M., Vapnik, V. N.: in Haussler, D. (Ed.), *5th Annual ACM Workshop on COLT*, ACM Press, Pittsburgh, PA, 1992, pp. 144–152. A training algorithm for optimal margin classifiers.

Breiman, L., Friedman, J. H., Olshen, R. H., Stone, C. J.: *Classification and Regression Trees*. Wadsworth Belmont, CA, 1984.

Brereton, R. G. (Ed.): *Multivariate Pattern Recognition in Chemometrics, Illustrated by Case Studies*. Elsevier, Amsterdam, the Netherlands, 1992.

Brereton, R. G.: *Chemometrics—Data Analysis for the Laboratory and Chemical Plant*. Wiley, Chichester, United Kingdom, 2006.

Brereton, R. G.: *Applied Chemometrics for Scientists*. Wiley, Chichester, United Kingdom, 2007.

Christianini, N., Shawe-Taylor, J.: *An Introduction to Support Vector Machines and Other Kernel-Based Learning Methods*. Cambridge University Press, Cambridge, NY, 2000.

Crawford, L. R., Morrison, J. D.: *Anal. Chem.* 40, 1968, 1469–1474. Computer methods in analytical mass spectrometry. Empirical identification of molecular class.

Croux, C., Dehon, C.: *Can. J. Stat.* 29, 2001, 473–492. Robust linear discriminant analysis using S-estimators.

Davies, A. N.: in Gauglitz, G., Vo-Dinh, T. (Ed.), *Handbook of Spectroscopy*, Vol. 2, Wiley-VCH, Weinheim, Germany, 2003, pp. 488–502. Mass spectrometry.

Drablos, F.: *Anal. Chim. Acta* 256, 1992, 145–151. Transformation of mass spectra.

Elyashberg, M. E., Blinov, K. A., Molodtsov, S. G., Smurnyi, E. D.: *J. Anal. Chem.* 63, 2008, 13–20. New computer-assisted methods for the elucidation of molecular structures from 2-D spectra.

Eriksson, L., Johansson, E., Kettaneh-Wold, N., Trygg, J., Wikström, C., Wold, S.: *Multi- and Megavariate Data Anaylsis*. Umetrics AB, Umea, Sweden, 2006.

Esposito Vinci, V., Tenenhaus, M.: in Esposito Vinci, V., Lauro, C., Morineau, A., Tenenhaus, M. (Ed.), PLS and related methods. *Proceedings of the PLS'01 International Symposium*, CISIA-CERESTA, Paris, France, 2001, pp. 117–130. PLS Logistic Regression.

Fisher, R. A.: *Ann. Eugenic*. 8, 1938, 376–386. The statistical utilization of multiple measurements.

Forina, M., Armanino, C., Leardi, R., Drava, G.: *J. Chemom*. 5, 1991, 435–453. A class-modelling technique based on potential functions.

Funatsu, K., Sasaki, S. I.: *J. Chem. Inf. Comput. Sci*. 36, 1996, 190–204. Recent advances in the automated structure elucidation system, CHEMICS. Utilization of two-dimensional NMR spectral information and development of peripheral functions for examination of candidates.

Gray, N. A. B.: *Computer-Assisted Structure Elucidation*. Wiley, New York, 1986.

Hastie, T., Tibshirani, R. J.: *J. Royal Stat. Soc. B*. 58, 1996, 155–176. Discriminant analysis by Gaussian mixtures.

Hastie, T., Tibshirani, R. J., Friedman, J.: *The Elements of Statistical Learning*. Springer, New York, 2001.

He, X., Fung, W. K.: *J. Multivariate Anal*. 72, 2000, 151–162. High breakdown estimation for multiple populations with applications to discriminant analysis.

Hubert, M., Van Driessen, K.: *Computat. Stat. Data Anal*. 45, 2004, 301–320. Fast and robust discriminant analysis.

Huberty, C. J.: *Applied Discriminant Analysis*. Wiley, New York, 1994.

Isenhour, T. L., Jurs, P. C.: in Mark, H. B. Jr., Mattson, J. S., Macdonald, H. C. Jr. (Eds.), Computer Fundamentals for Chemists, Marcel Dekker, New York, 1973, pp. 285–330. Learning machines.

Ivanciuc, O.: *Rev. Computat. Chem*. 23, 2007, 291–400. Applications of support vector machines in chemistry.

Johnson, R. A., Wichern, D. W.: *Applied Multivariate Statistical Analysis*. Prentice Hall, Upper Saddle River, NJ, 2002.

Jurs, P. C., Isenhour, T. L.: *Chemical Applications Pattern Recognition*. Wiley, New York, 1975.

Jurs, P. C., Kowalski, B. R., Isenhour, T. L.: *Anal. Chem*. 41, 1969, 21–27. Computerized learning machines applied to chemical problems. Molecular formula determination from low resolution mass spectrometry.

Kleinbaum, D. G., Klein, M.: *Logistic Regression*. Springer, New York, 2002.

Kowalski, B. R., Bender, C. F.: *J. Am. Chem. Soc*. 94, 1972, 5632–5639. Pattern recognition. A powerful approach to interpreting chemical data.

McLachlan, G. J., Pee, D.: *Finite Mixture Models*. Wiley, New York, 2000.

Meisel, W. S.: *Computer-Oriented Approaches to Pattern Recognition*. Academic Press, New York, 1972.

Meyer, D., Leisch, F., Hornik, K.: *Neurocomputing* 55, 2003, 169–186. The support vector machine under test.

Minsky, M. L., Papert, S. A.: *Perceptrons*. MIT Press, Cambridge, MA, 1969.

Munk, M. E.: *J. Chem. Inf. Comput. Sci*. 38, 1998, 997–1009. Computer-based structure determination: Then and now.

Naes, T., Isaksson, T., Fearn, T., Davies, T.: *A User-Friendly Guide to Multivariate Calibration and Classification*. NIR Publications, Chichester, United Kingdom, 2004.

Neudert, R., Penk, M.: *J. Chem. Inf. Comput. Sci*. 36, 1996, 244–248. Enhanced structure elucidation.

Nilsson, N. J.: *Learning Machines*. McGraw Hill, New York, 1965.

NIST: *Mass Spectral Database 98*. National Institute of Standards and Technology, www.nist.gov/srd/nist1a.htm, Gaithersburg, MD, 1998.

Penchev, P. N., Andreev, G. N., Varmuza, K.: *Anal. Chim. Acta* 388, 1999, 145–159. Automatic classification of infrared spectra using a set of improved expert-based features.

Preuss, D. R., Jurs, P. C.: *Anal. Chem.* 46, 1974, 520–525. Pattern recognition techniques applied to the interpretation of infrared spectra.

Rao, C. R.: *J. Royal Stat. Soc., Series B.* 10, 1948, 159–203. The utilization of multiple measurements in problems of biological classification.

Robien, W.: in Gauglitz, G., Vo-Dinh, T. (Ed.), *Handbook of Spectroscopy*, Vol. 2, Wiley-VCH, Weinheim, Germany, 2003, pp. 469–487. Nuclear magnetic resonance spectroscopy.

Rosenblatt, F.: *Proc. IRE* 48, 1960, 301–309. Perceptron simulation experiments.

Steinwart, I., Christmann, A.: *Support Vector Machines*. Springer, New York, 2008.

Thiele, S., Salzer, R.: in Gauglitz, G., Vo-Dinh, T. (Ed.), *Handbook of Spectroscopy*, Vol. 2, Wiley-VCH, Weinheim, Germany, 2003, pp. 441–468. Optical spetcroscopy.

Thissen, U., Pepers, M., Üstün, B., Melssen, W. J., Buydens, L. C. M.: *Chemom. Intell. Lab. Syst.* 73, 2004, 169–179. Comparing support vector machines to PLS for spectral regression applications.

Vandeginste, B. G. M., Massart, D. L., Buydens, L. C. M., De Jong, S., Smeyers-Verbeke, J.: *Handbook of Chemometrics and Qualimetrics: Part B*. Elsevier, Amsterdam, the Netherlands, 1998.

Vanden Branden, K., Hubert, M.: *Chemom. Intell. Lab. Syst.* 79, 2005, 10–21. Robust classification in high dimensions based on the SIMCA method.

Vapnik, V.: *The Nature of Statistical Learning Theory*. Springer, New York, 1995.

Varmuza, K.: *Pattern Recognition in Chemistry*. Springer, Berlin, Germany, 1980.

Varmuza, K.: in Lindon, J. C., Tranter, G. E., Holmes, J. L. (Ed.), *Encyclopedia of Spectroscopy and Spectrometry*, Academic Press, London, United Kingdom, 2000, pp. 232–243. Chemical structure information from mass spectrometry.

Varmuza, K., Werther, W.: *J. Chem. Inf. Comput. Sci.* 36, 1996, 323–333. Mass spectral classifiers for supporting systematic structure elucidation.

Verboven, S., Hubert, M.: *Chemom. Intell. Lab. Syst.* 75, 2005, 127–136. LIBRA: A MATLAB library for robust analysis.

Werther, W., Demuth, W., Krueger, F. R., Kissel, J., Schmid, E. R., Varmuza, K.: *J. Chemom.* 16, 2002, 99–110. Evaluation of mass spectra from organic compounds assumed to be present in cometary grains. Exploratory data analysis.

Will, M., Fachinger, W., Richert, J. R.: *J. Chem. Inf. Comput. Sci.* 36, 1996, 221–227. Fully automated structure elucidation—a spectroscopist's dream comes true.

Wold, S.: *Pattern Recogn.* 8, 1976, 127–139. Pattern recognition by means of disjoint principal component models.

Xu, Y., Dixon, S. J., Brereton, R. G., Soini, H. A., Novotny, M. V., Trebesius, K., Bergmaier, I., Oberzaucher, E., Grammer, K., Penn, D. J.: *Metabolomics* 3, 2007, 427–437. Comparison of human axillary odour profiles obtained by gas chromatography/mass spectrometry and skin microbial profiles obtained by denaturing gradient gel electrophoresis using multivariate pattern recognition.

Xu, Y., Zomer, S., Brereton, R. G.: *Crit. Rev. Anal. Chem.* 34, 2006, 177–188. Support vector machines: A recent method for classification in chemometrics.

6 Cluster Analysis

6.1 CONCEPTS

The term "cluster" has the meaning of "concentrated" group. It usually refers to the objects (in the variable space), but is also used for variables (in the space of the objects), or for both, variables and objects simultaneously. Speaking in terms of the objects, cluster analysis tries to identify concentrated groups (i.e., clusters) of objects, while no information about any group membership is available, and usually not even the number of clusters is known. In other words, cluster analysis tries to find groups containing similar objects (Everitt 1974; Gordon 1999; Kaufmann and Rousseeuw 1990; Massart and Kaufmann 1983; Ripley 1996). It is thus a method for UNSUPERVISED LEARNING, while in Chapter 5 (Classification) we treat methods for SUPERVISED LEARNING that require known group memberships at least for a training data set. In the following, we will focus on cluster analysis with the goal of identifying groups of objects.

The task of identifying concentrated groups of objects presumes that such a group structure is inherent in the data. It does, however, in general not assume that an object belongs to only one group, but it could be part of two or even more groups. Thus, clustering methods that perform a partitioning of the objects into separated groups will not always give the desired solution. For this reason many clustering algorithms have been proposed in the literature that do not only perform differently, but that even work on different principles. Further problems arise because the shape and size of clusters may be very different (Figure 6.1). The most important methods are

- PARTITIONING METHODS: Each object is assigned to exactly one group (Section 6.3).
- HIERARCHICAL METHODS: Objects and partitions are arranged in a hierarchy. An appropriate graphical representation of the result is a tree-like dendrogram (Section 6.4). It allows to determine manually the "optimal" number of clusters as well as to see the hierarchical relations between different groups of objects.
- FUZZY CLUSTERING METHODS: Each object is assigned by a membership coefficient to each of the found clusters. Usually, the membership coefficients are normalized to fall in the interval [0, 1], and thus they can be interpreted as probability that an object is assigned to anyone of the clusters (Section 6.5).
- MODEL-BASED CLUSTERING: The different clusters are supposed to follow a certain model, like a multivariate normal distribution with a certain mean and covariance (Section 6.6).

FIGURE 6.1 Different shapes of two-dimensional clusters: spherical, ellipsoidal, linear, crescent, ring, and spiral.

All these methods follow the idea of classifying the objects with respect to their similarity. Such a classification, however, can also be done "by eye," if an appropriate graphical representation of the data is available.

- PRINCIPAL COMPONENT ANALYSIS (PCA) (Chapter 3) and FACTOR ANALYSIS (Section 3.8.1): The first few principal components or factors represent a relevant part of the total data variance. Thus, when plotting pairs of principal component scores or factors, the data structure can be visually inspected in two dimensions in order to identify groups of objects. This approach works fine as long as objects of different groups are sufficiently different in the variable space, and the multidimensional space can be well represented by a projection (low intrinsic dimensionality). In practice, this is often the case.
- KOHONEN MAPPING (Section 3.8.3) and SAMMON'S NONLINEAR MAPPING (Section 3.8.4): These nonlinear methods can be seen as clustering algorithms because the distance information between the objects is represented in a condensed form for graphical inspection. Only the assignment of the objects to the clusters is not done by the algorithm and thus needs to be done "manually."
- CHERNOFF FACES: It is one of the many approaches to represent multivariate objects by ICON PLOTS that can be easily recognized and clustered by humans, and therefore can also be used for classification of objects. Various multivariate graphical representations have been suggested, such as stars, pie charts, polygons, flowers, castles (Everitt 1978) to visualize object vectors by visual patterns. Because humans are well trained to recognize faces Chernoff (1973) proposed to transform a vector into a cartoon face with characteristic features of the face (shape, length of nose, width of mouth, size of eyes, etc.) being controlled by the values of the vector components. Small sets of variables ($m = 5, \ldots, 15$) can be easily considered for parameterized faces; however, the variables must be carefully selected to appropriate face features—for instance, to obtain a smiling face if clinical medical data indicate no disease (Honda and Aida 1982). This selection is thus rather subjective, and a different assignment of the variables to face features could result in a completely different appearance of the symbols. A number of variations of the original Chernoff faces have been proposed (Flury and Riedwyl 1988; Otto 2007), and relationships among the chemical elements have been demonstrated by faces (Larsen 1986). Not only visual icons can be used to represent multivariate data of an object in a "human way," for

instance, AUDIO REPRESENTATIONS have been suggested for chemical analytical data (Sweeley et al. 1987; Yeung 1980), and for mass spectra (Varmuza 1986) (Figure 6.2), other human senses may follow in the future.

The outcome of any cluster analysis procedure are assignments of the objects to the clusters, where objects within a cluster are supposed to be similar to each other, and objects from different clusters are supposed to be dissimilar. This raises further issues.

- Closeness of objects needs to be measured, which is mostly done by a distance or similarity measure (Section 6.2).
- Since the "correct" number of clusters is unknown, a cluster validity measure needs to be consulted for the evaluation of the clustering solution (Section 6.7).

Usually one cannot expect a unique solution for cluster analysis. The result depends on the used distance measure, the cluster algorithm, and the chosen parameters; often

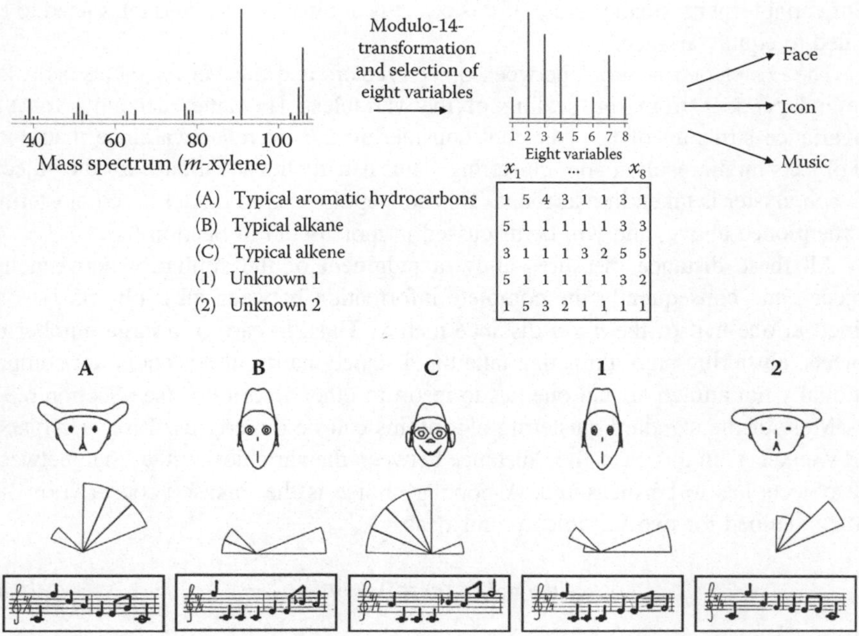

FIGURE 6.2 Representation of multivariate data by icons, faces, and music for human cluster analysis and classification in a demo example with mass spectra. Mass spectra have first been transformed by modulo-14 summation (see Section 7.4.4) and from the resulting 14 variables, 8 variables with maximum variance have been selected and scaled to integer values between 1 and 5. **A**, typical pattern for aromatic hydrocarbons; **B**, typical pattern for alkanes; **C**, typical pattern for alkenes; **1** and **2**, "unknowns" (2-methyl-heptane and meta-xylene). The 5×8 data matrix has been used to draw faces (by function "faces" in the ℝ-library "TeachingDemos"), segment icons (by ℝ-function "stars"), and to create small melodies (Varmuza 1986). Both unknowns can be easily assigned to the correct class by all three representations.

also from initial conditions. Application of complementary methods is advised, for instance, PCA, hierarchical cluster analysis, and fuzzy cluster analysis. The final success of a cluster analysis is determined whether the found clusters can be assigned to problem-relevant groups or conditions, or not. Application of unsupervised methods is often a recommendable starting step in data evaluation in order to obtain an insight into the data structure (to detect clusters or outliers) which may be important for a following development of classification or calibration models.

6.2 DISTANCE AND SIMILARITY MEASURES

Distance measures were already discussed in Section 2.4. The most widely used distance measure for cluster analysis is the EUCLIDEAN DISTANCE. The MANHATTAN DISTANCE would be less dominated by far outlying objects since it is based on absolute rather than squared differences. The MINKOWSKI DISTANCE is a generalization of both measures, and it allows adjusting the power of the distances along the coordinates. All these distance measures are not scale invariant. This means that variables with higher scale will have more influence to the distance measure than variables with smaller scale. If this effect is not wanted, the variables need to be scaled to equal variance.

The COSINE OF THE ANGLE between object vectors and the MAHALANOBIS DISTANCE are independent from the scaling of the variables. The latter accounts for the covariance structure of the data, but considering the overall covariance matrix of all objects might result in poor clustering. Thus usually the covariance for the objects in each cluster is taken into account. This concept is used in model-based clustering as mentioned above, and will be discussed in more detail in Section 6.6.

All these distance measures allow a judgment of the similarity between the objects, and consequently the complete information between all n objects is contained in one-half of the $n \times n$ distance matrix. Thus, in case of a large number of objects, clustering algorithms that take the distance matrix into account are computationally not attractive, and one has to resort to other algorithms (see Section 6.3).

Most of the standard clustering algorithms can be directly used for CLUSTERING THE VARIABLES. In this case, the "distance between the variables" rather than between the objects has to be measured. A popular choice is the PEARSON CORRELATION DISTANCE, defined for two variables x_j and x_k as

$$d_{\text{CORR}}(x_j, x_k) = 1 - |r_{jk}| \tag{6.1}$$

where r_{jk} is the Pearson correlation coefficient between variables x_j and x_k (see Section 2.3.2). A more robust distance measure could be achieved by replacing the Pearson correlation in Equation 6.1 by the Spearman rank correlation or by Kendall's tau correlation, see Section 2.3.2.

An alternative clustering method for variables is to use the correlation coefficient matrix in which each variable is considered as an object, characterized by the correlation coefficients to all other variables. PCA and other unsupervised methods can be applied to this matrix to obtain an insight into the similarities between the original variables.

Chemical structures are often characterized by BINARY VECTORS in which each vector component (with value 0 or 1) indicates absence or presence of a certain substructure (BINARY SUBSTRUCTURE DESCRIPTORS). An appropriate and widely used similarity measure for such binary vectors is the TANIMOTO INDEX (Willett 1987), also called JACCARD SIMILARITY COEFFICIENT (Vandeginste et al. 1998). Let $\boldsymbol{x}_{\mathrm{A}}$ and $\boldsymbol{x}_{\mathrm{B}}$ be binary vectors with m components for two chemical structures A and B, respectively. The Tanimoto index t_{AB} is given by

$$t_{\mathrm{AB}} = \frac{\sum_j \mathrm{AND}(x_{\mathrm{A}j}, x_{\mathrm{B}j})}{\sum_j \mathrm{OR}(x_{\mathrm{A}j}, x_{\mathrm{B}j})} \tag{6.2}$$

where

$\sum_j \mathrm{AND}(x_{\mathrm{A}j}, x_{\mathrm{B}j})$ is for the number of variables with a "1" in both vectors (logical AND)

$\sum_j \mathrm{OR}(x_{\mathrm{A}j}, x_{\mathrm{B}j})$ is the number of variables with a "1" in at least one of the vectors (logical OR)

In vector notation, Equation 6.2 can be written as

$$t_{\mathrm{AB}} = \frac{\boldsymbol{x}_{\mathrm{A}}^{\mathrm{T}} \cdot \boldsymbol{x}_{\mathrm{B}}}{\boldsymbol{x}_{\mathrm{A}}^{\mathrm{T}} \cdot \mathbf{1} + \boldsymbol{x}_{\mathrm{B}}^{\mathrm{T}} \cdot \mathbf{1} - \boldsymbol{x}_{\mathrm{A}}^{\mathrm{T}} \cdot \boldsymbol{x}_{\mathrm{B}}} \tag{6.3}$$

where $\mathbf{1}$ denotes a vector with ones of the same length as $\boldsymbol{x}_{\mathrm{A}}$ and $\boldsymbol{x}_{\mathrm{B}}$. The Tanimoto index is in the range of 0–1; the value 1 is obtained if all descriptors are pairwise equal. The Tanimoto index considers the fact that usually most of the descriptors are zero—simply because the high diversity of chemical structures requires a large number of different substructures for characterization. It is also reasonable that the presence of a substructure in both structures indicates similarity, while the absence in both structures has no meaning. For such binary vectors the Tanimoto index is therefore better suited than the HAMMING DISTANCE

$$d_{\mathrm{HAMMING}} = \sum_j \mathrm{XOR}(x_{\mathrm{A}j}, x_{\mathrm{B}j}) \tag{6.4}$$

with $\sum_j \mathrm{XOR}(x_{\mathrm{A}j}, x_{\mathrm{B}j})$ for the number of variables with different values in both vectors (logical exclusive OR). Note that the Hamming distance is equivalent to the squared Euclidean distance and equivalent to the Manhattan distance.

The Tanimoto index is a similarity measure; a corresponding distance measure d_{TANI}—also called ASYMMETRIC BINARY DISTANCE—is

$$d_{\mathrm{TANI}} = 1 - t_{\mathrm{AB}} = \frac{\sum_j \mathrm{XOR}(x_{\mathrm{A}j}, x_{\mathrm{B}j})}{\sum_j \mathrm{OR}(x_{\mathrm{A}j}, x_{\mathrm{B}j})} \tag{6.5}$$

which can be used for a cluster analysis of chemical structures. In $\mathbb{R}$ the distance matrix $\boldsymbol{D}$ containing d_{TANI} can be calculated from a binary descriptor matrix $\boldsymbol{X}$ by

```
ℝ: D <- dist(X, method="binary")  # lower triangle matrix
   D <- dist(X, method="binary", upper=TRUE, diag=TRUE)
                                        # complete distance matrix
```

The distribution of Tanimoto indices for randomly selected (or all) pairs of structures characterizes the diversity of a chemical structure database (Demuth et al. 2004; Scsibrany et al. 2003). For structure similarity searches, a number of other similarity measures have been suggested (Gasteiger and Engel 2003; Willett 1987).

To demonstrate the use of binary substructure descriptors and Tanimoto indices for cluster analysis of chemical structures we consider the 20 STANDARD AMINO ACIDS (Figure 6.3) and characterize each molecular structure by eight binary variables describing presence/absence of eight substructures (Figure 6.4). Note that in most practical applications—for instance, evaluation of results from searches in structure databases—more diverse molecular structures have to be handled and usually several hundred different substructures are considered. Table 6.1 contains the binary substructure descriptors (variables) with value "0" if the substructure is absent and "1" if the substructure is present in the amino acid; these numbers form the ***X***-matrix. Binary substructure descriptors have been calculated by the software SubMat (Scsibrany and Varmuza 2004), which requires as input the molecular structures in one file and the substructures in another file, all structures are in Molfile format (Gasteiger and Engel 2003); output is an ASCII file with the binary descriptors.

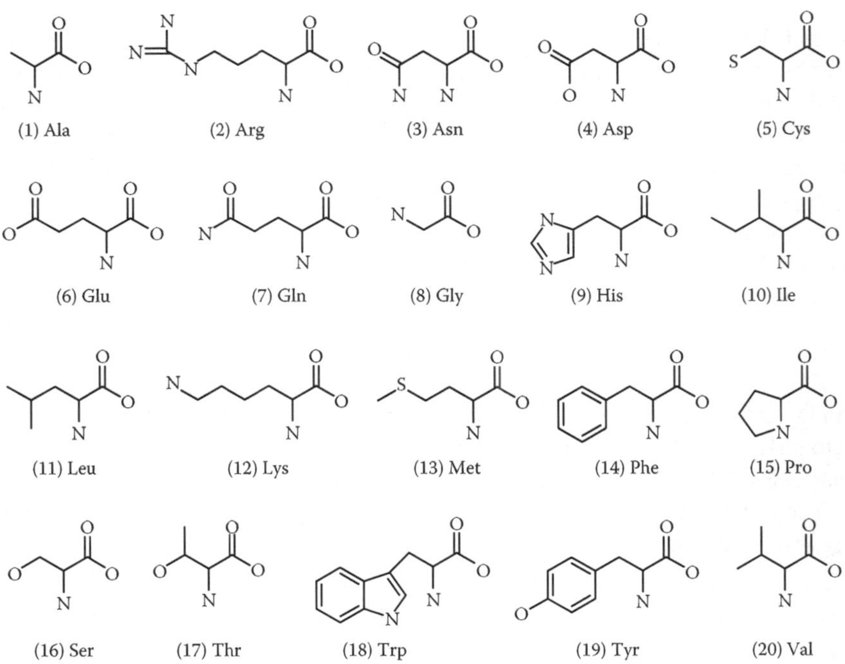

FIGURE 6.3 Twenty standard amino acids used for cluster analysis of chemical structures. For the three-letter codes, see Table 6.1.

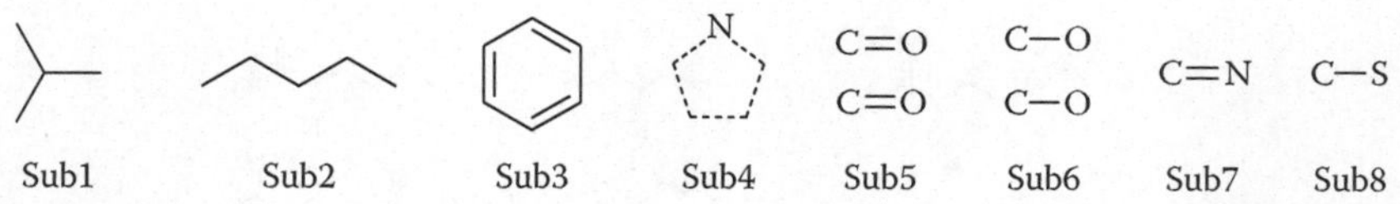

FIGURE 6.4 Eight substructures used for the generation of binary substructure descriptors to characterize the 20 standard amino acids. Sub1, tertiary C-atom; Sub2, C_5-chain; Sub3, benzene ring; Sub4, any 5-membered ring with at least one N-atom; Sub5, additional carbonyl group; Sub6, additional C–OH group; Sub7, C=N; Sub8, C–S.

A cluster analysis of the amino acid structures by PCA of the *X*-matrix is shown in Figure 6.5a; note that PCA optimally represents the Euclidean distances. The score plot for the first two principal components (preserving 27.1% and 20.5% of the total variance) shows some clustering of similar structures. Four structure pairs have identical variables: 1 (Ala) and 8 (Gly), 5 (Cys) and 13 (Met), 10 (Ile) and 11 (Leu), and 16 (Ser) and 17 (Thr). Objects with identical variables of course have identical scores, but for a better visibility the pairs have been artificially

TABLE 6.1
Twenty Standard Amino Acids Characterized by Eight Binary Substructure Descriptors

			Substructure Descriptors							
No	**Amino Acid**	**Code**	**Sub1**	**Sub2**	**Sub3**	**Sub4**	**Sub5**	**Sub6**	**Sub7**	**Sub8**
1	Alanine	Ala	0	0	0	0	0	0	0	0
2	Arginine	Arg	0	1	0	0	0	0	1	0
3	Asparagine	Asn	0	0	0	0	1	0	0	0
4	Aspartic acid	Asp	0	0	0	0	1	1	0	0
5	Cysteine	Cys	0	0	0	0	0	0	0	1
6	Glutamic acid	Glu	0	1	0	0	1	1	0	0
7	Glutamine	Gln	0	1	0	0	1	0	0	0
8	Glycine	Gly	0	0	0	0	0	0	0	0
9	Histidine	His	0	0	0	1	0	0	1	0
10	Isoleucine	Ile	1	1	0	0	0	0	0	0
11	Leucine	Leu	1	1	0	0	0	0	0	0
12	Lysine	Lys	0	1	0	0	0	0	0	0
13	Methionine	Met	0	0	0	0	0	0	0	1
14	Phenylalanine	Phe	0	0	1	0	0	0	0	0
15	Proline	Pro	0	1	0	1	0	0	0	0
16	Serine	Ser	0	0	0	0	0	1	0	0
17	Threonine	Thr	0	0	0	0	0	1	0	0
18	Tryptophan	Trp	0	1	1	1	0	0	0	0
19	Tyrosine	Tyr	0	0	1	0	0	1	0	0
20	Valine	Val	1	0	0	0	0	0	0	0

Note: For chemical structures, see Figures 6.3 and 6.4.

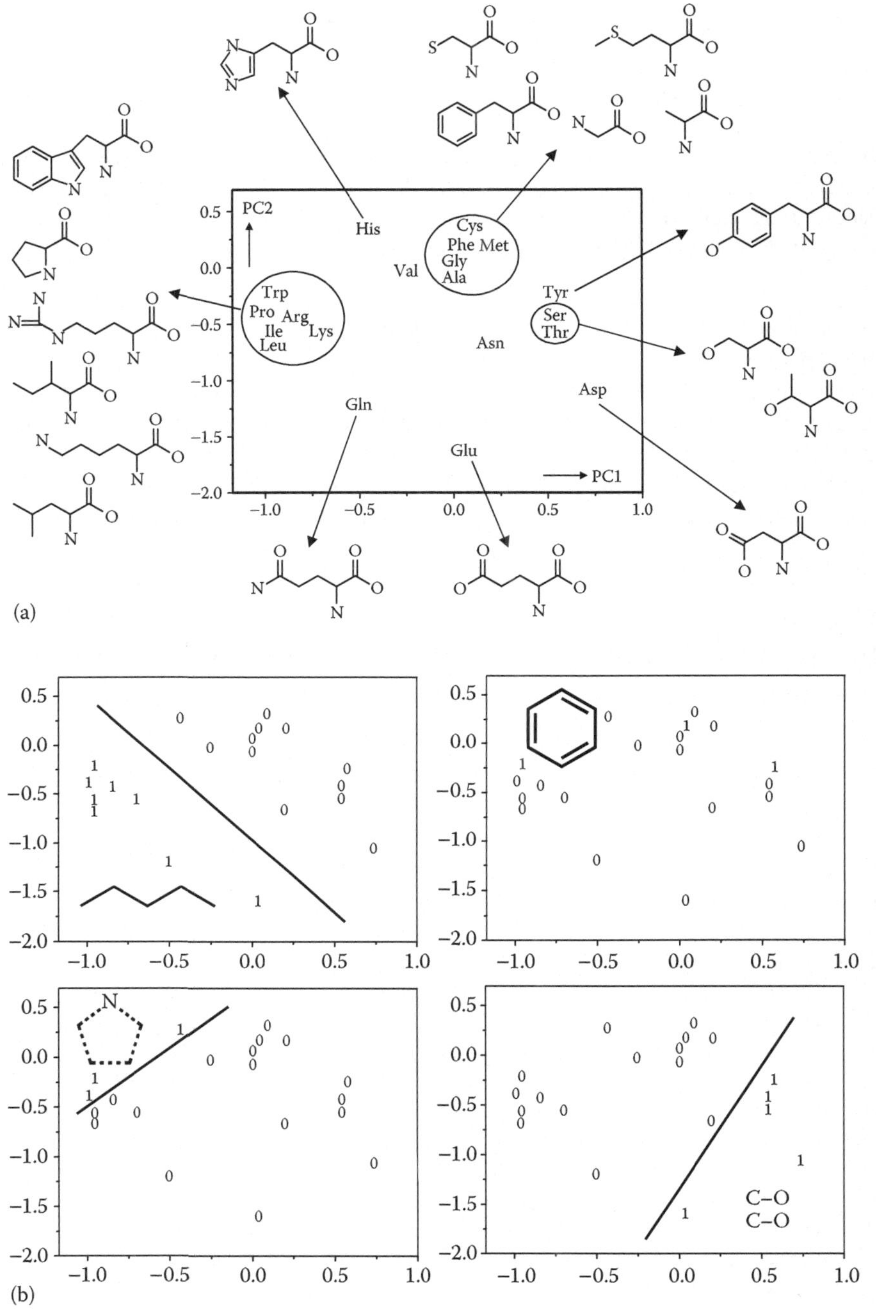

FIGURE 6.5 PCA score plot (a) of $n = 20$ standard amino acid structures characterized by $m = 8$ binary descriptors (27.1% and 20.5% of the total variance preserved in PC1 and PC2). In the lower plots (b) presence/absence of selected four substructures is indicated.

separated in the plot. The four plots in Figure 6.5b show that amino acids containing the substructures 2 (C_5-chain), 4 (5-membered ring with nitrogen), or 6 (an additional C-OH) are well separated from the amino acids not containing these substructures; however, amino acids with a benzene ring do not form a cluster in this PCA score plot.

Hierarchical cluster analysis (Section 6.4)—with the result represented by a dendrogram—is a complementary, nonlinear, and widely used method for cluster analysis. The distance measure used was d_{TANI} (Equation 6.5), and the cluster mode was "average linkage." The dendrogram in Figure 6.6 (without the chemical structures) was obtained from the descriptor matrix $\boldsymbol{X}$ by

```
ℝ: D <- dist(X, method= "binary")
   res <- hclust(D, method= "average")
   plot(res)
```

Four pairs of structures with identical descriptors merge at a distance of zero. From the chemist's point of view clustering appears more satisfying than the linear projection method PCA (with only 47.6% of the total variance preserved by the first two PCA scores). A number of different clustering algorithms have been applied to the 20 standard amino acids by Willet (1987).

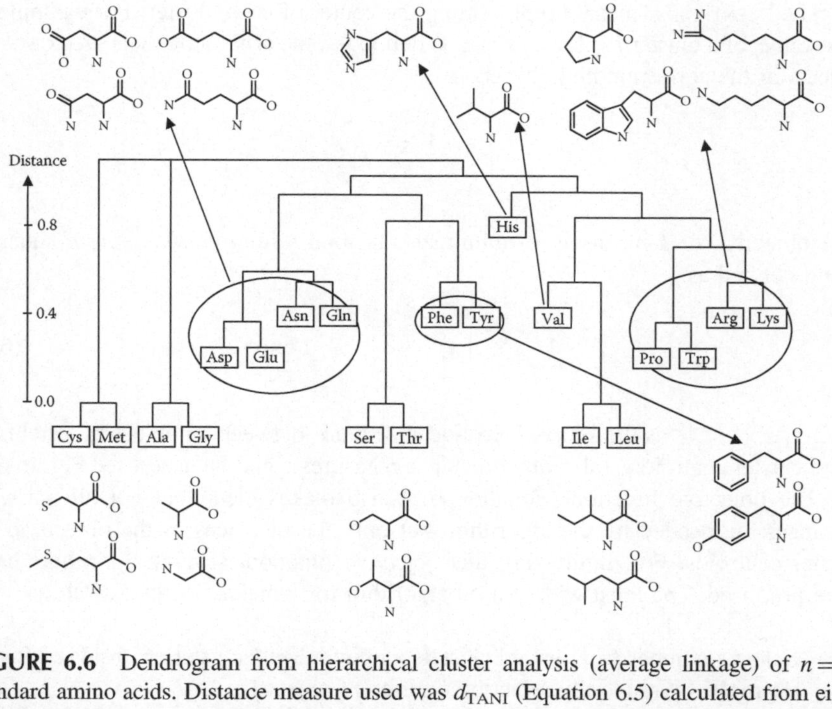

FIGURE 6.6 Dendrogram from hierarchical cluster analysis (average linkage) of $n = 20$ standard amino acids. Distance measure used was d_{TANI} (Equation 6.5) calculated from eight binary substructure descriptors. Four structure pairs with identical descriptors merge at a distance of zero. Clustering widely corresponds to the chemist's point of view.

6.3 PARTITIONING METHODS

Given objects $\boldsymbol{x}_1, \ldots, \boldsymbol{x}_n$ which have been measured for m variables, partitioning methods aim at assigning each object to one of k different clusters. A cluster must consist of at least one object, and can include at most n objects—in this case only one cluster with all objects exists. We will use the notation $\boldsymbol{x}_i^{(j)}$ if an observation i (for $i=1, \ldots, n$) has been assigned to a cluster j (for $j=1, \ldots, k$). Furthermore, n_j denotes the number of objects that have been assigned to cluster j. Thus the clustered objects are denoted by $\boldsymbol{x}_1^{(1)}, \ldots, \boldsymbol{x}_{n_1}^{(1)}, \boldsymbol{x}_1^{(2)}, \ldots, \boldsymbol{x}_{n_2}^{(2)}, \ldots, \boldsymbol{x}_1^{(k)}, \ldots, \boldsymbol{x}_{n_k}^{(k)}$. For partitioning methods we have $n_1 + \cdots + n_k = n$.

Many different algorithms have been proposed for partitioning the objects. Often, the algorithms are adapted to the type or geometry of the data. For example, for the recognition of contours of images, the clustering method should be able to identify long-stretched linear and curved clusters, while for measurements of features on persons, elliptically shaped clusters may be more appropriate. Most algorithms use the data matrix, $\boldsymbol{X}$, and the desired number k of clusters as input information. The result—n numbers with the assignments of the objects to the k clusters—depends on random initializations used in the clustering algorithm; however, a fast clustering algorithm can be run many (say 100) times, and the most frequent outcome can be taken as the final result.

The most widely known algorithm for partitioning is the k-MEANS ALGORITHM (Hartigan 1975). It uses pairwise distances between the objects, and requires the input of the desired number k of clusters. Internally, the k-means algorithm uses the so-called CENTROIDS (means) representing the center of each cluster. For example, a centroid $\boldsymbol{c}_j$ of a cluster $j=1, \ldots, k$ can be defined as the arithmetic mean vector of all objects of the corresponding cluster, i.e.,

$$\boldsymbol{c}_j = \frac{1}{n_j} \sum_{i=1}^{n_j} \boldsymbol{x}_i^{(j)} \tag{6.6}$$

The objective of k-means is to minimize the total within-cluster sum-of-squares, being defined as

$$\sum_{j=1}^{k} \sum_{i=1}^{n_j} ||\boldsymbol{x}_i^{(j)} - \boldsymbol{c}_j||^2 \rightarrow \min \tag{6.7}$$

where $||\boldsymbol{x}_i^{(j)} - \boldsymbol{c}_j||^2$ is the squared Euclidean distance between an object $\boldsymbol{x}_i^{(j)}$ of cluster j and cluster centroid $\boldsymbol{c}_j$ (also other distance measures could be used), see Figure 6.7.

The objective function (Equation 6.7) also makes clear that not all pairwise distances are needed by the algorithm, but only the distances of the objects to all cluster centroids. For minimizing this objective function, several algorithms have been proposed. The most widely used algorithm for k-means works as follows:

1. Select a number k of desired clusters and initialize k cluster centroids $\boldsymbol{c}_j$, for example, by randomly selecting k different objects.
2. Assign each object to the cluster with the closest centroid, i.e., compute for each object the distance $||\boldsymbol{x}_i - \boldsymbol{c}_j||$ for $j=1, \ldots, k$ and assign $\boldsymbol{x}_i$ to the cluster where the minimum distance to the cluster centroid appears.

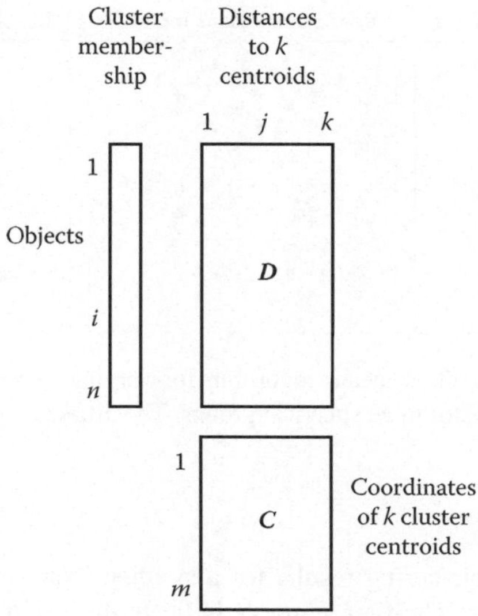

FIGURE 6.7 Matrices for *k*-means clustering of *n* objects with *m* variables.

3. Recompute the cluster centroids using Equation 6.6 for the new object assignments.
4. Repeat steps 2 and 3 until the centroids become stable.

The algorithm usually always converges; however, it does not necessarily find the global minimum of the objective function (Equation 6.7). The outcome of the *k*-means algorithm also depends on the initialization of the cluster centroids in step 1. As a possible solution, the algorithm can be run several times to reduce this drawback.

The number *k* of clusters being inherent in the data set is usually unknown, but it is needed as an input of the *k*-means algorithm. Since the algorithm is very fast, it can be run for a range of different numbers of clusters, and the "best" result can be selected. Here, "best" refers to an evaluation of the results by cluster validity measures (see Section 6.7).

In chemoinformatics and bioinformatics JARVIS–PATRICK CLUSTERING is a popular method for cluster analysis (Willett 1987). This method (Jarvis and Patrick 1973) uses a nearest-neighbor table of size $n \times g$ containing the *g*-nearest neighbors of all *n* objects (any usual distance measure can be applied). Two objects are placed into the same cluster if they are near neighbors and share a minimum number, *z*, of other objects as near neighbors. The first parameter, *g*, must be at least 2; a larger value guides to a small number of clusters and clusters may form chains. The other parameter, *z*, must be between 1 and *g*; low values guide to more compact clusters. The algorithm partitions the objects into nonoverlapping, nonhierarchical groups, and is recommended for clusters with complicated shapes.

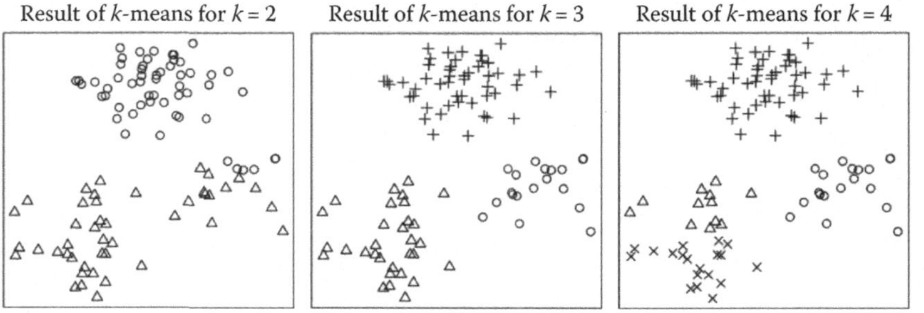

FIGURE 6.8 Results of the *k*-means algorithm for varying number of clusters, *k*, for an artificial data set consisting of three spherical groups. The different symbols correspond to the cluster results.

Figure 6.8 shows clustering results for a synthetic data example in two dimensions with three groups. The plot symbols indicate the result of *k*-means clustering for $k=2$ (left), $k=3$ (middle), and $k=4$ (right). Here it is obvious that the choice $k=3$ gave the best result since it directly corresponds to the visually evident data groups.

Figure 6.9 also shows clustering results for another synthetic data example, again with three groups, but they are not spherical like in Figure 6.8 but two of them are elliptical. The *k*-means algorithm has difficulties with this configuration, and it tends to find spherical clusters. This results from the objective function (Equation 6.7) which accounts for distances symmetrically around a centroid. Methods treated later in this chapter will be able to handle such a data configuration.

```
R: res <- kmeans(X,3) # k-means clustering of X with 3 clusters
```

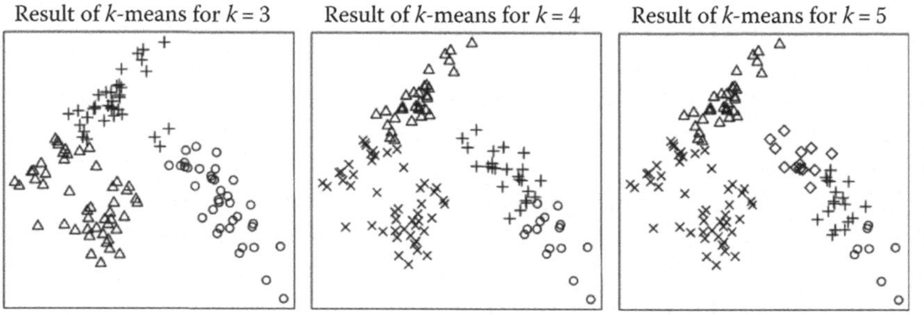

FIGURE 6.9 Results of the *k*-means algorithm for varying number of clusters, *k*, for an artificial data set consisting of one spherical and two elliptical groups. The different symbols correspond to the cluster results. None of the results identifies the three groups correctly because *k*-means tends to find spherical clusters.

6.4 HIERARCHICAL CLUSTERING METHODS

The standard hierarchical clustering algorithms produce a whole set of cluster solutions, namely a partitioning of the objects into $k=1,\ldots, n$ clusters. The partitions are ordered hierarchically, and there are two possible procedures:

- AGGLOMERATIVE METHODS: In the first level of the hierarchy, each of the n objects forms a separate cluster, resulting in n clusters. In the next level the two closest clusters are merged, and so on, until finally all objects are in one single cluster.
- DIVISIVE METHODS: In the first level of the hierarchy, all n objects are in one single cluster. In the next level, this cluster is split into two smaller clusters. The next level again splits one of the clusters into two smaller clusters, and so on, until finally each object forms a separate cluster.

Splitting a cluster is computationally more demanding than merging two clusters, because not only the cluster to be split has to be found, but also the objects that will form two new clusters have to be identified. Therefore, divisive methods are not very commonly used.

The basic information for splitting or merging clusters is the similarity or distance between the clusters. In Section 6.2 we only mentioned methods for measuring the distance between objects, but not for groups of objects forming clusters. Let us denote $\boldsymbol{x}_i^{(j)}$ and $\boldsymbol{x}_i^{(l)}$ for $i=1,\ldots, n$ as all observations that have been assigned to cluster j and l, respectively, with the cluster sizes n_j and n_l. Then the DISTANCE BETWEEN the TWO CLUSTERS with index j and l can be determined by various methods, where the following are most frequently used:

- COMPLETE LINKAGE: $\max_i ||\boldsymbol{x}_i^{(j)} - \boldsymbol{x}_i^{(l)}||$
- SINGLE LINKAGE: $\min_i ||\boldsymbol{x}_i^{(j)} - \boldsymbol{x}_i^{(l)}||$
- AVERAGE LINKAGE: $\text{average}_i ||\boldsymbol{x}_i^{(j)} - \boldsymbol{x}_i^{(l)}||$
- CENTROID METHOD: $||\boldsymbol{c}_i - \boldsymbol{c}_j||$
- WARD'S METHOD: $||\boldsymbol{c}_i - \boldsymbol{c}_j|| \cdot \frac{\sqrt{2n_jn_l}}{\sqrt{n_j+n_l}}$

We used the Euclidean distance in these definitions, but also other distance measures can be considered (Section 6.2). Complete linkage takes the maximum of all pairwise distances between those objects that belong to cluster j and those that belong to cluster l. So, if even only two objects in both clusters are far away, the distance between both clusters will be large. In contrast, single linkage uses the minimum of the pairwise distances. Thus, the distance between two clusters will be small as soon as there exists one object in one cluster that is close to one or another object in the other cluster, independent on how far the remaining objects are. Average linkage computes the average distance between all pairs of objects in the two clusters, i.e., it computes the sum of all pairwise cluster distances and divides by the number of pairs $n_j \cdot n_l$. The centroid method uses the distance between the cluster centroids $\boldsymbol{c}_j$ and $\boldsymbol{c}_l$. This, however, does not lead to strictly increasing distances within the clustering

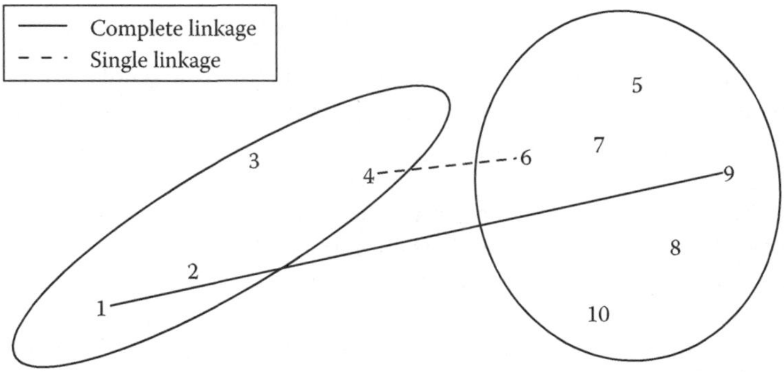

FIGURE 6.10 Distance measure between two clusters, and the concepts of complete and single linkage are shown. Complete linkage takes the maximum of all pairwise distances between the objects of the two clusters, while single linkage uses the minimum. The remaining methods discussed in the text are based on averages of all pairwise distances, or on distances between the cluster centroids.

procedure, and thus a visualization of the results is difficult. This problem is corrected by the factor used in Ward's method. An illustration of complete and single linkage is shown in Figure 6.10. The other methods are based on averages or on distances between the cluster centroids, and are thus somewhere in between single and complete linkage.

Coming back to agglomerative clustering, we can now outline an algorithm:

1. Define each object as a separate cluster and compute all pairwise distances.
2. Merge those clusters (=objects) with the smallest distance into a new cluster.
3. Compute the distances between all clusters using complete linkage, single linkage, average linkage, or other methods.
4. Merge those clusters with smallest distance from step 3.
5. Proceed with steps 3 and 4 until only one cluster remains.

It can be seen that this algorithm is computationally expensive and memory demanding if the number of objects is large. The computation of the distances in subsequent steps is also based on this distance matrix. Complete linkage usually results in homogeneous clusters in the early stages of the agglomeration, but the resulting clusters will be small. Single linkage is known for the chaining effect, because even quite inhomogeneous clusters can be linked just by chance as soon as two objects are very similar.

The strategy of complete linkage leads for the example shown in Figure 6.10 to the hierarchy displayed in Figure 6.11. The algorithm starts with linking pairs of objects (visualized by the connecting lines) into new clusters of size 2. Then single objects are merged to the clusters, and finally clusters are merged. This merging is indicated by the ellipses, where the increasing line thickness corresponds to higher levels in the hierarchy.

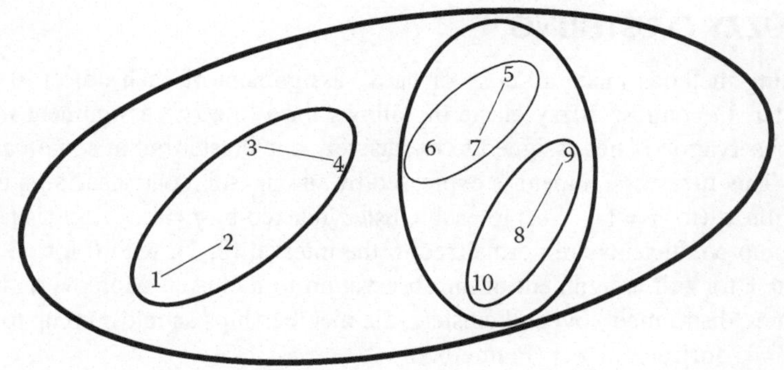

FIGURE 6.11 Resulting hierarchy from complete linkage applied to the data of Figure 6.10. Thicker lines correspond to higher levels in the hierarchy, and thus to larger clusters.

The results of hierarchical clustering are usually displayed by a DENDROGRAM. For the example used in Figure 6.10 the resulting dendrograms for complete and single linkage are shown in Figure 6.12. This tree-like diagram shows the objects as "leaves" at the bottom, and branches merge according to the order given by the algorithm. This presentation is appropriate because the objects in clusters of a lower level of the hierarchy are always a subset of the objects in clusters of a higher level of the hierarchy. The dendrogram shows the evolution of the cluster tree from bottom to top, in the scale of the cluster distance measure. It can be used for identifying the number of clusters that is most likely inherent in the data structure. If such a number, say k, exists, we would expect k clear branches of the complete tree, i.e., when merging the k clusters to $k - 1$ clusters, the cluster distance (height) will increase considerably. A given number of clusters can be achieved by cutting the tree at a certain height.

```
ℝ: res<-hclust(dist(X),method="complete")  #hierarchical
                        # cluster analysis using complete linkage
   plot(res)            # plot results with dendrogram
```

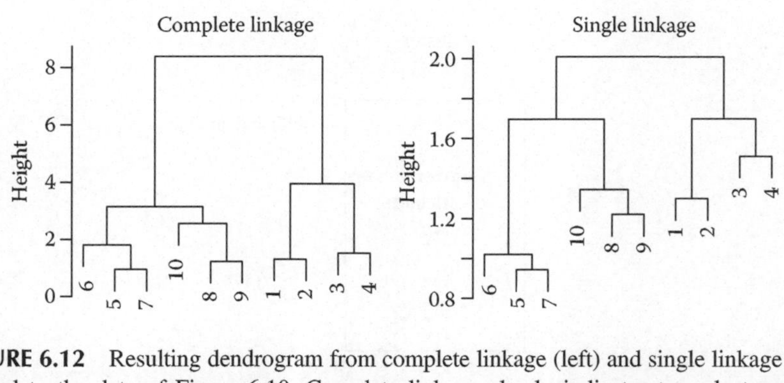

FIGURE 6.12 Resulting dendrogram from complete linkage (left) and single linkage (right) applied to the data of Figure 6.10. Complete linkage clearly indicates two clusters, single linkage two to four clusters.

6.5 FUZZY CLUSTERING

Partitioning methods make a "crisp or hard" assignment of each object to exactly one cluster. In contrast, fuzzy clustering allows for a "fuzzy" assignment meaning that an observation is not assigned to exclusively one cluster but at some part to all clusters. This fuzzy assignment is expressed by MEMBERSHIP COEFFICIENTS u_{ij} for each observation $\boldsymbol{x}_i$ (for $i = 1, \ldots, n$) to each cluster indexed by $j = 1, \ldots, k$. Usually, the membership coefficients are normalized to the interval [0, 1], with 0 for no assignment and 1 for full assignment of an observation to a cluster. Moreover, since the objects are "distributed" over all clusters, the memberships should sum up to 1, thus $\sum_{j=1}^{k} u_{ij} = 1$ for each object (Figure 6.13).

Also for fuzzy clustering different algorithms exist being designed for different types of applications and needs. The first and most prominent algorithm is the FUZZY C-MEANS ALGORITHM (Bandemer and Näther 1992; Bezdek 1981; Dunn 1973). The objective function is similar to k-means:

$$\sum_{j=1}^{k} \sum_{i=1}^{n_j} u_{ij}^2 \cdot ||\boldsymbol{x}_i^{(j)} - \boldsymbol{c}_j||^2 \rightarrow \min \tag{6.8}$$

Instead of using the Euclidean distance, also other distance measures can be considered. Moreover, another power than 2 could be used for the membership coefficients, which will change the characteristics of the procedure (degree of fuzzification). Similar to k-means, the number of clusters k has to be provided as an input, and the algorithm also uses cluster centroids $\boldsymbol{c}_j$ which are now computed by

$$\boldsymbol{c}_j = \frac{\sum\limits_i u_{ij}^2 \cdot \boldsymbol{x}_i}{\sum\limits_i u_{ij}^2} \tag{6.9}$$

So, the cluster centroids are weighted averages of all observations, with weights based on the membership coefficients of all observations to the corresponding cluster. When using only memberships of 0 and 1, this algorithm reduces to k-means.

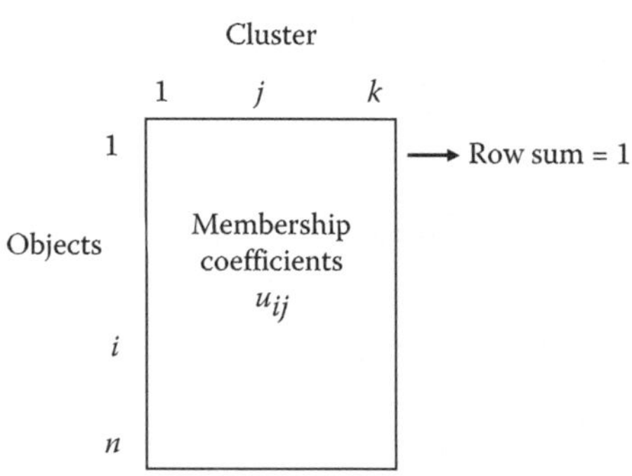

FIGURE 6.13 Fuzzy clustering uses membership coefficients to assign each object with varying probabilities to all k clusters.

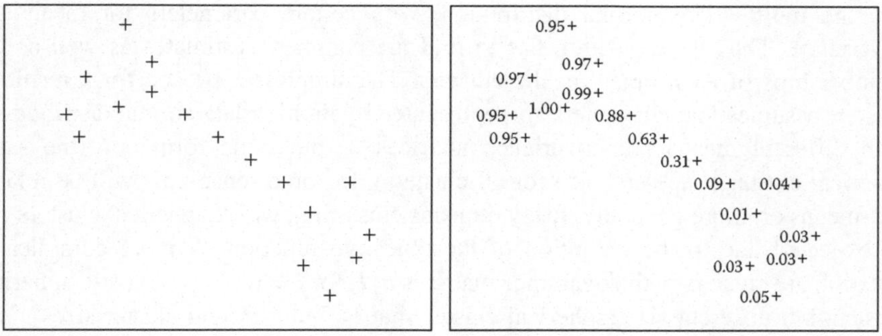

FIGURE 6.14 Example with two groups connected by a bridge of some objects (left) and resulting membership coefficients from fuzzy clustering for the left-hand side cluster.

Minimization of the objective function (Equation 6.8) is done by an iterative optimization. In each iteration step the membership coefficients are updated by

$$u_{ij} = \frac{1}{\sum_{l=1}^{k} \left(\|\boldsymbol{x}_i - \boldsymbol{c}_j\| / \|\boldsymbol{x}_i - \boldsymbol{c}_l\| \right)^2} \tag{6.10}$$

and the cluster centroids are recalculated by Equation 6.9. The iteration is stopped if the membership coefficients, or alternatively the cluster centers, change only marginally. Similar to the k-means algorithm, fuzzy c-means can be run for several values k of the number of clusters, and the "best" result can be chosen (see Section 6.7).

Figure 6.14 shows a demo example of two clusters that are connected by some points (left). In the right picture the membership coefficients resulting from the fuzzy c-means algorithm are displayed for the left-hand side cluster (note that the membership coefficients for the right-hand side cluster are the differences to 1). While the cluster memberships are close to 1 or 0 for objects belonging to the constructed groups, the memberships become intermediate in between. If a crisp membership to the clusters is needed, the objects can be assigned to one of the clusters by using a threshold of 0.5, corresponding to the reciprocal of the number of clusters.

```
ℝ:  library(e1071)
    res <- cmeans(X,3)   # fuzzy clustering with 3 clusters
```

6.6 MODEL-BASED CLUSTERING

While the previously discussed clustering methods are free of any assumptions about the data (except that the data should have an inherent grouping structure), model-based clustering assumes a statistical model of the clusters (Fraley and Raftery 1998). Each cluster is supposed to be represented by a statistical distribution, like the multivariate normal distribution, with certain parameters for mean and covariance. Thus the algorithm has to find the parameter estimates as well as the memberships of each object to the clusters. The simplest approach for k resulting clusters assumes that all clusters are represented by multivariate normal distributions with different means but covariance matrices of the same form $\sigma^2 \boldsymbol{I}$ (the same spherical cluster shape and size for all clusters). In some sense, this will be related to k-means or, more generally, fuzzy c-means clustering, where spherical clusters can be expected due to the definition of the objective function. A more complicated situation are clusters with covariance matrices $\sigma_j^2 \boldsymbol{I}$, for $j = 1, \ldots, k$, i.e., still spherical clusters but different size (spherical cluster shapes, but different cluster sizes). In a third type of cluster models the covariance matrix has not a diagonal form (as in the previous two model types) which is required to model elliptically symmetric clusters. Thus, the most general form are clusters with different covariance matrices $\boldsymbol{\Sigma}_j$.

Figure 6.15 visualizes the different cluster models mentioned above. The left picture is the result of using a model with the same form $\sigma^2 \boldsymbol{I}$ for all clusters. The middle picture changes the cluster size with $\sigma_j^2 \boldsymbol{I}$. The right picture shows the most general cluster model, each having a different covariance matrix $\boldsymbol{\Sigma}_j$. Clearly, there exist several more different possible model classes.

The solution for model-based clustering is based on the EXPECTATION MAXIMIZATION (EM) ALGORITHM. It uses the likelihood function and iterates between the expectation step (where the group memberships are estimated) and the maximization step (where the parameters are estimated). As a result, each object receives a membership to each cluster like in fuzzy clustering. The overall cluster result can be evaluated by the value of the negative likelihood function which should be as small as possible. This allows judging which model for the clusters is best suited (spherical clusters, elliptical clusters) and which number of clusters, k, is most appropriate.

The example used in Figure 6.9 is analyzed by model-based clustering; this synthetic data set contains two elliptical and one spherical group of objects. The different types of cluster models are evaluated for a number of clusters varying from 2 to 6. Figure 6.16 shows the evaluation of the various models (denoted by

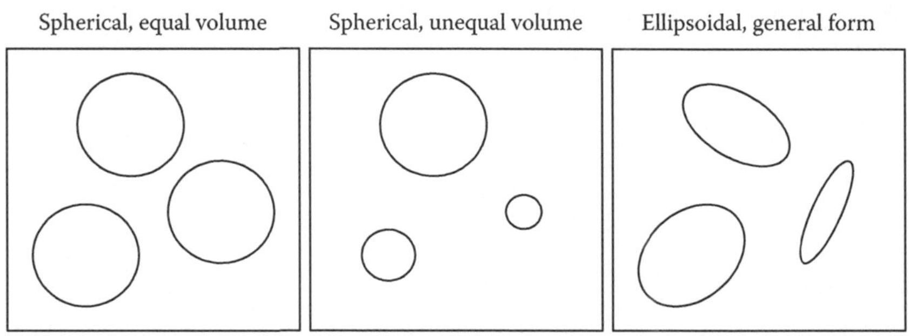

FIGURE 6.15 Three different models for model-based clustering: equal and diagonal covariance (left), diagonal covariance of different size (middle), and different covariance matrix for each cluster (right).

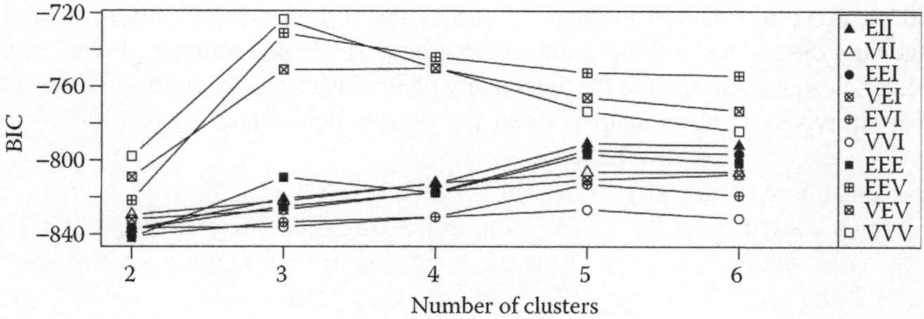

FIGURE 6.16 Cluster evaluation for model-based clustering (example from Figure 6.9) using different types of models (see text and legend) and different numbers of clusters. That model and that number of clusters with the largest BIC value will be taken: three clusters with different volume, shape, and orientation (type "VVV").

three-letter codes) with the BIC measure (see Section 4.2.4). For example, model "EII" uses spherical clusters with equal volume, model "VII" uses spherical clusters with unequal volume, and the most general model with elliptical clusters of varying volume, shape, and orientation is the model "VVV" (Figure 6.16 legend). The BIC measure should be a maximum, and thus the model "VVV" with three clusters is preferred.

Figure 6.17 shows some diagnostic plots for the optimal model resulting from the analysis of the BIC values in Figure 6.16. The left picture visualizes the result of the classification, the middle picture shows the uncertainty of the classification by symbol size, and the right picture shows contour lines for the clusters. The result of model-based clustering fits much better to the visually evident groups than the result from *k*-means clustering (shown in Figure 6.9).

Although model-based clustering seems to be restrictive to elliptical cluster forms resulting from models of multivariate normal distributions, this method has several advantages. Model-based clustering does not require the choice of a distance measure, nor the choice of a cluster validity measure because the BIC measure can be

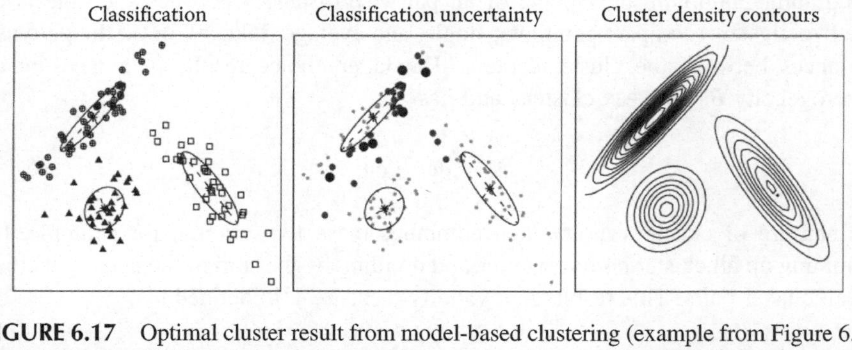

FIGURE 6.17 Optimal cluster result from model-based clustering (example from Figure 6.9, result in Figure 6.16); assignment of the objects to the clusters denoted by different symbols (left), symbol size according to uncertainty of the assignments (middle), and contours of the clusters (right).

used for this purpose. The number of clusters has to be provided only for a certain range, and the method automatically selects the best suitable number of clusters and types of cluster models. Note that also many other clustering algorithms are restricted to certain types of cluster shapes, usually even spherical shape.

```
R: library(mclust)
   res <- Mclust(X,2:6)  # model-based clustering with 2 to
          # 6 clusters; the best model can then be selected
   plot(res)             # diagnostic plots
```

6.7 CLUSTER VALIDITY AND CLUSTERING TENDENCY MEASURES

The crucial point for most clustering algorithms is the correct identification of the number k of clusters that is somehow inherent in the data. In most cases, there does not exist such a correct number, but there might exist an optimal number of clusters with respect to some quality criterion. As mentioned above, the goal of cluster analysis is to find clusters where the objects within the clusters are as similar as possible and objects between different clusters are as dissimilar as possible. This can be evaluated by defining a measure of HOMOGENEITY within the clusters and HETEROGENEITY between the clusters. Naturally, there exist probably as many measures for homogeneity and heterogeneity as clustering algorithms. Measures of homogeneity can be based on the maximum, minimum, or average of the distances between all objects of a cluster, or an average distance of the objects within a cluster to the cluster center. Thus, a possible choice for a measure of homogeneity W_j WITHIN a cluster j is

$$W_j = \sum_{i=1}^{n_j} ||\boldsymbol{x}_i^{(j)} - \boldsymbol{c}_j||^2 \tag{6.11}$$

Similarly, a measure of heterogeneity between two clusters can be based on the maximum, minimum, or average of all pairwise distances between the objects of the two clusters (compare complete, single, and average linkage), or on the pairwise distances between the cluster centers. The latter choice results in a measure of heterogeneity B_{jl} BETWEEN cluster j and l as

$$B_{jl} = ||\boldsymbol{c}_j - \boldsymbol{c}_l||^2 \tag{6.12}$$

A measure of cluster validity then combines these two criteria, for example, by summing up all cluster homogeneities and dividing by the sum of the heterogeneities of all cluster pairs. This results in a validity measure $V(k)$ defined as

$$V(k) = \frac{\sum_{j=1}^{k} W_j}{\sum_{j<l=1}^{k} B_{jl}} \tag{6.13}$$

which depends on the chosen number k of clusters. Since the homogeneities should be small and the heterogeneities large, this cluster validity measure should be small. A graph picturing the number of clusters versus the validity measure will often show a knee which indicates the optimal number of clusters: on the left of the knee the validity measure increases fast, whereas on the right of the knee only a marginal improvement of the validity measure can be made.

Figure 6.18 shows the outcome of the validity measure $V(k)$ when applied to the examples used in Figures 6.8 and 6.9. Although we know in advance that the optimal number of clusters is three for both examples, results have been computed for $k=2,\ldots, 10$ for the algorithms k-means, fuzzy c-means, and model-based clustering. The left picture is for the example of Figure 6.8 (three spherical clusters), and the knee at a value of 3 is rather obvious for all considered algorithms. In the right picture we used the example of Figure 6.9 (two elliptical clusters and one spherical cluster). Here it is not clear whether $k=3$ or $k=4$ or $k=5$ should be taken, although we know from model-based clustering that $k=3$ leads to the optimal solution (see Figure 6.17). Other validity measures may suggest a different optimal number of clusters and in that sense this kind of diagnostics should not be considered as "absolute truth" but rather as an indication.

```
R: library(chemometrics)
   clvalidity(X,2:10)  # generates the plot with cluster
                # validity measures and returns the validities
```

Since cluster analysis will only give sensible results if there is a grouping structure inherent in the data, methods have been developed that check the CLUSTERING TENDENCY, that is, they check for the existence of subgroups with higher

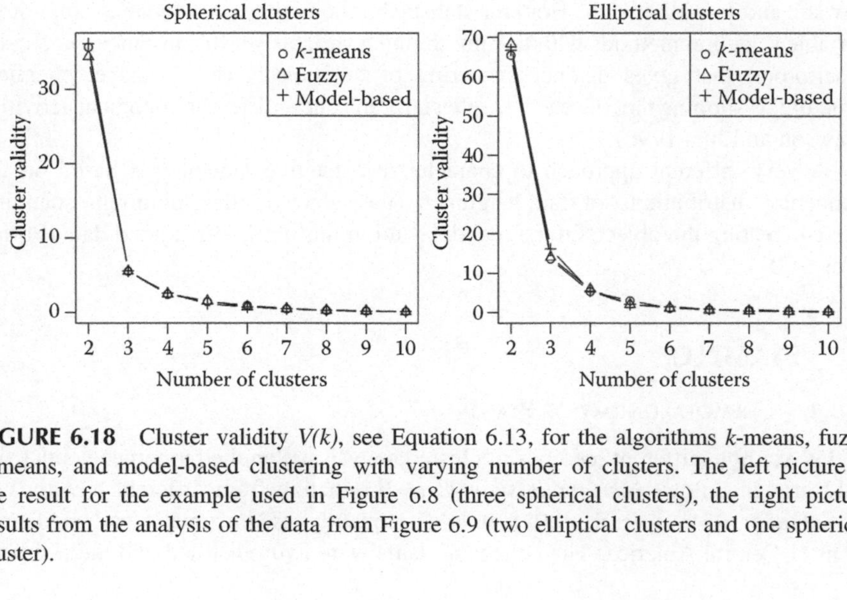

FIGURE 6.18 Cluster validity $V(k)$, see Equation 6.13, for the algorithms k-means, fuzzy c-means, and model-based clustering with varying number of clusters. The left picture is the result for the example used in Figure 6.8 (three spherical clusters), the right picture results from the analysis of the data from Figure 6.9 (two elliptical clusters and one spherical cluster).

homogeneity. A simple possibility would be to visually inspect the plots of the first few pairs of principal components. A more formal way to check for the clustering tendency is the HOPKINS STATISTIC (Hopkins and Skellam 1954) which is based on the null hypothesis that the data points are uniformly distributed in the data space. If this null hypothesis cannot be rejected, the result of any cluster analysis procedure will be only a random partitioning of the objects, depending on the actual algorithm used. The Hopkins statistic compares Euclidean distances

d_W of a data object to the nearest neighboring object
d_U of an arbitrary (artificial) point in the data space to the nearest neighboring object

A number n^* of data points and arbitrary points are randomly selected, and the resulting distances $d_W(i)$ and $d_U(i)$ for each considered point $i = 1, \ldots, n^*$ are computed. The Hopkins statistic is defined as

$$H = \frac{\sum_i d_U(i)}{\sum_i d_U(i) + \sum_i d_W(i)} \tag{6.14}$$

In presence of clustering tendency, $d_W(i)$ will tend to be smaller than $d_U(i)$, and thus H will be larger than 0.5 and at most be 1. Practically, the Hopkins statistic is computed for several random selections of points, and the average of all results for H is used for a decision: if this average is greater than 0.75 then the null hypothesis can be rejected with a high confidence.

Since the value of H depends on the choice of n^*, modifications of this procedure have been proposed (Fernandez Pierna and Massart 2000). Another modification of the Hopkins statistic—published in the chemometrics literature—concern the distributions of the values of the used variables (Hodes 1992; Jurs and Lawson 1991; Lawson and Jurs 1990). The Hopkins statistic has been suggested for an evaluation of variable selection methods with the aim to find a variable set (for instance, molecular descriptors) that gives distinct clustering of the objects (for instance, chemical structures)—hoping that the clusters reflect, for instance, different biological activities (Lawson and Jurs 1990).

A very different approach to characterize clustering tendency is based on the frequency distributions of the lengths of the edges in the minimum spanning tree connecting the objects in the real data and in uniformly distributed data (Forina et al. 2001).

6.8 EXAMPLES

6.8.1 CHEMOTAXONOMY OF PLANTS

In this example different methods of cluster analysis are applied to terpene data from wild growing, flowering *Hyptis suaveolens* originating from different locations in El Salvador (Grassi et al. 2005); it is a frequently used aromatic and medicinal plant of Central America. The collected plants were hydrodistilled and the resulting essential oils were analyzed by GC-MS. The original data set has been reduced for this example to 30 plants characterized by the concentrations (mass % in essential oil) of the seven terpenes sabiene, β-pinene, 1,8-cineole, γ-terpinene, fenchone, α-terpinolene, and fenchol. The geographical regions of the plants are denoted by "North," "South," and "East"; for the plants in the East, a distinction was made whether the plants grew at low or high altidude (>150 m above sea level). For cluster analysis only the terpene concentrations are used to recognize different chemotypes of the plant. The grouping information can then be used to verify if the resulting clusters are related to their geographical location.

This example belongs to CHEMOTAXONOMY, a discipline that tries to classify and identify organisms (usually plants, but also bacteria, and even insects) by the chemical or biochemical composition (e.g., fingerprint of concentrations of terpenes, phenolic compounds, fatty acids, peptides, or pyrolysis products) (Harborne and Turner 1984; Reynolds 2007; Waterman 2007). Data evaluation in this field is often performed by multivariate techniques.

As a first step the *Hyptis* data are visualized using PCA (Chapter 3). By reducing the data dimensionality to a few principal components, it is possible to graphically inspect the data for a grouping structure. Figure 6.19 shows the resulting plot of the first two principal components for the original unscaled data (left) and for the autoscaled data (right). As plot symbols, we coded the information of the geographical origin: North (N), South (S), East with high altitude (H, East-high) and east with low altitude (e, East-low). In both plots one can see groups of samples, although they are not clearly separated. Samples from the South form the most compact cluster, while samples from the East are split into several clusters. The PCA plot for the original data (Figure 6.19, left) allows a better division into the

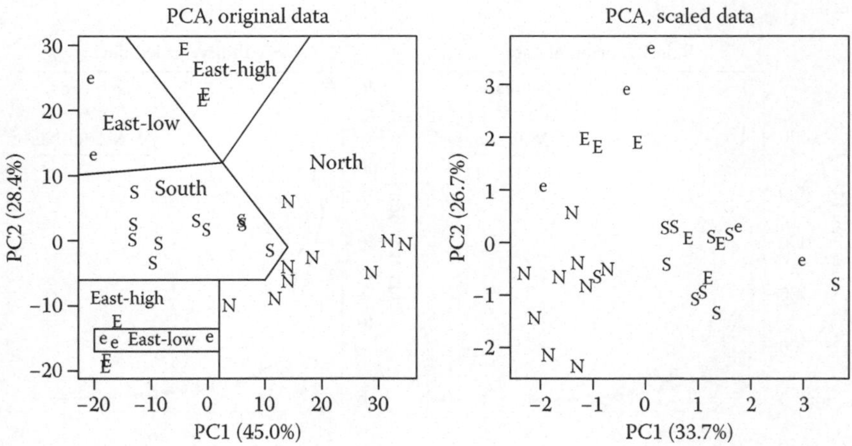

FIGURE 6.19 Projection of the *Hyptis* data on the first two principal components. Using the original data for PCA (left) results in a better distinction of the geographical regions than using the autoscaled data (right). The plot symbols (E—East-high, e—East-low, N—North, S—South) refer to the geographical origin.

geographical groups than the plot for the scaled data. Therefore, the PCA score plots from the original data will be used in the following to visualize the result of the different cluster analysis methods. Although the unscaled data lead to an informative PCA projection, they are not necessarily suitable for cluster analysis, because most of the clustering methods are not scale invariant. Therefore, we will also compare cluster results for the unscaled and for the scaled data.

By looking at the data one can observe right-skewed distributions for some of the variables. Thus an appropriate data transformation (e.g., the log-transformation) can improve the quality of the cluster results. However, it turned out that the results changed only marginally for the transformed data, and thus they will not be presented in the following.

In order to get an idea about the number of groups that is inherent in the data we compute the CLUSTER VALIDITIES for the range from two to nine clusters (see Section 6.7) using Equation 6.13. Figure 6.20 shows the resulting plot when using the original (left) and the autoscaled (right) data for the methods *k*-means clustering, fuzzy clustering, and model-based clustering. Due to the graphs the original data require a smaller number of clusters (about five) than the scaled data (between six and eight). Since we also want to compare the outcomes of different cluster analysis methods for the original and scaled data, the number of clusters will be chosen as six in both cases.

The results of *K*-MEANS CLUSTERING (Section 6.3) for six clusters are marked in the PCA projection of Figure 6.19 (left) in Figure 6.21. The left plot shows the results of *k*-means applied to the original data, and the right plot for the autoscaled data. The plot symbols represent the cluster assignments (but not the geographical regions). An almost perfect agreement with the geographical origins of the objects is seen in the result for the scaled data (Figure 6.21, right). The groups East-high and East-low

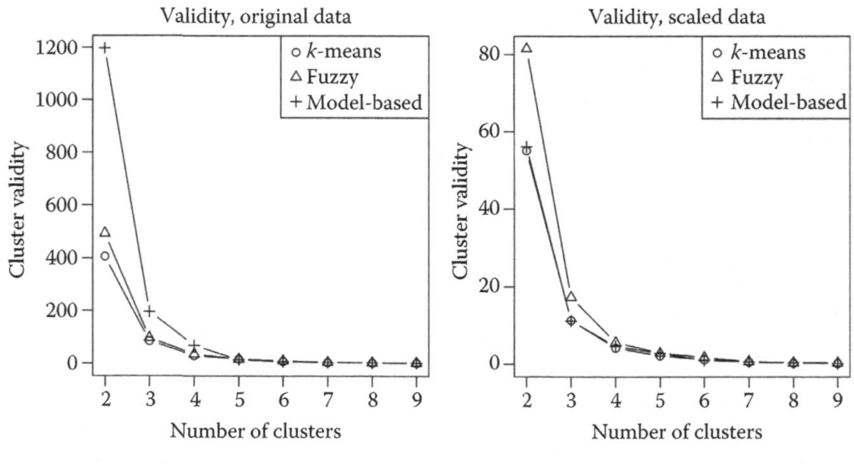

FIGURE 6.20 Cluster validities for the *Hyptis* data for two to nine clusters analyzed with the methods *k*-means clustering, fuzzy clustering, and model-based clustering. For the left plot the original data were used, for the right plot the data were autoscaled.

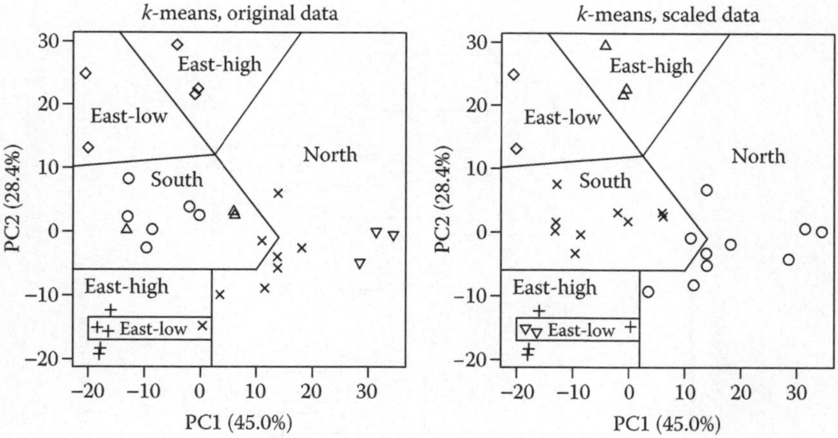

FIGURE 6.21 *k*-means clustering with six clusters for the original (left) and scaled (right) *Hyptis* data. The plot symbols correspond to the found clusters, and they are marked in the PCA projection obtained from the original data (Figure 6.19, left).

are split into two subclusters that correspond to the different locations in the PCA projection. In total, only two objects in the plot are located in a wrong geographical region. The results for the original data (Figure 6.21, left) are much worse which demonstrates the dependence of *k*-means clustering on the scale and thus on the cluster shapes. The larger clusters for North and South were subdivided into two clusters, but even when merging these clusters the result does not improve much. The algorithm cannot distinguish the groups from the East coming from different altitudes.

The results from HIERARCHICAL CLUSTERING with the method complete linkage based on the Euclidean distances (Section 6.4) are shown in Figure 6.22. The dendrogram when using the original data (left) is quite different from the dendrogram for the scaled data (right), where the structure corresponds well to the geographical regions. Both results reveal two unusual objects (one from South, one from East-low) that are quite dissimilar to the other objects. When cutting the dendrogram for the scaled data at a height defining six resulting clusters, these objects would form single-object clusters, and the remaining clusters would show several wrong assignments. Applying other hierarchical clustering methods (single linkage, average linkage, Ward's method) or using other distance measures does not lead to a significant improvement of the cluster results.

FUZZY CLUSTERING allows a fuzzy assignment of the objects to the clusters which is expressed by the membership coefficients (Section 6.5). Figure 6.23 shows the results of fuzzy clustering with six clusters for the original (left) and scaled (right) data. The plot symbols correspond to a hard assignment, by assigning each object to the cluster with the largest membership coefficient. In addition, the symbol size in this plot is proportional to this largest membership coefficient. Thus large symbols represent objects that were assigned to the corresponding group with high reliability. We would expect that the symbol size for objects that were assigned to the wrong

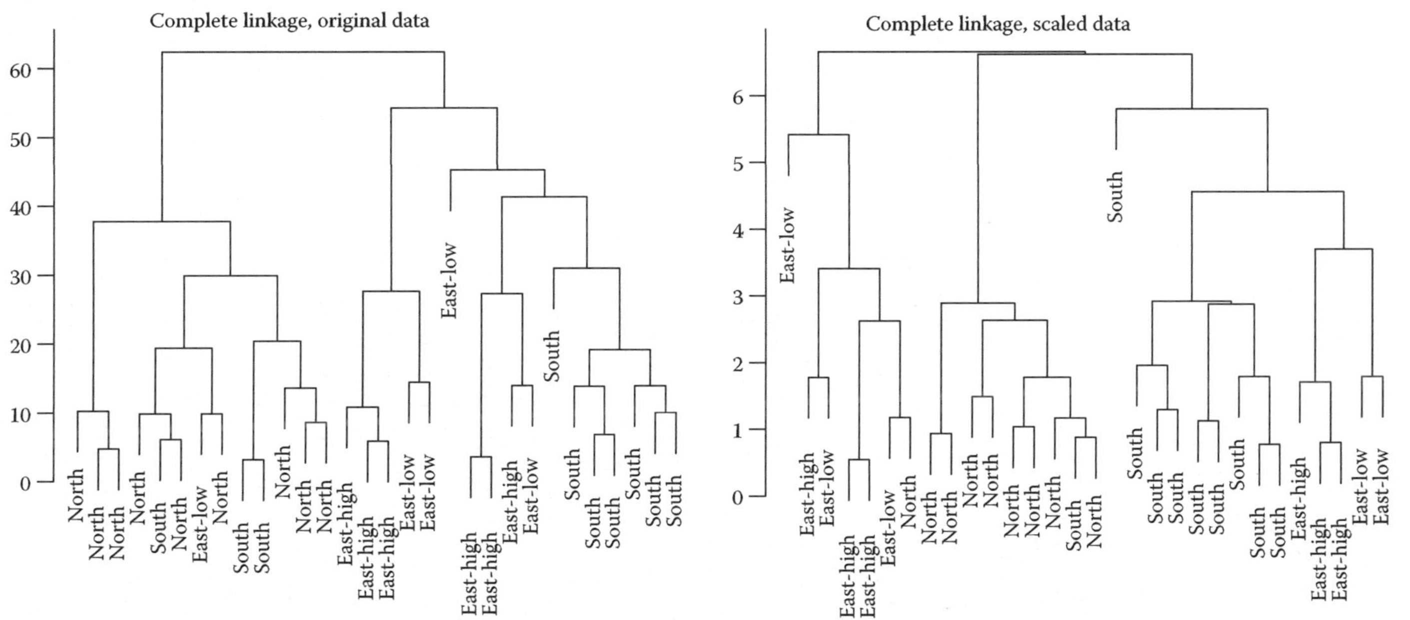

FIGURE 6.22 Hierarchical clustering based on complete linkage for the original (left) and scaled (right) *Hyptis* data.

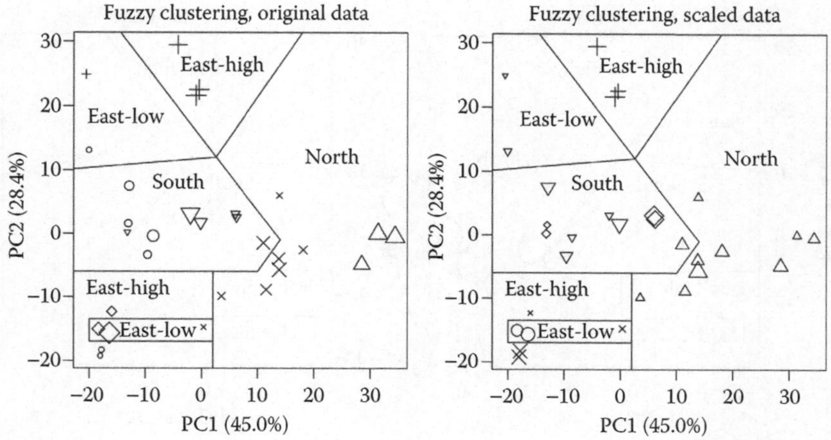

FIGURE 6.23 Fuzzy clustering with six clusters for the original (left) and scaled (right) *Hyptis* data. The plot symbols correspond to the found clusters with the largest membership coefficient, and their size is proportional to this coefficient. The results are presented in the PCA projection obtained from the original data (Figure 6.19, left).

geographical region would be smaller than that for correctly assigned objects. However, this is only true for some of the objects. In both plots, the objects from South were split into two clusters; in the result for the original data (left) also the objects from North were subdivided into two smaller clusters.

Finally, MODEL-BASED CLUSTERING (Section 6.6) is applied to the original (Figure 6.24, left) and scaled (Figure 6.24, right) data with six clusters. The symbols

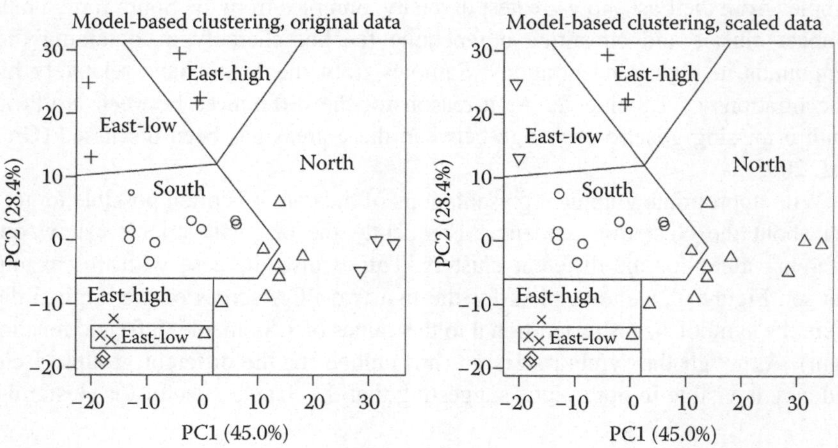

FIGURE 6.24 Model-based clustering with six clusters for the original (left) and scaled (right) *Hyptis* data. The plot symbols correspond to the found clusters, and the symbol sizes reflect the reliability of the assignments. The results are presented in the PCA projection obtained from the original data (Figure 6.19, left).

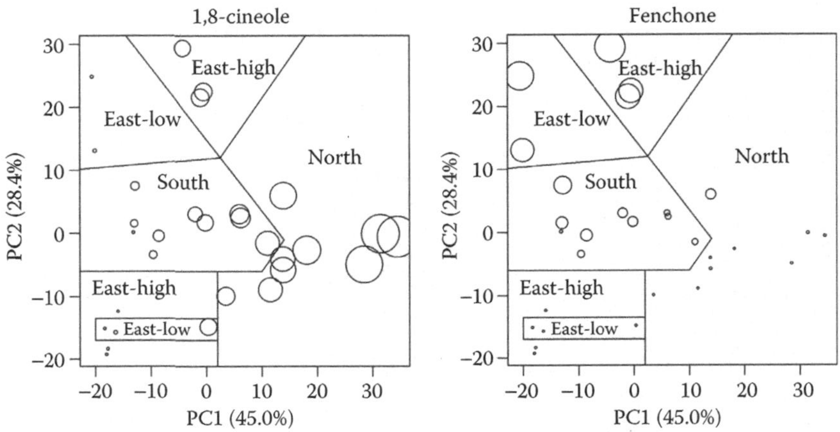

FIGURE 6.25 Plots of the first two PCA scores obtained from the original data (Figure 6.19, left). The symbol sizes are proportional to the concentrations of 1,8-cineole (left) and fenchone (right). Since the values are quite different for the groups, these variables are useful for cluster interpretation.

in the plot correspond to the resulting clusters, and the symbol sizes to the reliability that the objects are assigned to the clusters (inverse uncertainty). In this example, model-based clustering gives quite comparable results for the original and scaled data, and they correspond well to the geographical origin.

In summary, all clustering methods confirmed a separation of the samples into geographical regions, with model-based clustering and *k*-means clustering performing best. The altitude of the sample collection places is not reflected by clusters. Samples from the East show greatest diversity. Samples from the South form a rather compact cluster and constitute a fenchone–fenchol chemotype containing these compounds in high concentrations. Samples from the North have relatively high concentrations of 1,8-cineole. As a reason for the differences between North and South a missing genetic exchange between these areas has been discussed (Grassi et al. 2005).

With appropriate graphical presentations of the data it is often possible to get an idea about the clustering tendency. Frequently the original variables show quite different values for the different clusters. This is also the case with this example data set. Figure 6.25 shows plots for the first two PCA scores of the original data, where the symbol size is proportional to the values of 1,8-cineole (left) and fenchone (right). Although the symbol sizes are not unique for the different groups, a clear tendency is visible in both plots, suggesting that the data are useful for clustering.

6.8.2 Glass Samples

The glass vessels data have been used previously in this book (Janssen et al. 1998). They consist of four different types of glass vessels, and 180 objects have been measured for 13 variables. A projection of the data on the first two principal components was

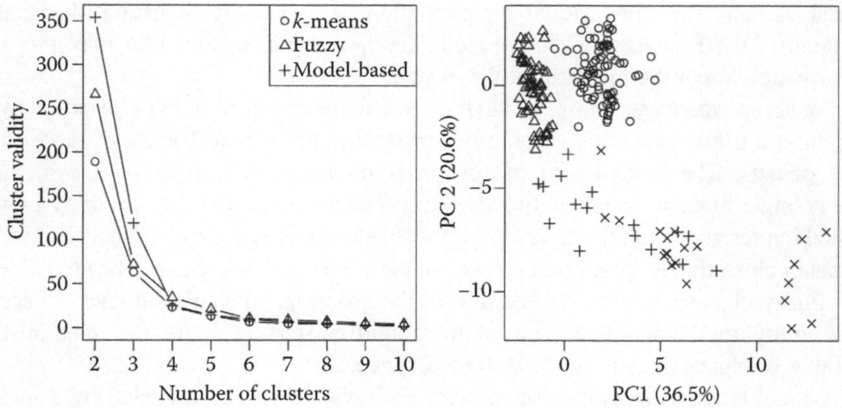

FIGURE 6.26 Cluster validity measures for the glass vessels data (left) and result from model-based clustering for $k = 4$ (right) as a projection on the first two robust principal components (compare Figure 3.10, right).

shown in Figure 3.10. Although there are only four different types of glass vessels, there can exist more than four groups in the multivariate data. This was also seen in the projection on the first pair of robust principal components (Figure 3.10, right) where the larger group 1 was visible as a composition of at least two (but probably even four) groups of objects.

Figure 6.26 (left) shows the plot with the cluster validity measures. Although a fast decay of the measure is visible, it is not evident for which number of clusters the curves flatten out, i.e., the knee in the plot is not very clear. For the final analysis we opted for model-based clustering and used the optimal choice from the BIC measure. The optimal BIC was attained for $k = 4$ for general elliptical models as described in Section 6.6 (see Figure 6.16). The result of four clusters corresponds to the result obtained from the cluster validity measure for model-based clustering. Figure 6.26 (right) shows the final cluster result as a projection on the first two robust principal components. Indeed, the larger group was subdivided into two smaller groups, well corresponding to the visual impression. Obviously, the clustering algorithm had difficulties to identify correctly other (quite small) groups (compare Figure 3.10, left).

6.9 SUMMARY

Cluster analysis of objects is often the first step in multivariate data analysis, and has the aim to obtain an insight into the data structure. In most cases at least ideas about object groups and the number of potential clusters are available in advance, and cluster analysis indicates if a supposed grouping is reflected in the data. Cluster analysis gives useful hints for a following development of classification models (for the recognition of predefined object groups), such as if the groups form compact single clusters, or are split into subgroups, and the presence of outliers.

For many data sets in chemistry, a PCA score plot will be the method of choice to obtain a first impression of clustering, especially in the case of highly correlating variables. The two-dimensional representation can be easily investigated visually (human is the best pattern/cluster recognizer), and the loading plot may provide information about the meaning of the clusters.

A complementary nonlinear method is a representation of the clustering by a dendrogram; however, this graphic becomes difficult to read for about more than 100 objects. The structure of the dendrogram depends strongly on the method used (single linkage, complete linkage, etc.). The dendrogram often provides a good visual impression about the inherent number of clusters. For a large number of objects, k-means clustering is appropriate; however, the number of clusters must be predefined.

Fuzzy clustering gives, in addition to the group membership for each object, a probability for belonging to one of the found clusters; also for this method the number of clusters has to be defined in advance.

Model-based clustering assumes that each cluster can be modeled by a multivariate normal distribution (with varying parameters). If the clusters can be well modeled in this way, the method is powerful, and can estimate an optimum number of clusters. Especially for higher-dimensional data it is computer time demanding.

For PCA, hierarchical cluster analysis, and model-based cluster analysis, the number of clusters can be selected by the user by inspecting the result. If the number of clusters has to be defined in advance, it is advisable to test different numbers of clusters, and to compare the results by measures of cluster validity (Equation 6.13). If different sets of variables are available, a clustering tendency measure (Equation 6.14) can help to find a good variable set. For the interpretation of the meaning of the clusters, the distribution of characteristic variables among the found clusters may be helpful (although it is a univariate approach), for instance, by including the variable as a third dimension into the PCA score plot, as used in Figure 6.25.

There is no one best cluster analysis method (see Table 6.2 for an overview). The variability in the methods and in the parameters demands some discipline of the user to avoid that cluster analysis is applied under different conditions until the desired result is obtained.

TABLE 6.2
Overview of Cluster Analysis Methods

	k-Means	Hierarchical Clustering	Fuzzy c-Means	Model-Based
Uses distance measure	Yes	Yes	Yes	No
Size of distance matrix	$n \times k$	$n \times n$	$n \times k$	—
Fix number of clusters in advance	Yes	No	Yes	No (range)
Provides information about cluster uncertainty	No	No	Yes	Yes
Optimal number of clusters chosen by the algorithm	No	No	No	Yes

Note: n, number of objects; k, number of clusters.

REFERENCES

Bandemer, H., Näther, W.: *Fuzzy Data Analysis*. Kluwer Academic, Dordrecht, the Netherlands, 1992.

Bezdek, J. C.: *Pattern Recognition with Fuzzy Objective Function Algorithms*. Plenum Press, New York, 1981.

Chernoff, H.: *J. Am. Stat. Assoc.* 68, 1973, 361–368. The use of faces to represent points in *k*-dimensional space graphically.

Demuth, W., Karlovits, M., Varmuza, K.: *Anal. Chim. Acta* 516, 2004, 75–85. Spectral similarity versus structural similarity: Mass spectrometry.

Dunn, J. C.: *J. Cybern.* 3, 1973, 32–57. A fuzzy relative of the ISODATA process and its use in detecting compact well-separated clusters.

Everitt, B.: *Cluster Analysis*. Heinemann Educational, London, United Kingdom, 1974.

Everitt, B.: *Graphical Techniques for Multivariate Data*. Heinemann Educational Books, London, United Kingdom, 1978.

Fernandez Pierna, J. A., Massart, D. L.: *Anal. Chim. Acta* 408, 2000, 13–20. Improved algorithm for clustering tendency.

Flury, B., Riedwyl, H.: *Multivariate Statistics: A Practical Approach*. Chapman & Hall, Boca Raton, FL, 1988.

Forina, M., Lanteri, S., Esteban Diez, I.: *Anal. Chim. Acta* 446, 2001, 59–70. New index for clustering tendency.

Fraley, C., Raftery, A.: *Comp. J.* 41, 1998, 578–588. How many clusters? Which clustering method? Answers via model-based cluster analysis.

Gasteiger, J., Engel, T.: *Chemoinformatics—A Textbook*. Wiley-VCH, Weinheim, Germany, 2003.

Gordon, A. D.: *Classification*. Chapman & Hall, CRC, Boca Raton, FL, 1999.

Grassi, P., Nunez, M. J., Varmuza, K., Franz, C.: *Flavour Fragr. J.* 20, 2005, 131–135. Chemical polymorphism of essential oils of *Hyptis suaveolens* from El Salvador.

Harborne, J. B., Turner, B. L.: *Plant Chemosystematics*. Academic Press, Orlando, FL, 1984.

Hartigan, J. A.: *Clustering Algorithms*. Wiley, New York, 1975.

Hodes, L.: *J. Chem. Inf. Comput. Sci.* 32, 1992, 157–166. Limits of classification. 2. Comment on Lawson and Jurs.

Honda, N., Aida, S.: *Pattern Recogn.* 15, 1982, 231–242. Analysis of multivariate medical data by face method.

Hopkins, B., Skellam, J. G.: *Ann. Bot.* 18, 1954, 213–227. A new method for determining the type of distribution of plant individuals.

Janssen, K. H. A., De Raedt, I., Schalm, O., Veeckman, J.: *Microchim. Acta* 15 (suppl.), 1998, 253–267. Compositions of 15th–17th century archaeological glass vessels excavated in Antwerp.

Jarvis, R. A., Patrick, E. A.: *IEEE Trans. Comput.* C-22, 1973, 1025–1034. Clustering using a similarity measure based on shared near neighbors.

Jurs, P. C., Lawson, R. G.: *Chemom. Intell. Lab. Syst.* 10, 1991, 81–83. Analysis of chemical structure—biological activity relationships using clustering methods.

Kaufmann, L., Rousseeuw, P. J.: *Finding Groups of Data*. Wiley, New York, 1990.

Larsen, R. D.: *J. Chem. Educ.* 63, 1986, 505–507. Features associated with chemical elements (faces).

Lawson, R. G., Jurs, P. C.: *J. Chem. Inf. Comput. Sci.* 30, 1990, 36–41. New index for clustering tendency and its application to chemical problems.

Massart, D. L., Kaufmann, L.: *The Interpretation of Analytical Chemical Data by the Use of Cluster Analysis*. Wiley, New York, 1983.

Otto, M.: *Chemometrics—Statistics and Computer Application in Analytical Chemistry*. Wiley-VCH, Weinheim, Germany, 2007.

Reynolds, T.: *Phytochemistry* 68, 2007, 2887–2895. The evolution of chemosystematics.

Ripley, B. D.: *Pattern Recognition and Neural Networks*. Cambridge University Press, Cambridge, NY, 1996.

Scsibrany, H., Varmuza K.: Software Submat. Laboratory for Chemometrics, Vienna University of Technology, www.lcm.tuwien.ac.at., Vienna, Austria, 2004.

Scsibrany, H., Karlovits, M., Demuth, W., Müller, F., Varmuza, K.: *Chemom. Intell. Lab. Syst.* 67, 2003, 95–108. Clustering and similarity of chemical structures represented by binary substructure descriptors.

Sweeley, C. C., Holland, J. F., Towson, D. S., Chamberlin, B. A.: *J. Chromatogr. A* 399, 1987, 173–181. Interactive and multi-sensory analysis of complex mixtures by an automated gas chromatography system.

Vandeginste, B. G. M., Massart, D. L., Buydens, L. C. M., De Jong, S., Smeyers-Verbeke, J.: *Handbook of Chemometrics and Qualimetrics: Part B*. Elsevier, Amsterdam, the Netherlands, 1998.

Varmuza, K.: Chemometrics: The computer as a partner in chemical data interpretation. *Lecture at German-Austrian Chemist's Meeting*, 14 May 1986, music played by J. Jaklin on a diatonic accordion, Innsbruck, Austria, 1986.

Waterman, P. G.: *Phytochemistry* 68, 2007, 2896–2903. The current status of chemical systematics.

Willett, P.: *Similarity and Clustering in Chemical Information Systems*. Research Studies Press, Letchworth, United Kingdom, 1987.

Yeung, E. S.: *Anal. Chem.* 52, 1980, 1120–1123. Pattern recognition by audio representation of multivariate analytical data.

7 Preprocessing

7.1 CONCEPTS

The development of useful models requires appropriate methods but even more important are suitable data. For instance in a first trial, it may not be essential whether the regression method principal component regression or partial least-squares or an artificial neural network is applied to make a calibration model, but it is essential that the used x-data have a strong relationship with the modeled y-data. Choice of x-data is often limited, because of the specific chemical–physical properties of the samples (objects), the availability of instruments, and also because we often only suppose that some x-data may be related to the desired y-data. Available x-data can be improved by appropriate transformation or extension—in general by an appropriate preprocessing. Some preprocessing methods are solely based on mathematical concepts, others are inspired by the chemical–physical background of the data and the problem. Selected preprocessing methods that are important in chemometrics are briefly described in this chapter. A book has been dedicated to wavelet transforms (Chau et al. 2004), other chemistry-specific data transformations are described in Brereton (2006), Smilde et al. (2004), and Vandeginste et al. (1998).

Basic preprocessing methods have already been described in Section 2.2. If the DISTRIBUTION of a variable is highly skewed, data transformations like log-transformation, power transformation, Box–Cox transformation or logit transformation are useful (Section 2.2.1). Column-wise transformations of a matrix $\boldsymbol{X}$ are mean-centering and scaling—the most used method is autoscaling (all variables will have zero mean and a standard deviation of one, thereby they get equal statistical weight, Section 2.2.2). The use of robust measures (median, MAD) instead of classical measures (mean, standard deviation) should be considered. Row-wise transformations of a matrix $\boldsymbol{X}$ are normalization to a constant sum of the variables or to a constant maximum or constant vector length (Section 2.2.3). Attention must be given to transformations of compositional data to avoid "closing" and not to produce artificial correlations between the variables; various versions of the logratio transformation are helpful in such cases (Section 2.2.4).

7.2 SMOOTHING AND DIFFERENTIATION

If the x-data of an object are time-series or digitized data from a continuous spectrum (infrared, IR; near infrared, NIR) then smoothing and/or transformation to first or second derivative may be appropriate preprocessing techniques. Smoothing tries to reduce random noise and thus removes narrow spikes in a spectrum. Differentiation extracts relevant information (but increases noise). In the first derivative an additive baseline is removed and therefore spectra that are shifted in parallel to other absorbance values will have identical first derivative spectra. A second derivative removes a constant and a linear baseline. Smoothing and differentiation is performed separately for each object vector $\boldsymbol{x}_i$.

The most used technique in chemistry for smoothing and differentiation is the so-called SAVITZKY–GOLAY METHOD, which is a LOCAL POLYNOMIAL REGRESSION requiring equidistant and exact x-values. Mathematically, for each point j with value x_j a weighted sum of the neighboring values is calculated (a linear combination). The weights determine whether a smoothing is performed or a derivative is calculated. The number of neighbors and the degree of the polynomial control the strength of smoothing. A vector component x_j is transformed by

$$x_j^* = \frac{1}{N} \sum_{h=-k}^{k} c_h x_{j+h} \tag{7.1}$$

where

x_j^* is the new value (of a smoothed curve or a derivative)
N is the normalizing constant
k is the number of neighboring values at each side of j
c_h are the coefficients that depend on the used polynomial degree and on the goal (smoothing, first or second derivative)

For instance, if we fit a second-order polynomial through a window of five points ($k=2$) the coefficients c_{-2}, c_{-1}, c_0, c_1, c_2 are for smoothing -3, 12, 17, 12, -3, for the first derivative -2, -1, 0, 1, 2, and for the second derivative 2, -1, -2, -1, 2; for details see Savitzky and Golay (1964). Note that this simple and fast scheme is an exact solution. Coefficients for other window sizes and for cubic polynomials are, for instance, given in Brereton (2006); the originally published coefficients (Savitzky and Golay 1964) have been corrected several times (Steinier et al. 1972). Of course for the first and the last k points (vector components) no x^* can be calculated. Window size and polynomial degree should be chosen in accordance with the shape of the curve and the desired noise reduction. Figures 7.1 and 7.2 show an example.

7.3 MULTIPLICATIVE SIGNAL CORRECTION

This method is specifically applied to NIR data obtained by reflectance or transmission measurements on diffuse samples. The method has originally been developed to reduce the disturbing effect of light scattering (small particles scatter light more than larger particles) and is also called MULTIPLICATIVE SCATTER CORRECTION (MSC) (Geladi et al. 1985; Naes et al. 1990). For a comprehensive treatment of the subject see Naes et al. (2004). In contrary to derivatives of the spectra, multiplicative signal correction retains the spectral shape.

Let matrix $\boldsymbol{X}(n \times m)$ contain the n spectra from a calibration set with each spectrum i comprising m absorbance values x_{ij}. The spectra are represented by the row vectors $\boldsymbol{x}_i^{\mathrm{T}}$. In calibration experiments the chemical differences of the samples

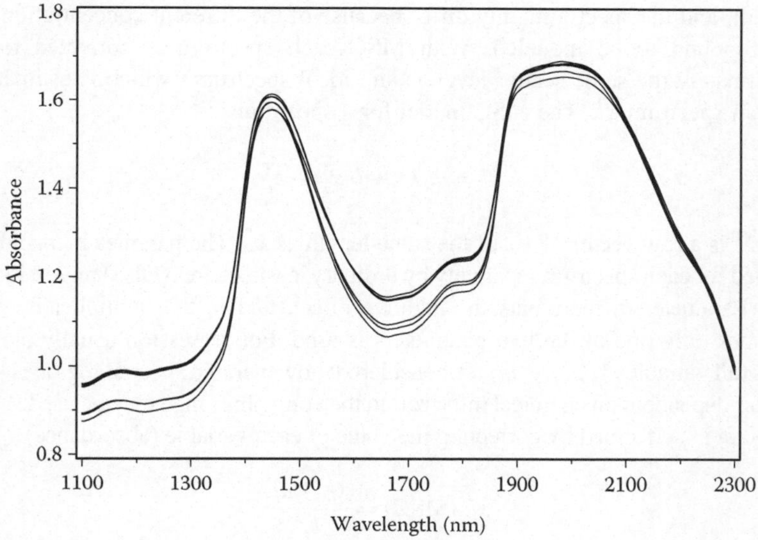

FIGURE 7.1 NIR spectra from different fermentations of rye with yeast. The seven spectra have been measured in reflection mode in opaque mashes. Wavelength interval is 1100–2300 nm; the number of data points is 241. The samples differ in the ethanol contents (62.2–84.1 g/L).

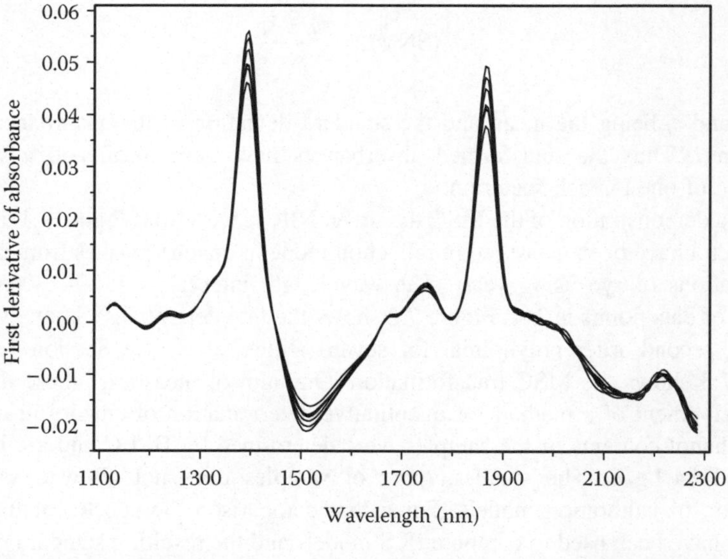

FIGURE 7.2 First derivatives of the seven NIR spectra from Figure 7.1. The Savitzky–Golay method was applied with a second-order polynomial for seven points.

are small, and the spectra mainly differ because of the different concentrations y_i of the compound to be modeled. With MSC each spectrum is corrected to have approximately the same scatter level as an "ideal spectrum" which is estimated by the mean spectrum $\bar{\boldsymbol{x}}^{\mathrm{T}}$. The MSC model for a spectrum i is

$$\boldsymbol{x}_i^{\mathrm{T}} = a_i \mathbf{1}^{\mathrm{T}} + b_i \bar{\boldsymbol{x}}^{\mathrm{T}} + \boldsymbol{e}_i^{\mathrm{T}} \tag{7.2}$$

where $\mathbf{1}^{\mathrm{T}}$ is a row vector of 1's of the same length as $\boldsymbol{x}_i^{\mathrm{T}}$. The parameters a_i and b_i are estimated for each spectrum separately by ordinary least-squares (OLS) regression of $\boldsymbol{x}_i^{\mathrm{T}}$ on $\bar{\boldsymbol{x}}^{\mathrm{T}}$. Parameter a_i represents the additive effect, and b_i the multiplicative effect; sometimes only one of the two parameters is used. For regression usually all wavelengths (all variables $1, 2, \ldots, m$) are considered; however, if a spectral region is known to be less dependent on chemical information then only this range may be used. MSC of the spectra is performed by correcting the value of each variable (absorbance) x_{ij} by

$$x_{ij}\ (\mathrm{MSC}) = \frac{x_{ij} - a_i}{b_i} \tag{7.3}$$

For transformation of future spectra the mean spectrum must be stored; first a_i and b_i have to be calculated by OLS regression and then Equation 7.3 is applied. It has been shown that calibration models from MSC spectra often require less (PLS-)components and have smaller prediction errors than models obtained from the original spectra (Naes et al. 2004).

STANDARD NORMAL VARIATE (SNV) transformation is closely related to MSC (Barnes et al. 1989, 1993; Helland et al. 1995). SNV treats each spectrum separately by autoscaling the absorbance values (row-wise) by

$$x_{ij}\ (\mathrm{SNV}) = \frac{x_{ij} - \bar{x}_i}{s_i} \tag{7.4}$$

with $\bar{x}_i$ and s_i being the mean and the standard deviation of the absorbances x_{ij} in spectrum i. Thus the transformed absorbances have zero mean and a standard deviation of one in each spectrum.

For a demonstration of the MSC the seven NIR spectra from Figure 7.1 are used. The spectra have been measured in reflection mode in opaque mashes from different fermentations of rye using yeast. The wavelength interval is 1100–2300 nm; the number of data points is 241. Figure 7.2 shows the first derivatives (Savitzky–Golay method, second-order polynomial for seven points, $k = 3$, see Section 7.2), and Figure 7.3 shows the MSC transformation. One aim of measuring these data was the development of a method for quantitative determination of ethanol in mash by NIR. Ethanol contents of the samples were determined by HPLC and are between 62.2 and 84.1 g/L. The small number of samples does not allow a satisfying evaluation of calibration models. For a first comparison the spectra of the seven samples have been used to compute PLS models and the resulting standard errors of calibration (SEC, see Section 4.2.3) are as follows (Unscrambler 2004): 3.6 g/L (two PLS components) for the original spectra, 1.1 g/L (three PLS components) for the

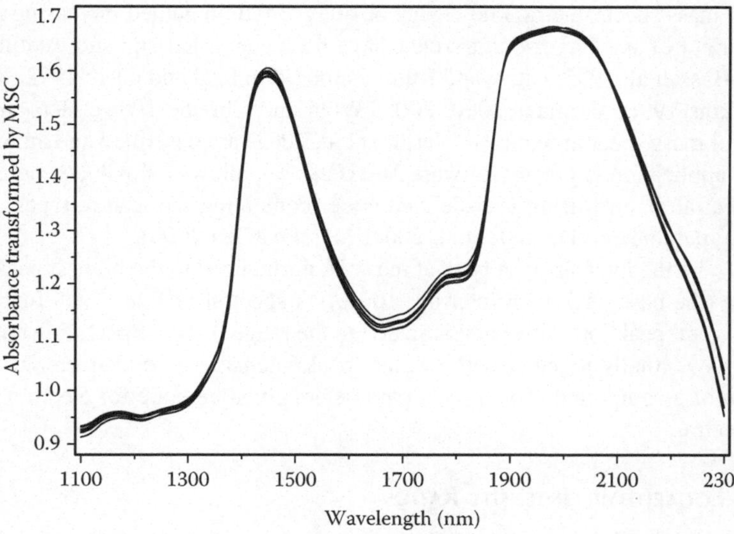

FIGURE 7.3 MSC of the seven NIR spectra from Figure 7.1.

first derivative spectra, and 0.7 g/L (four PLS components) for the MSC spectra. In this example, MSC transformation and the first derivative give better models than the original spectra; the number of PLS components is similar for all used data sets.

7.4 MASS SPECTRAL FEATURES

The investigation of relationships between low resolution mass spectra and chemical structures of organic compounds has for long been a challenging task in chemometrics and spectroscopy. Because a low resolution mass spectrum with integer mass numbers and peak intensities is a vector by definition, the direct use of such vectors is easy and tempting. However, application of multivariate classification methods for the recognition of substance classes or substructures had only very limited success (Jurs and Isenhour 1975; Varmuza 1980). Remarkably, in one of the first multivariate approaches to spectra–structure problems not the peak intensities themselves but variables derived from the peak intensities have been used successfully for assigning organic compounds to one of 12 selected substance classes (Crawford and Morrison 1968). The authors used the so-called REDUCED MASS SPECTRA consisting of exactly 14 "peaks": The first is the sum of the peak intensities at masses 1, 15, 29, ... $(1+14z$, with z being 0, 1, 2, ...), the second peak is the sum of the peak intensities at masses 2, 16, 30, ... $(2+14z)$, and so on until the 14th peak is computed from masses $(14+14z)$. Mass spectroscopists knew since long that reduced mass spectra show patterns that are characteristic for some substance classes. The spectroscopic reason is simple: peaks that are characteristic for a substance class appear in a homologous series of compounds at masses that differ by a multiple of 14 mass units (corresponding to CH_2 groups). In analogy to the modulo function such mass spectral variables are called modulo-14 features. The idea of transforming mass spectral peak lists by applying mass spectrometric knowledge or rules of thumb gained increasing interest and a variety of MASS SPECTRAL FEATURES have been suggested and successfully used (Cabrol-Bass et al. 1995; Curry and Rumelhardt 1990; Erni and Clerc 1972; Klawun and Wilkins 1996; Varmuza 2000, 2005; Wold and Christie 1984). Here the most successful mass spectral features (Werther et al. 2002) are described and an example for their application is given. Software *MassFeatGen* allows a flexible generation of mass spectral features from user-defined codes containing the feature types and the necessary parameters (Demuth et al. 2004; MassFeatGen 2006).

Let I_M be the intensity of a peak at mass M, normalized to the highest peak in the spectrum (the base peak with intensity 100%). A spectral feature x_j is a function of one or several peak intensities and is scaled to the range 0–100. Spectral features can be used—eventually together with original peak intensities—to characterize a mass spectrum of a compound. The typical parameters given here are for electron impact mass spectra.

7.4.1 Logarithmic Intensity Ratios

The logarithm of the ratio of a peak intensity I_M and the intensity $I_{M+\Delta M}$ of a neighboring peak is calculated under the restrictions $I_M = \max(I_M, 1)$ and $I_{M+\Delta M} = \max(I_{M+\Delta M}, 1)$ by

$$x = 100 \frac{\ln(I_M / I_{M+\Delta M}) + \ln 100}{2 \ln 100} \tag{7.5}$$

These spectral features reflect competing fragmentation reactions. Masses M used are typically 39–150, and ΔM is 1 and 2.

7.4.2 Averaged Intensities of Mass Intervals

These features reflect the varying distribution of peaks in the lower and higher mass range and are capable to discriminate some compound classes (e.g., aliphatic and aromatic compounds). Typical mass ranges $M1$ to $M2$ are 33–50, 51–70, 71–100, 101–150.

$$x = \sum_{M=M1}^{M2} \frac{I_M}{M2 - M1 + 1} \tag{7.6}$$

7.4.3 Intensities Normalized to Local Intensity Sum

These features emphasize isolated peaks in the spectrum even if they have only low intensity. The local intensity sum is the sum of peak intensities in a mass interval $\pm\Delta M$ (typically ± 3) around a considered mass M (typically between 33 and 150)

$$x = 100 \frac{I_M}{\sum_{Q=M-\Delta M}^{M+\Delta M} I_Q} \tag{7.7}$$

7.4.4 Modulo-14 Summation

As discussed above, these 14 features are characteristic for some substance classes. First, the 14 possible sums S_j ($j = 1, 2, \ldots, 14$) are calculated by

$$S_j = \sum_{L=1}^{14} I_{L+14K}, \quad K = 1, 2, 3, \ldots \tag{7.8}$$

with the summation eventually restricted to selected mass ranges, e.g., 30–120, and 121–800. Then the sums are scaled to a maximum value of 100 by

$$S_{\mathrm{MAX}} = \max\,(S_1, S_2, \ldots, S_{14}) \tag{7.9}$$

$$x_j = 100 S_j / S_{\mathrm{MAX}} \tag{7.10}$$

7.4.5 Autocorrelation

Characteristic mass differences result, e.g., from losses of small stable molecules or atoms from ions. The appearing periodicity in the spectrum can be described by autocorrelation features. Mass differences ΔM are typically 1, 2, 14–60, and the sums, e.g., can be calculated for the ranges $M1$ to $M2$ of 31–120 and 121–800.

$$x = 100 \frac{\sum_{M=M1}^{M2} I_M I_{M+\Delta M}}{\sum_{M=M1}^{M2} I_M I_M} \tag{7.11}$$

7.4.6 Spectra Type

These features characterize the distribution of the peaks across the mass range. Feature x_{DUST} indicates the relative amount of peak intensities in the low mass range up to mass 78. Feature x_{BASE} is the base peak intensity in percentage of the total sum of the peak intensities, I_{ALL}. The relative peak intensities at even masses are described by feature x_{EVEN}.

$$x_{\mathrm{DUST}} = 100 \sum_{M=1}^{78} \frac{I_M}{I_{\mathrm{ALL}}} \tag{7.12}$$

$$x_{\mathrm{BASE}} = 100 \times 100 / I_{\mathrm{ALL}} \tag{7.13}$$

$$x_{\mathrm{EVEN}} = 100 \sum_{Q=1}^{400} \frac{I_{2Q}}{I_{\mathrm{ALL}}} \tag{7.14}$$

7.4.7 Example

As an example for the use of mass spectral features we look at a MASS SPECTRAL–STRUCTURE RELATIONSHIP. It is known that for some substance classes, the degree of

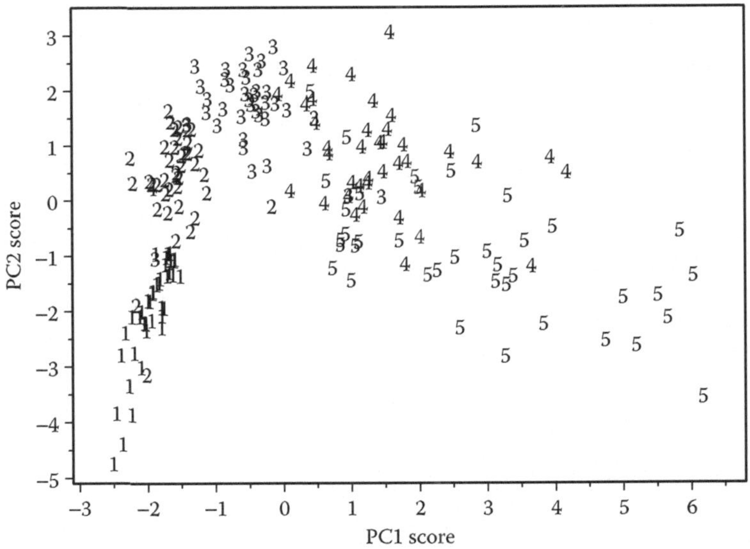

FIGURE 7.4 The PCA score plot of mass spectral data from ketones shows clustering by the number of double bond equivalents (indicated by the numbers 1–5). Variables used are 14 autoscaled modulo-14 features.

unsaturation in the molecule (number of double bond equivalents, DBE) is reflected in the peak pattern. The data set used consists of the low resolution mass spectral peak lists of 200 ketones of molecular formula $C_nH_mO_1$, randomly selected from the NIST MS Database (NIST 1998; Varmuza 2001). The compounds are from five groups with DBE = 1, 2, 3, 4, and 5, respectively, each group containing 40 compounds. The mass spectra have been transformed into 14 modulo-14 features (Equations 7.8 through 7.10).

PCA was performed with the autoscaled variables, and the score plot in Figure 7.4 shows a good clustering of the DBE groups with some overlap. The variances preserved by PC1 and PC2 are only 29.4% and 18.5%; thus the PCA projection may be only an approximate picture of the distances in the 14-dimensional variable space.

For the determination of the DBE group from mass spectral data, *k*-nearest neighbor (*k*-NN) classification was applied and the results are summarized in Table 7.1. Five different variable sets have been used and full cross validation (leave-one-out) has been applied. The group with DBE = 1 is classified best—corresponding to the good separation of this group in the PCA score plot. The larger the DBE becomes the more overlap appears in the PCA score plot and also *k*-NN produces more wrong assignments. Using the 14 modulo-14 variables in *k*-NN classification gives much better results (89.5% correct) than using the peak intensities at 14 mass numbers with highest variance of the peak intensities (72.5% correct). However, the distances in the two- or three-dimensional PCA space yield lower success rates than the 14-dimensional variable space. Best results with 93% correct assignments have been obtained with an extended variable set containing 50 mass spectral features (Varmuza 2001).

TABLE 7.1
***k*-NN Classification of the Number of DBE in Ketones from Mass Spectral Data**

		Number of Correct Prediction				
		$m=14$	$m=2$	$m=3$	$m=14$	$m=50$
DBE	n	Modulo-14	PCA Scores	PCA Scores	Intensities	MS Features
1	40	40	40	40	39	40
2	40	36	36	37	29	39
3	40	34	33	34	31	34
4	40	35	24	29	20	37
5	40	34	29	28	26	36
% correct		89.5	81.0	84.0	72.5	93.0

Note: The used variable sets are: 14 modulo-14 features (autoscaled); 2 and 3 PCA scores calculated from the autoscaled modulo-14 features; peak intensities at 14 selected mass numbers (with maximum variances of the peak intensities); 50 mass spectral features. The numbers of correct predictions are from a leave-one-out test; n is the number of spectra in the five DBE groups

The importance of an appropriate transformation of mass spectra has also been shown for relationships between the similarity of spectra and the corresponding chemical structures. If a spectra similarity search in a spectral library is performed with spectral features (instead of the original peak intensities), the first hits (the reference spectra that are most similar to the spectrum of a query compound) have chemical structures that are highly similar to the query structure (Demuth et al. 2004). Thus, spectral library search for query compounds—not present in the database—can produce useful structure information if compounds with similar structures are present.

Nature is sometimes difficult but never insidious. Fortunately, the SIMILARITY PRINCIPLE—compounds with similar chemical structures often possess similar properties or activities—is valid in many cases, and thus gives multivariate models a chance.

REFERENCES

Barnes, R. J., Dhanoa, M. S., Lister, S. J.: *Appl. Spectrosc.* 43, 1989, 772–777. Standard normal variate transformation and de-trending of near-infrared diffuse reflectance spectra.

Barnes, R. J., Dhanoa, M. S., Lister, S. J.: *J. Near Infrared Spectrosc.* 1, 1993, 185–186. Correction of the description of standard normal variate (SNV) and De-Trend transformations in practical spectroscopy with applications in food and beverage analysis.

Brereton, R. G.: *Chemometrics—Data Analysis for the Laboratory and Chemical Plant.* Wiley, Chichester, United Kingdom, 2006.

Cabrol-Bass, D., Cachet, C., Cleva, C., Eghbaldar, A., Forrest, T. P.: *Can. J. Chem.* 73, 1995, 1412–1426. Application pratique des reseaux neuro mimetiques aux donnees spectroscopiques (infrarouge et masse) en vue de l'elucidation structurale.

Chau, F. T., Liang, Y. Z., Gao, J., Shao, X. G.: *Chemometrics—From Basics to Wavelet Transform.* Wiley, Hoboken, NJ, 2004.

Crawford, L. R., Morrison, J. D.: *Anal. Chem.* 40, 1968, 1469–1474. Computer methods in analytical mass spectrometry. Empirical identification of molecular class.

Curry, B., Rumelhardt, D. E.: *Tetrahedron Comput. Methodol.* 3, 1990, 213–237. MSnet: A neural network which classifies mass spectra.

Demuth, W., Karlovits, M., Varmuza, K.: *Anal. Chim. Acta* 516, 2004, 75–85. Spectral similarity versus structural similarity: Mass spectrometry.

Erni, F., Clerc, J. T.: *Helv. Chim. Acta* 55, 1972, 489–500. Strukturaufklärung organischer Verbindungen durch computerunterstützten Vergleich spektraler Daten.

Geladi, P., MacDougall, D., Martens, H.: *Appl. Spectrosc.* 39, 1985, 491–500. Linearization and scatter-correction for NIR reflectance spectra of meat.

Helland, I. S., Naes, T., Isaksson, T.: *Chemom. Intell. Lab. Syst.* 29, 1995, 233–241. Related versions of the multiplicative scatter correction method for preprocessing spectroscopic data.

Jurs, P. C., Isenhour, T. L.: *Chemical Applications of Pattern Recognition*. Wiley, New York, 1975.

Klawun, C., Wilkins, C. L.: *J. Chem. Inf. Comput. Sci.* 36, 1996, 249–257. Joint neural network interpretation of infrared and mass spectra.

MassFeatGen: Software for generation of mass spectral features. Demuth W., Varmuza K., Laboratory for Chemometrics, Vienna University of Technology, www.lcm.tuwien.ac.at, Vienna, Austria, 2006.

Naes, T., Isaksson, T., Kowalski, B. R.: *Anal. Chem.* 62, 1990, 664–673. Locally weighted regression and scatter correction for near-infrared reflectance data.

Naes, T., Isaksson, T., Fearn, T., Davies, T.: *A User-Friendly Guide to Multivariate Calibration and Classification*. NIR Publications, Chichester, United Kingdom, 2004.

NIST: Mass Spectral Database 98. National Institute of Standards and Technology, www.nist.gov/srd/nist1a.htm, Gaithersburg, MD, 1998.

Savitzky, A., Golay, M. J. E.: *Anal. Chem.* 36, 1964, 1627–1639. Smoothing and differentiation of data by simplified least squares procedure.

Smilde, A., Bro, R., Geladi, P.: *Multi-Way Analysis with Applications in the Chemical Sciences*. Wiley, Chichester, United Kingdom, 2004.

Steinier, J., Termonia, Y., Deltour, J.: *Anal. Chem.* 44, 1972, 1906–1909. Comments on smoothing and differentiation of data by simplified least square procedure.

Unscrambler: Software. Camo Process AS, www.camo.no, Oslo, Norway, 2004.

Vandeginste, B. G. M., Massart, D. L., Buydens, L. C. M., De Jong, S., Smeyers-Verbeke, J.: *Handbook of Chemometrics and Qualimetrics: Part B*. Elsevier, Amsterdam, the Netherlands, 1998.

Varmuza, K.: *Pattern Recognition in Chemistry*. Springer, Berlin, Germany, 1980.

Varmuza, K.: in Lindon, J. C., Tranter, G. E., Holmes, J. L. (Eds.), *Encyclopedia of Spectroscopy and Spectrometry*, Vol., Academic Press, London, United Kingdom, 2000, pp. 232–243. Chemical structure information from mass spectrometry.

Varmuza, K.: *Analytical Sciences* 17(suppl.), 2001, i467–i470. Recognition of relationships between mass spectral data and chemical structures by multivariate data analysis.

Varmuza, K.: in Pomerantsev, A. L. (Ed.), *Progress in Chemometrics Research*, Vol., Nova Science Publishers, New York, NY, 2005, pp. 67–87. Global and local chemometric models of spectra–structure relationships.

Werther, W., Demuth, W., Krueger, F. R., Kissel, J., Schmid, E. R., Varmuza, K.: *J. Chemom.* 16, 2002, 99–110. Evaluation of mass spectra from organic compounds assumed to be present in cometary grains. Exploratory data analysis.

Wold, H., Christie, O. H. J.: *Anal. Chim. Acta* 165, 1984, 51–59. Extraction of mass spectral information by a combination of autocorrelation and principal components models.

Appendix 1 Symbols and Abbreviations

VARIABLES

Only the more important and more frequently used variables are listed here. Chemometrics literature sometimes uses different symbols as statistics or economics. In some cases, the same symbol has to be used for different variables; however, the particular meaning should be always clear from the definition.

a	Number of components used, for instance in PCA or in PLS.
$\boldsymbol{B}$	Loading matrix with m rows (variables, features) and l columns (components). b_{jl} is the loading of variable j for component l. b_j are also used for regression coefficients. In chemometrics not $\boldsymbol{B}$ but $\boldsymbol{P}$ is often used for the x-loading matrix in PCA and PLS, and $\boldsymbol{Q}$ for the y-loading matrix.
c_{jk}	Covariance between variables j and k; $\boldsymbol{C}$ is the covariance matrix.
d	Distance between two objects in multivariate variable space; for instance Euclidean distance, d(Euclid), city block distance, d(city), or Mahalanobis distance, d(Mahalanobis).
e	Error (in prediction of dependent variable y) or residual; e_i is the error for object i; $\boldsymbol{E}$ is the x-residual matrix, $\boldsymbol{F}$ is the y-residual matrix.
λ	Eigenvalue.
m	Number of variables (features, descriptors).
MSE	Mean of squared errors, MSE_{CV} for cross validation, MSE_{TEST} for an independent test set.
n	Number of objects (samples, patterns).
$\boldsymbol{P}$	Loadings matrix with m rows (variables, features) and a columns (components). p_{jl} is the loading of variable j for component l. As usual in chemometrics, $\boldsymbol{P}$ is used for x-loadings in PCA and PLS.
PRESS	Predicted residual error sum of squares (sum of squared prediction errors).
r_{jk}	(Pearson) correlation coefficient between variables j and k; r^2 is the squared (Pearson) correlation coefficient; for instance, between experimental y and predicted y.
RMSE	Root of mean squared errors (square root of MSE).
RSS	Residual sum of squares (sum of squared residuals/errors).
s	Standard deviation.
SEC	Standard error of calibration.
SEP	Standard error of prediction, SEP_{CV} for cross validation, SEP_{TEST} for an independent test set.

$\boldsymbol{T}$	Score matrix with n rows (objects, samples) and a columns (components). t_{il} is the score of component l for object i. As usual in chemometrics, $\boldsymbol{T}$ is used for x-scores in PCA and PLS.
$\boldsymbol{U}$	Score matrix with n rows (objects, samples) and a columns (components). u_{il} is the score of component l for object i. In chemometrics not $\boldsymbol{U}$ but $\boldsymbol{T}$ is often used for the x-score matrix in PCA and PLS. $\boldsymbol{U}$ is used in chemometrics especially for the y-score matrix.
v	Variance.
$\boldsymbol{W}$	Loading weight matrix in PLS.
$\boldsymbol{X}$	Variable (feature) matrix with n rows (objects, samples) and m columns (variables, features); matrix element x_{ij} is the value of variable j for object i. x_j are the independent variables in regression models. $\bar{x}$ is the arithmetic mean.
$\boldsymbol{y}$	Property vector for n objects (samples). The property may be continuous (for instance boiling point) or categorical (for instance class membership). y_i is the property value for object i. In some examples, a matrix $\boldsymbol{Y}$ (n rows, q columns) is used for a combined treatment of more than one property. y is the dependent variable in regression models; $\hat{y}$ is a predicted y-value.

ABBREVIATIONS

For better readability, a few abbreviations—widely used in chemometrics, chemoinformatics, and chemistry—are used.

Multivariate Data Analysis and Statistics

ANN	Artificial neural network
CCA	Canonical correlation analysis
CV	Cross validation
GA	Genetic algorithm
k-NN	*k*-Nearest neighbor
LDA	Linear discriminant analysis
MLR	Multiple linear regression (used as general term for all linear regression methods like OLS, PLS, PCR)
MSC	Multiplicative signal/scatter correction (preprocessing of NIR data)
NIPALS	Nonlinear iterative partial least-squares
NLM	Nonlinear mapping
OLS	Ordinary least-squares (regression)
PC	Principal component (in PCA or PLS, for instance PC1, PC2)
PCA	Principal component analysis
PCR	Principal component regression
PLS	Partial least-squares (regression)
RBF	Radial basis function
SIMCA	Soft independent modeling of class analogies

SOM	Self-organizing map
SVD	Singular value decomposition
SVM	Support vector machine

Chemistry

GC	Gas chromatography
IR	Infrared spectroscopy/spectrum
MS	Mass spectrometry/spectrum
NIR	Near infrared spectroscopy/spectrum
NMR	Nuclear magnetic resonance spectroscopy/spectrum
PAT	Process analytical technology
QSAR	Quantitative structure–activity relationship
QSPR	Quantitative structure–property relationship
UV/VIS	Ultraviolet and visible spectroscopy/spectrum

Appendix 2 Matrix Algebra

A.2.1 DEFINITIONS

Basic understanding and efficient use of multivariate data analysis methods require some familiarity with matrix notation. The user of such methods, however, needs only elementary experience; it is for instance not necessary to know computational details about matrix inversion or eigenvector calculation; but the prerequisites and the meaning of such procedures should be evident. Important is a good understanding of matrix multiplication. A very short summary of basic matrix operations is presented in this section. Introductions to matrix algebra have been published elsewhere (Healy 2000; Manly 2000; Searle 2006).

Most of the user-friendly software packages in chemometrics hide all underlying matrix algebra, on the other hand, powerful software tools are available that allow calculations with matrices almost like doing simple calculations with a pocket calculator. The development of chemometric methods was mainly performed by using the programming environment MATLAB (Matlab 2000; Martinez and Martinez 2002). Recently, also the free software product and programming environment R—denoted here by character $\mathbb{R}$—(Chambers 2007; R 2008) is increasingly used (as throughout this book), as well as the free product OCTAVE (Octave 1998; Alsberg and Hagen 2006).

Multivariate data are represented by one or several matrices. Variables (scalars, vectors, matrices) are written in italic characters; SCALARS in lower or upper case (examples: n, A), VECTORS in bold face lower case (example: $\boldsymbol{b}$). Vectors are always column vectors; row vectors are written as transposed vectors (example: $\boldsymbol{b}^{\mathrm{T}}$). MATRICES are written in bold face upper case characters (example: $\boldsymbol{X}$). The first index of a matrix element denotes the row, the second the column. Examples: x_{ij} or $x(i, j)$ is an element of matrix $\boldsymbol{X}$, located in row i and column j; $\boldsymbol{x}_i^{\mathrm{T}}$ is the vector of row i; $\boldsymbol{x}_j$ is the vector of column j.

Figure A.2.1 summarizes this notation and shows some special matrices. In a ZERO MATRIX, all elements are zero. In a QUADRATIC MATRIX, the number of rows, n, is equal to the number of columns, m; the (main) diagonal is from element (1, 1) to element (n, n). A DIAGONAL MATRIX is a square matrix with all nondiagonal elements zero. The IDENTITY MATRIX is a diagonal matrix with all diagonal elements equal to 1. A SYMMETRIC MATRIX is a square matrix with each element (i, j) equal to the mirrored element (j, i).

The vectors and matrices in Figure A.2.1 can be defined in $\mathbb{R}$ as follows:

```
ℝ:  x <- c(3,1,4)                  # input of a vector
    xt <- t(x)                     # transpose of vector
    A <- cbind(x,c(0,2,7))         # column binding of two vectors
    At <- t(A)                     # transpose of matrix
    matrix(0,nrow=3,ncol=2)        # zero matrix
    diag(c(3,2,8))                 # diagonal matrix
```

$$\boldsymbol{x} = \begin{bmatrix} 3 \\ 1 \\ 4 \end{bmatrix} \quad \boldsymbol{x}^{\mathrm{T}} = \begin{bmatrix} 3 & 1 & 4 \end{bmatrix} \quad A = \begin{bmatrix} 3 & 0 \\ 1 & 2 \\ 4 & 7 \end{bmatrix} \quad A^{\mathrm{T}} = \begin{bmatrix} 3 & 1 & 4 \\ 0 & 2 & 7 \end{bmatrix}$$

Zero matrix, Diagonal matrix, Identity matrix, Symmetric matrix

$$\begin{bmatrix} 0 & 0 \\ 0 & 0 \\ 0 & 0 \end{bmatrix} \quad \begin{bmatrix} 3 & 0 & 0 \\ 0 & 2 & 0 \\ 0 & 0 & 8 \end{bmatrix} \quad \begin{bmatrix} 1 & 0 & 0 \\ 0 & 1 & 0 \\ 0 & 0 & 1 \end{bmatrix} \quad \begin{bmatrix} 3 & 5 & 9 \\ 5 & 2 & 4 \\ 9 & 4 & 8 \end{bmatrix}$$

FIGURE A.2.1 Matrix transpose and special types of matrices.

```
diag(3)                             # identity matrix
matrix(c(3,5,9,5,2,4,9,4,8),ncol=3)  # symmetric matrix
X <- read.csv(file="matrix.csv")   # import of a csv-file,
                                   # e.g. from Excel
```

A vector $\boldsymbol{x} = (x_1, x_2, \ldots, x_m)^{\mathrm{T}}$ can be considered as a point in an m-dimensional space with the coordinates given by the vector components, or as a straight line (vector) from the origin to this point. A matrix $\boldsymbol{X}$ of dimension $(n \times m)$ can be considered as a point cloud in an m-dimensional space.

A.2.2 ADDITION AND SUBTRACTION OF MATRICES

Addition and subtraction of matrices are performed element-wise; the matrices must have the same size (see Figure A.2.2). Multiplication of a vector or a matrix with a scalar is also performed element-wise; in the case of a vector the resulting vector has the same direction but a different length. Vectors $\boldsymbol{a}$ and $\boldsymbol{b} = -\boldsymbol{a}$ have the same length but reverse direction.

The calculations shown in Figure A.2.2 can be performed in ℝ as follows:

```
ℝ: A <- matrix(c(3,0,1,2,4,7),ncol=2,byrow=TRUE)    # matrix A
   B <- matrix(c(-1,3,0,2,5,1),ncol=2,byrow=TRUE)   # matrix B
   C <- A+B  # sum of A and B
   X <- matrix(c(2,3,1,0,1,2),ncol=2,byrow=TRUE)    # matrix X
   Y <- X*3  # each element multiplied by 3
```

A.2.3 MULTIPLICATION OF VECTORS

Two types of vector multiplication exist:

1. The SCALAR PRODUCT (DOT PRODUCT, INNER PRODUCT) requires two vectors of the same length; the result is a scalar obtained by pair-wise multiplication of

$$\underset{A}{\begin{bmatrix} 3 & 0 \\ 1 & 2 \\ 4 & 7 \end{bmatrix}} + \underset{B}{\begin{bmatrix} -1 & 3 \\ 0 & 2 \\ 5 & 1 \end{bmatrix}} = \underset{C}{\begin{bmatrix} 2 & 3 \\ 1 & 4 \\ 9 & 8 \end{bmatrix}} \qquad \underset{X}{\begin{bmatrix} 2 & 3 \\ 1 & 0 \\ 1 & 2 \end{bmatrix}} \times 3 = \underset{Y}{\begin{bmatrix} 6 & 9 \\ 3 & 0 \\ 3 & 6 \end{bmatrix}}$$

FIGURE A.2.2 Addition of matrices and multiplication of a matrix with a scalar.

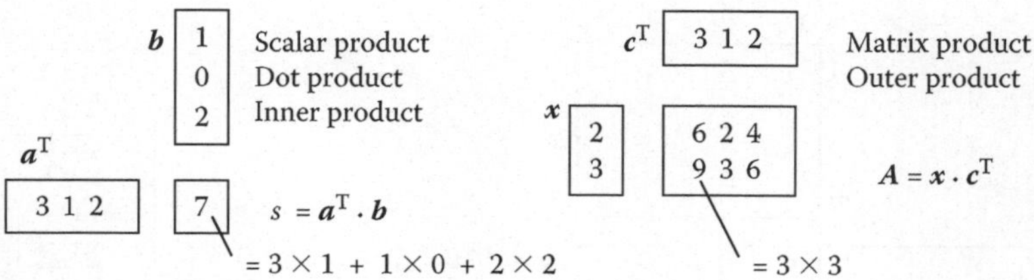

FIGURE A.2.3 Multiplication of vectors.

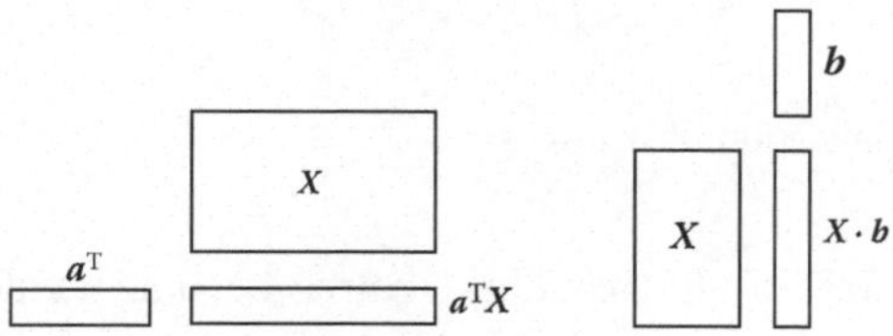

FIGURE A.2.4 Multiplication of a vector and a matrix.

corresponding vector elements. In a scalar product, for instance $\boldsymbol{a}^{\mathrm{T}}\boldsymbol{b}$, the first vector must be a row vector, and the second a column vector (see Figure A.2.3). Note that the scalar product of orthogonal vectors is zero. The length of a vector $\boldsymbol{a}$—also called the Euclidean norm, $||\boldsymbol{a}||$—is the square root of $\boldsymbol{a}^{\mathrm{T}}\boldsymbol{a}$. A vector can be scaled to unit length by dividing each component by $||\boldsymbol{a}||$.

2. The MATRIX PRODUCT (OUTER PRODUCT) of two vectors is a matrix (example: $\boldsymbol{a}\boldsymbol{b}^{\mathrm{T}}$. The first vector must be a column vector, and the second a row vector (see Figure A.2.3).

   The two possibilities of multiplying a vector and a matrix are illustrated in Figure A.2.4.

The multiplications shown in Figure A.2.3 can be performed in $\mathbb{R}$ as follows:

```
R:  a <- c(3,1,2)           # vector a
    b <- c(1,0,2)           # vector b
    s <- drop(t(a)%*%b)     # scalar product, "drop" produces a
                            # number rather than a matrix
    x <- c(2,3)             # vector x
    c <- c(3,1,2)           # vector c
    A <- x%*%t(c)           # matrix product
```

A.2.4 MULTIPLICATION OF MATRICES

Multiplication of two matrices is the most important operation in multivariate data analysis. It is not performed element-wise, but each element of the resulting matrix is a scalar product (see Section A.2.3). A matrix $\boldsymbol{A}$ and a matrix $\boldsymbol{B}$ can be multiplied by $\boldsymbol{A} \cdot \boldsymbol{B}$ only if the number of columns in $\boldsymbol{A}$ is equal to the number of rows in $\boldsymbol{B}$; this

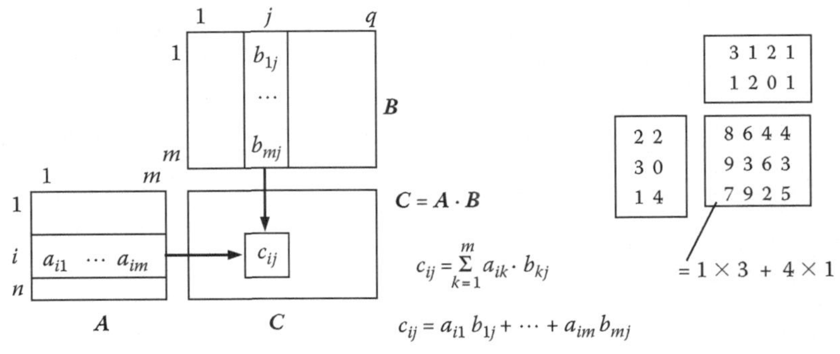

FIGURE A.2.5 Multiplication of matrices.

condition and the size of the resulting matrix become evident when using an arrangement as shown in Figure A.2.5. Each element of the result matrix is the scalar product of the corresponding row vector in $\boldsymbol{A}$ and the corresponding column vector in $\boldsymbol{B}$. In general $\boldsymbol{A} \cdot \boldsymbol{B}$ is not equal to $\boldsymbol{B} \cdot \boldsymbol{A}$ and both forms only exist if $\boldsymbol{A}$ and $\boldsymbol{B}$ are quadratic. The point symbol for multiplication is often omitted and the product is for instance written as $\boldsymbol{AB}$.

The example for matrix multiplication in Figure A.2.5 can be computed in ℝ as

```
R:  A <- matrix(c(2,2,3,0,1,4),ncol=2,byrow=TRUE)   # matrix A
    B <- matrix(c(3,1,2,1,1,2,0,1),ncol=4,byrow=TRUE)
                                                    # matrix B
    C <- A%*%B                                      # matrix product
```

A.2.5 MATRIX INVERSION

Matrix inversion is analogous to division. Multiplication of $\boldsymbol{A}$ with its inverse $\boldsymbol{A}^{-1}$ gives an identity matrix, $\boldsymbol{I}$ (see Figure A.2.6). The inverse is only defined for square matrices that are not singular. A matrix is SINGULAR if at least one row (or column) contains equal numbers, or at least one column (or row) is a linear combination of

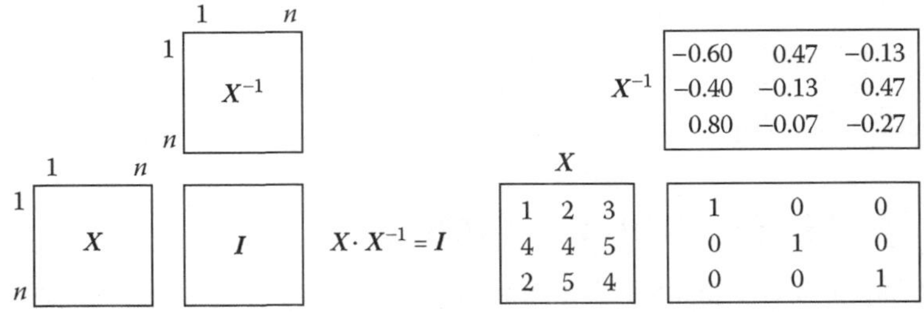

FIGURE A.2.6 Inversion of a matrix.

other columns (rows). Data from chemistry are often highly correlated and give a (near) singular covariance matrix; consequently, methods are required that do not need inversion of the covariance matrix. Matrix inversion is performed by an iterative procedure.

The example in Figure A.2.6 can be computed in ℝ as

```
ℝ:  X <- matrix(c(1,2,3,4,4,5,2,5,4),ncol=3,byrow=TRUE)
                                    # matrix X
    Xinv <- solve(X)                # X inverse
    X%*%Xinv-diag(3)                # results in matrix with zeros
```

A.2.6 EIGENVECTORS

Eigenvectors of square matrices are frequently used in multivariate methods. Each eigenvector is connected with an EIGENVALUE, λ, and in most applications the eigenvectors are ordered by decreasing λ and are normalized to length 1 (see Figure A.2.7).

A nonsingular matrix $\boldsymbol{A}$ of size $(m \times m)$ has m eigenvectors with eigenvalues that are all greater than zero. Any two eigenvectors are orthogonal. If $\boldsymbol{A}$ is singular then less than m eigenvectors with eigenvalue > 0 exist.

Calculation of eigenvectors requires an iterative procedure. The traditional method for the calculation of eigenvectors is JACOBI ROTATION (Section 3.6.2). Another method—easy to program—is the NIPALS algorithm (Section 3.6.4). In most software products, singular value decomposition (SVD), see Sections A.2.7 and 3.6.3, is applied. The example in Figure A.2.7 can be performed in ℝ as follows:

```
ℝ:  A <- matrix(c(2,3,4,3,7,5,4,5,1),ncol=3,byrow=TRUE)
                                                  # matrix A
    res <- eigen(A)      # eigenvectors and eigenvalues of A
    B <- res$vectors     # matrix with eigenvectors of A
    L <- res$values      # vector with eigenvalues of A
    A%*%B[,1]            # A multiplied with first eigenvector
    B[,1]*L[1]           # gives the same result
```

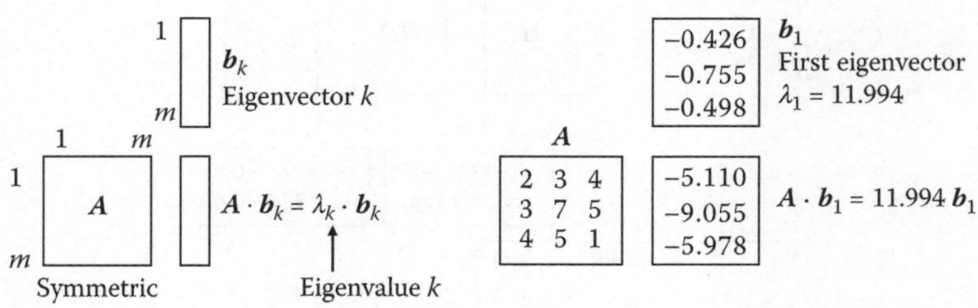

FIGURE A.2.7 Eigenvectors and eigenvalues of a matrix.

A.2.7 SINGULAR VALUE DECOMPOSITION

SVD performs a decomposition of a matrix $\boldsymbol{X}$ of size $(n \times m)$ into $\boldsymbol{X} = \boldsymbol{U} \cdot \boldsymbol{D} \cdot \boldsymbol{V}^{\mathrm{T}}$ where

- $\boldsymbol{U}$ is an orthogonal matrix of dimension $(n \times n)$.
- $\boldsymbol{D}$ is of dimension $(n \times m)$ and has entries $d_i \geq 0$ at the diagonal positions. (i, i) for $i = 1, \ldots, \min(n, m)$; all other elements are zero.
- $\boldsymbol{V}$ is an orthogonal matrix of dimension $(m \times m)$.

If r is the rank of $\boldsymbol{X}$, then $d_1, \ldots, d_r$ are positive, and they are called SINGULAR VALUES of $\boldsymbol{X}$. Usually the columns of $\boldsymbol{U}$ and $\boldsymbol{V}$ are normalized to length 1, and the first r columns are arranged according to a descending order of the singular values $d_1 \geq \cdots \geq d_r > 0$. Figure A.2.8 illustrates SVD for a simple example.

Note that some programming environments (e.g., $\mathbb{R}$) use a different convention for the matrix dimensions in order to save memory. If $q = \min(n, m)$, then $\boldsymbol{U}$ is $(n \times q)$, $\boldsymbol{D}$ is $(q \times q)$, and $\boldsymbol{V}$ is $(m \times q)$.

The SVD can also be expressed in terms of eigenvectors and eigenvalues. Let r denote the rank of $\boldsymbol{X}$. Then the first r columns of the matrix $\boldsymbol{U}$ are the eigenvectors of $\boldsymbol{X} \cdot \boldsymbol{X}^{\mathrm{T}}$ to the eigenvalues $d_1^2, \ldots, d_r^2$. The first r columns of the matrix $\boldsymbol{V}$ are the eigenvectors of $\boldsymbol{X}^{\mathrm{T}} \cdot \boldsymbol{X}$ to the same eigenvalues $d_1^2, \ldots, d_r^2$. For the application of SVD to principal component analysis, see Section 3.6.3.

The example in Figure A.2.8 can be performed in $\mathbb{R}$ as follows:

```
R: X <- matrix(c(1,0,0,-1,2,4),ncol=2,byrow=TRUE)  # matrix X
   res <- svd(X)        # SVD of X, res contains U, D, and V
   U <- res$u           # matrix U
   d <- res$d           # diagonal elements of D
   D <- diag(d)         # matrix D
   V <- res$v           # matrix V
   X - U%*%D%*%t(V)     # results in a matrix with zeros
```

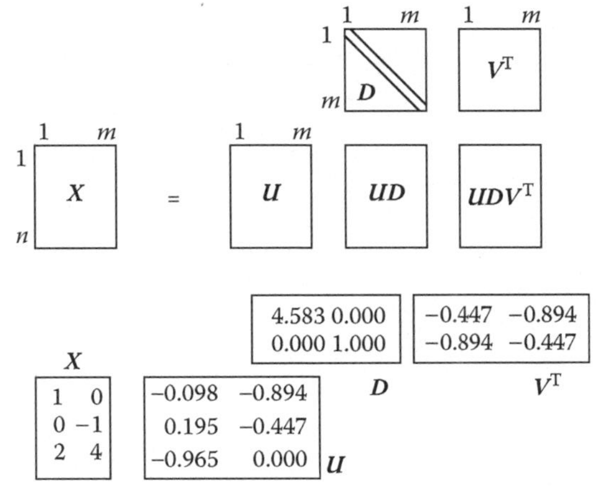

FIGURE A.2.8 SVD of a matrix $\boldsymbol{X}$.

REFERENCES

Alsberg, B. K., Hagen, O. J.: *Chemom. Intell. Lab. Syst.* 84, 2006, 195–200. How Octave can replace Matlab in chemometrics.

Chambers, J. M.: *Software for Data Analysis: Programming with R.* Springer, New York, 2007.

Healy, M. J. R.: *Matrices for Statistics.* Oxford University Press, New York, 2000.

Manly, B. F. J.: *Multivariate Statistical Methods: A Primer.* Chapman & Hall, London, United Kingdom, 2000.

Martinez, W. L., Martinez, A. R.: *Computational Statistics Handbook with Matlab.* Chapman & Hall and CRC, Boca Raton, FL, 2002.

Matlab: Software. *The Mathworks Inc.*, www.mathworks.com, Natick, MA, 2000.

Octave: Software. J. W. Eaton, University of Wisconsin, www.gnu.org/software/octave, Madison, WI, 1998.

R: Software, a language and environment for statistical computing. R Development Core Team, Foundation for Statistical Computing, www.r-project.org, Vienna, Austria, 2008.

Searle, S. R.: *Matrix Algebra Useful for Statistics*, Wiley-Interscience, Hoboken, NJ, 2006.

Appendix 3 Introduction to ℝ

A.3.1 GENERAL INFORMATION ON ℝ

ℝ is a software environment mainly intended for statistical computing. It is an open source implementation of the S language, which has mainly been written by John Chambers from Bell Laboratories (Becker et al. 1988). ℝ has originally been developed by Ross Ihaka and Robert Gentleman in 1994 at the University of Auckland (Ihaka and Gentleman 1996). Besides an ℝ Development Core Team (ℝ 2008), researchers and people from practice are continuously working on a further development. More than 1500 contributed packages are available now to any user, including latest implementations of statistical methods as well as other software implementations.

ℝ is freely available under the GNU General Public License. It offers high computational speed, a useful help system, a huge variety of statistical methods, and the possibility of extension by own developed methods. ℝ requires basic knowledge in programming and statistics.

Online manuals and Wiki pages for ℝ can be found at the ℝ homepage:

http://www.r-project.org

There exist various books on ℝ, as well as books on how to use ℝ for statistical computing. Recent books are for instance Braun and Murdoch 2007, Chambers 2007, Sakar 2007, and Spector 2008.

A.3.2 INSTALLING ℝ

ℝ can be installed via the CRAN (Comprehensive ℝ Archive Network) at

http://cran.r-project.org

It is available for different operating systems (Windows, Linux, Mac). The base system needs to be installed first. Under Windows, the base system is installed via the executable ".exe" file at http://cran.r-project.org/bin/windows/base/.

A.3.3 STARTING ℝ

Under Windows, ℝ is started either via the ℝ start icon, or via the file *Rgui.exe*. A window called *RGui* opens, containing a window called *R Console*, a simple graphical user interface.

Under Linux or Mac, one has to type R in a terminal window. A prompt sign ">" appears and ℝ commands can be submitted.

A.3.4 WORKING DIRECTORY

It is recommended to define an own WORKING DIRECTORY when working with ℝ on a specific project. The reason is that ℝ optionally saves all objects (data, functions, results, etc.) in the working directory, and there could be confusion if objects from different projects are saved in the same directory.

Under Windows, the working directory can be selected either via the *RGui* window (*File → Change directory*) or by typing the command setwd("*fullpathname*"). For other systems, the last option can be used, or one executes ℝ already in the desired directory. With the command getwd(), one can check for the directory.

A.3.5 LOADING AND SAVING DATA

ℝ offers the possibility to load data from various data formats. The following commands create an object "dat" in ℝ with the data table.

dat = read.table("*filename.txt*", header = TRUE)
: loads data tables which are available in a text format.

dat = read.csv2("*filename.csv*")
: loads data in tables which are available in csv format (e.g. from MS-Excel).

dat = source("*filename.txt*")
: loads a text file that might have been saved by dump().

dat = load("*filename.RData*")
: loads data that have been saved by save() in ℝ binary format.

For saving data objects that are available during the current ℝ session, one can use the following commands:

write.table(*object*,"*filename.txt*", header = TRUE)
: saves an ℝ object into the text file "*filename.txt*".

write.csv2(*object*,"*filename.csv*")
: saves an ℝ object into "*filename.csv*" that can be read, e.g., by Excel.

dump("*object*","*filename.txt*")
: saves an ℝ object as text file "*filename.txt*".

save(*object*,"*filename.RData*")
: saves an ℝ object into an ℝ binary format.

A.3.6 IMPORTANT ℝ FUNCTIONS

For any ℝ function, one can either look at the code by typing the name of the function (e.g., sd), or one can execute the function by using brackets () after the function name (e.g., sd()). Arguments can be submitted inside the brackets.

Some important basic functions are

q()	Ends the current session. A question follows whether the current work space (i.e., all objects that are made available in the session) should be saved. If this is wanted, they are saved in the file named "*.RData*".
help(*functionName*)	Provides help for the function *functionName*.
help.search("*topic*")	Searches for the topic *topic* in all installed ℝ packages.
help.start()	Starts the ℝ help system with online manuals; interactive help, search possibilities, and many other functionalities are available.
ls()	Lists the objects that are available in the current session.
save.image()	Saves all objects that are available in the current session to the file "*.RData*".
example(*function*)	Executes the examples that are available at the help page to the function *function*.
str(*object*)	Shows the structure of the ℝ object *object*.
summary(*object*)	Provides a short summary of the ℝ object *object*.
library(*packName*)	Loads the contributed package *packName*. All available contributed packages are listed and downloadable at the CRAN.
#	All text (in the line) after this symbol is ignored. This is typically used for comments in functions.
x <- 3	Assigns the value 3 to the object *x*. Here one could also use x = 3, but in some cases this can be misleading. One can assign any number, character, vector, matrix or a more general object (see below), or function to another object.

A.3.7 OPERATORS AND BASIC FUNCTIONS

MATHEMATICAL AND LOGICAL OPERATORS, COMPARISON

<table>
<tr><td>+</td><td>Addition</td><td>!=</td><td>Unequal</td></tr>
<tr><td>−</td><td>Subtraction</td><td>></td><td>Larger</td></tr>
<tr><td>*</td><td>Multiplication</td><td>>=</td><td>Larger or equal</td></tr>
<tr><td>/</td><td>Division</td><td><</td><td>Less</td></tr>
<tr><td>^</td><td>Power</td><td><=</td><td>Less or equal</td></tr>
<tr><td>%*%</td><td>Matrix multiplication</td><td>!</td><td>Negation</td></tr>
<tr><td>&&</td><td>Logical AND</td><td>||</td><td>Logical OR</td></tr>
<tr><td>&</td><td>Logical AND (elementwise)</td><td>&</td><td>Logical OR (elementwise)</td></tr>
<tr><td>==</td><td>Equal</td><td></td><td></td></tr>
</table>

Special Elements

TRUE	True
FALSE	False
Inf	Infinity
–Inf	Minus infinity
NULL	Null (empty) object
NA	Not available (missing value)
NaN	Not a number (not defined)

Mathematical Functions

The following functions are applied to an object *x*:

sqrt(*x*)	Square root
min(*x*), max(*x*)	Minimum, maximum
abs(*x*)	Absolute values
sum(*x*)	Sum of all values
log(*x*), log10(*x*)	Natural logarithm and base 10 logarithm
exp(*x*)	Exponential function
sin(*x*), cos(*x*), tan(*x*)	Trigonometric functions
eigen(*x*)	Eigenvectors and eigenvalues of *x*
svd(*x*)	Singular value decomposition of *x*

Matrix Manipulation

rbind, cbind	Combine rows (columns)
t(*x*)	Matrix transpose of *x*
solve(*x*)	Matrix inverse of *x*
rowSums(*x*), colSums(*x*)	Row sums or column sums of *x*
rowMeans(*x*), colMeans(*x*)	Row means or column means of *x*
apply(*x*,1,*function*), apply(*x*,2,*function*)	Apply the function *function* (e.g. *mean*) to the rows (code 1) or columns (code 2) of *x*

Statistical Functions

mean(*x*), median(*x*)	Mean, median
sd(*x*), mad(*x*)	Standard deviation, median absolute deviation
var(*x*), cov(*x*)	Variance and covariance of the elements of *x*
var(*x, y*), cov(*x, y*)	Covariance between *x* and *y*
cor(*x*)	Correlation matrix of *x*
cor(*x, y*)	Correlation between *x* and *y*
scale(*x*)	Scaling and centering of a matrix
quantile(*x,probs*)	Quantile of *x* to probabilities *probs*
dnorm, pnorm, qnorm, rnorm	Density, distribution function, quantile function, and random generation for the normal distribution

Density, distribution function, quantile function, and random generation are also available for various other distributions. Instead of dnorm (equivalently for other versions), the functions are named dt, dunif, dchisq, dexp, dbinom, dpois, dgamma, dbeta, dlnorm, etc.

A.3.8 DATA TYPES

logical	Logical value, either TRUE (T) or FALSE (F)
integer	Integer (number without decimals) value (e.g., 1283)
double	A real numeric value (e.g., 2.98)
character	Character expression, e.g., "aabbc"

Conversion of x to a different data type (if appropriate) can be done by

as.logical(*x*), as.numeric(*x*), as.integer(*x*), as.double(*x*), as.character(*x*)

The commands

is.logical(*x*), is.numeric(*x*), is.integer(*x*), is.double(*x*), is.character(*x*)

check whether x is the specified data type.

Missing Values

The command is.na(*x*) checks for each row of x if it contains missing values (NA).

na.omit(*x*) excludes entire rows of x that contain missing values.

A.3.9 DATA STRUCTURES

vector	Consists of a desired number of elements which are all of the same data type, e.g., logical, numeric, or character
factor	Contains information about categorical data, it contains very few different levels (characters, numbers, or combinations thereof)
matrix	Matrix with rows and columns; all elements must consist of the same data type
array	Array with rows, columns, slices, etc.; all elements must consist of the same data type
data.frame	A data frame can combine different data types, e.g., a matrix where factors, numeric, and character variables can occur in different columns.
list	It is a general data object combining vectors, factors, matrices data frames, and lists

A.3.10 SELECTION AND EXTRACTION FROM DATA OBJECTS

Examples for Creating Vectors

x <- 5:10	Gives a vector *x* with elements	5 6 7 8 9 10
x <- c(1,3,5:10)	Gives a vector *x* with elements	1 3 5 6 7 8 9 10
x <- seq(from = 2,to = 4, by = 0.5)	Gives a vector *x* with elements	2.0 2.5 3.0 3.5 4.0
x <- rep(3,times = 4)	Gives a vector *x* with elements	3 3 3 3

Generally speaking, if the arguments are not specified in a function call, the parameters are taken in the same sequence as the arguments are defined. For example, the function seq(*from,to*) takes as first argument from and as second argument to. Thus, seq(from = 5,to = 10) gives the same as seq(5,10), but seq(10,5) would result in 10 9 8 7 6 5. Moreover, the arguments can be abbreviated and provided in a different order in the call; they only need to be unique. For example, seq(t = 10,f = 5) gives the same as seq(from = 5,to = 10).

Examples for Selecting Elements from a Vector or Factor

Consider the vector x <- c("a","b","c","d","e").

x[2:4]	Gives	"b" "c" "d"
x[c(1,3,5)]	Gives	"a" "c" "e"
x[-c(1,5)]	Gives	"b" "c" "d" # without elements 1 and 5
x[x! = "a"]	Gives	"b" "c" "d" "e" # without element "a"

Examples for Selecting Elements from a Matrix, Array, or Data Frame

As for vectors, the selection is specified within square brackets []. For selecting rows and columns of a matrix, two arguments have to be provided, separated by a comma, e.g. [*rows,columns*]. Analogously, for arrays, the selections in the different ways of the array are specified as separate arguments, e.g. [*way1,way2,way3*].

x[2,4]	Select element in 2nd row and 4th column
x[2,]	Select entire 2nd row
x[,4]	Select entire 4th column
x[,1:4]	Select first 4 columns
x[,"*VarName*"]	Select column with name *VarName*
x$*VarName*	Select column with name *VarName* (only if *x* is a data frame)

Examples for Selecting Elements from a List

x[[3]]	Select 3rd list element
x$*ListName*	Select list element with name *ListName*

A.3.11 GENERATING AND SAVING GRAPHICS

Functions Relevant for Graphics

plot(*x*)	Plot the values of a column (vector) *x* versus the index
plot(*x*,*y*)	Plot the elements in the column (vector) *x* against those in *y*
points(*x*, *y*)	Add points to a plot
lines(*x*, *y*)	Add lines to a plot
text(*x*, *y*,"*end*")	Place the text "end" at the location specified by *x* and *y*
legend(*x*, *y*, *leg*)	Place legend given by character vector *leg* at location *x* and *y*

Relevant Plot Parameters

main = "*title*"	Add title *title* to the plot
xlab, ylab	Character for *x*- (or *y*-) axis labels
xlim, ylim	Vector with minimum and maximum for *x*- (or *y*-) axis
pch	Number (e.g., pch = 3) of the plot symbol (plot character)
col	Name (e.g., col = "red") or number (e.g., col = 2) of the symbol color
cex	Scaling for text size and symbol size

Statistical Graphics

hist(*x*)	Plot a histogram of the frequencies of *x*
plot(density(*x*))	Plot a density function of *x*
boxplot(*x*)	Boxplot of *x*
qqnorm(*x*)	QQ-plot of *x*

Saving Graphic Output

When generating a graphic, a new window called *R Graphics Device* opens with the plot. Under Windows, the graphic can be directly saved via *File* → *Save as* in the desired format. The graphic output can also be directly sent to a file. Depending on the desired file format, this is done by the commands postscript, pdf, png, or jpeg with the file name (and optionally the size of the graphic) as argument. After specifying the file name, the graphic can be generated and the information will be redirected to the file rather than to the *R Graphics Device*. Finally, the file needs to be closed with dev.off(). An example is

```
pdf(file = "FileName.pdf",width = 6,height = 4)
boxplot(x)
dev.off()
```

The pdf-file *FileName.pdf* is generated with the size 6 in. times 4 in. containing a boxplot of *x*.

REFERENCES

Becker, R. A., Chambers, J. M., Wilks, A. R.: *The New S Language: A Programming Environment for Data Analysis and Graphics*. Chapman & Hall, London, United Kingdom, 1988.

Braun, W. J., Murdoch, D. J.: *A First Course in Statistical Programming with R*. Springer, New York, 2007.

Chambers, J. M.: *Software for Data Analysis: Programming with R*. Springer, New York, 2007.

Ihaka, R., Gentleman, R.: *J. Computat. Graph. Stat*. 5, 1996, 299–314. R: A language for data analysis and graphics.

ℝ: Software, a language and environment for statistical computing. R Development Core Team, Foundation for Statistical Computing, www.r-project.org, Vienna, Austria, 2008.

Sakar, D.: *Lattice Multivariate Data Visualization with R*. Springer, New York, 2007.

Spector, P.: *Data Manipulation with R*. Springer, New York, 2008.

Index

D

E

F

G

H

I

J

K

L

M

N

Q

R

U

V

W

Z